# *Audio IC Users' Handbook*

*Newnes Circuits Manual Series*

**Audio IC Circuits Manual**   R. M. Marston
**Modern CMOS Circuits Manual**   R. M. Marston
**Diode, Transistor & FET Circuits Manual**   R. M. Marston
**Electronic Alarm Circuits Manual**   R. M. Marston
**Instrumentation & Test Gear Circuits Manual**   R. M. Marston
**Modern TTL Circuits Manual**   R. M. Marston
**Op-amp & OTA Circuits Manual**   R. M. Marston
**Optoelectronics Circuits Manual**   R. M. Marston
**Power Control Circuits Manual**   R. M. Marston
**Timer/Generator Circuits Manual**   R. M. Marston

# Audio IC Users' Handbook

R. M. MARSTON

*To my Lady Esther, with love.*

Newnes
An imprint of Butterworth-Heinemann
Linacre House, Jordan Hill, Oxford OX2 8DP
A division of Reed Educational and Professional Publishing Ltd

⅋ A member of the Reed Elsevier plc group

OXFORD  BOSTON  JOHANNESBURG
MELBOURNE  NEW DELHI  SINGAPORE

First published 1997

© R. M. Marston 1997

All rights reserved. No part of this publication
may be reproduced in any material form (including
photocopying or storing in any medium by electronic
means and whether or not transiently or incidentally
to some other use of this publication) without the
written permission of the copyright holder except
in accordance with the provisions of the Copyright,
Designs and Patents Act 1988 or under the terms of a
licence issued by the Copyright Licensing Agency Ltd,
90 Tottenham Court Road, London, England W1P 9HE.
Applications for the copyright holder's written permission
to reproduce any part of this publication should be addressed
to the publishers

**British Library Cataloguing in Publication Data**
A catalogue record for this book is available from the British Library

ISBN 0 7506 3006 X

**Library of Congress Cataloguing in Publication Data**
A catalogue record for this book is available from the Library of Congress

Composition by Scribe Design, Gillingham, Kent, UK
Printed in Great Britain by
Biddles Ltd, Guildford and King's Lynn

# Contents

| | | |
|---|---|---|
| **Preface** | | vii |
| 1 | Audio basics | 1 |
| 2 | Op-amp audio processing circuits | 20 |
| 3 | Dedicated audio processing IC circuits | 58 |
| 4 | Audio pre-amplifier circuits | 101 |
| 5 | Audio power amplifier circuits | 116 |
| 6 | High-power audio amplifiers | 150 |
| 7 | LED bar-graph displays | 177 |
| 8 | Audio delay-line systems and circuits | 201 |
| 9 | Power supply circuits | 258 |
| 10 | End notes | 271 |
| **Index** | | 285 |

# *Preface*

A vast range of audio and audio-associated ICs are readily available for use by amateur and professional design engineers and constructors. This new wide-ranging 64 000-word handbook explains – with the aid of almost 400 illustrations – the operating principles of the most important of these devices and serves as a practical users guide to seventy-three of the most popular and useful of the currently available audio ICs. It deals with individual audio power amplifier ICs with power outputs ranging from a few milliwatts to 68W per channel, and with audio devices ranging from simple linear amplifiers to complex dynamic range compressors, electronic tone/volume control ICs, special noise reduction ICs, analogue and digital delay lines, and with audio-associated ICs such as dot- and bar-graph display driver ICs and voltage regulators.

The handbook is split into ten chapters. The opening chapter gives a concise description of 'audio' and audio-system basic principles and theory. The next five chapters deal with the theoretical and practical aspects of audio signal processing circuits, audio pre-amplifiers, low-power audio amplifiers, and high-power audio amplifiers. Chapters 7 and 8 deal with the audio-associated subjects of LED dot- and bar-graph displays (which can give a visual indication of signal levels, etc.) and analogue and digital delay-line ICs and systems (which can be used to give special sound effects such as echo and reverberation). Chapter 9 deals with power supply circuits for use in audio systems. The final chapter presents a useful collection of information, dealing with matters such as loudspeaker selection, heatsink size calculation, power supply requirements, and audio power amplifier basic design techniques.

The handbook, though aimed specifically at all practical design engineers, technicians and experimenters, will doubtless also be of great interest to all amateurs and students of electronics. It deals with its subject in an easy-to-read, down-to-earth, mainly non-mathematical but very comprehensive manner. Each chapter starts off by explaining the basic principles of its subject and then, where appropriate, goes on to present the reader with a great mass of practical circuits and data, all of which have been fully evaluated and/or verified by the author.

Throughout the volume, great emphasis is placed on practical 'user' information and circuitry, and the book abounds with useful circuits and graphs. Most of the ICs and other devices used in the practical circuits are modestly priced and readily available types, with universally recognized type numbers.

R. M. Marston, 1997

# 1
# Audio basics

## Audio signals

An audio *signal* is a waveform with a fundamental frequency that falls within the range 16Hz to 20kHz, which is the recognized span of human hearing. The audio signal may be acoustic (a sound wave) or electric (a voltage or current). *Figure 1.1* shows some important features of the audio spectrum. Thus, the human voice generates acoustic waves within the range 100Hz to 4.5kHz, human hearing has a sensitivity peak at about 3.5kHz, and most music waveforms fall within the spectrum 50Hz to 16kHz.

Humans with unimpaired hearing have an aural response curve that varies with frequency, loudness, and with age. *Figure 1.2* shows typical response curves for 18 to 25 year olds, at both 'very low' and 'moderate to loud' sound levels. The aural response falls off sharply (typically at a 10 to 14dB/octave rate) below 200Hz, and drops dramatically at frequencies above 16kHz.

Above the age of 25 years, the ear's relative sensitivity to signal frequencies greater than 500Hz falls off (when compared to a typical 25 year old) in proportion to both frequency and the age of the listener, in the basic manner shown in *Figure 1.3*. Thus, to hear a 10kHz signal with the same clarity as a

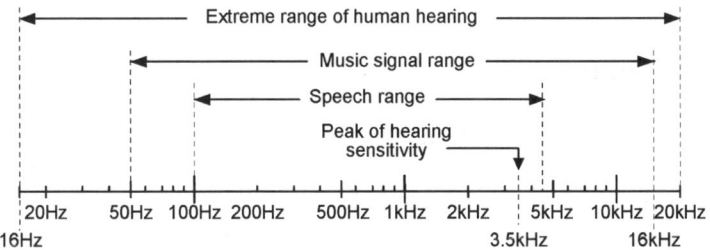

**Figure 1.1** *Important features of the audio spectrum*

## 2  Audio basics

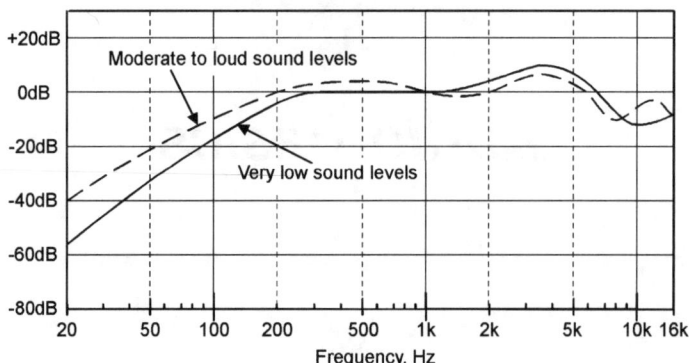

**Figure 1.2**  *Typical aural frequency response curves for healthy 18 to 25 year olds*

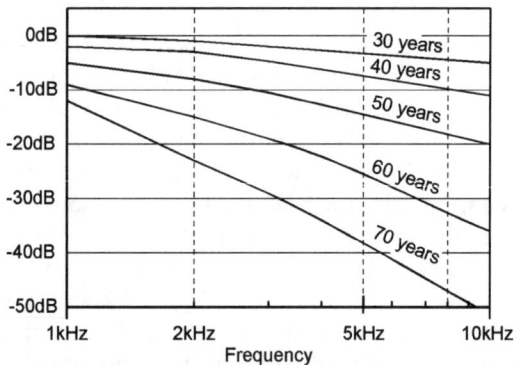

**Figure 1.3**  *The ear's relative sensitivity to signal frequencies above 500Hz deteriorates with age, typically in the manner shown here*

25 year old, a listener needs to give the signal 5dB of boost at the age of 30, 11dB at the age of 40, 20dB at 50, 35dB at the age of 60, and a hefty 55dB at the age of 70.

## Audio systems

An audio *system* is one which enables an acoustic audio signal to be conveyed from a *source* point to a *destination* point by artificial means. Prior to the invention of electronic amplifying devices, all short-range audio systems consisted of simple mechanical devices such as megaphones, hearing horns, or speaking tubes, and the most widely used long-range systems were the telephone and the hand-cranked gramaphone.

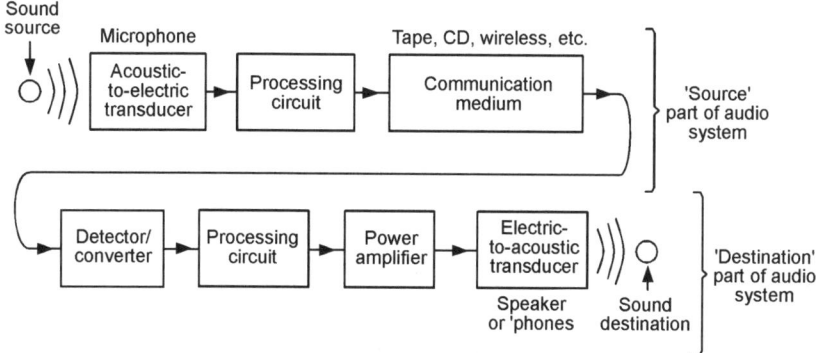

**Figure 1.4** *Basic elements of all modern electronic audio systems*

All modern electronic audio systems consist of the basic elements shown in *Figure 1.4*. At the *source* end of the system, the sound is picked up by a microphone (an acoustic-to-electric transducer) and the resulting electrical output signal is passed to an electronic processing circuit, where the signal may be amplified and/or filtered, etc., before being applied to the system's communication medium, which usually takes the form of a tape, a disc (record), a CD, a length of cable, or some form of wireless link.

At the *destination* end of the system, the basic audio signal is extracted from the communication medium via a suitable detector/converter circuit and is then changed back to its original form via another signal processing circuit. The resulting audio signal is then passed through a power amplifier before being connected to an electric-to-acoustic transducer such as a loudspeaker or headphone, which converts the signal back into its original audible form.

## Ausio system fidelity

In audio system jargon, 'fidelity' is a rough measure of the ability of a system (or major sub-system) to reproduce an accurate copy of an original audio input signal, and is expressed in terms of either 'hi-fi', 'medium-fi', or 'low-fi'. A system's 'fi' requirement depends on the specific application, as shown in *Figure 1.5* and indicated by the following notes.

A hi-fi system is one that gives a good to superb quality of sound reproduction over the full audio frequency spectrum. A hi-fi performance is required in all good music reproduction systems. Most sophisticated modern home entertainment audio units give a hi-fi performance.

A medium-fi system is one that gives a moderate to good quality of sound reproduction over most or all of the audio frequency spectrum. Such a performance is adequate for use in portable radios and portable cassette or

4  Audio basics

| TYPE OF UNIT | 'FI' TYPE | NOTES |
|---|---|---|
| Sophisticated home entertainment unit | hi-fi | High-performance sterio unit |
| Portable radio | medium-fi | Typically operate over the 20Hz to 12kHz audio spectrum and have very simple controls |
| Portable cassette/CD unit | medium-fi | |
| Simple Karaoke unit | medium-fi | |
| Loud hailer | low-fi | Operate mainly over the 100Hz to 4.5kHz speech frequency spectrum |
| Public address system | low-fi | |
| Intercom | low-fi | |
| Mobile phone | low-fi | |

**Figure 1.5** *Examples of fidelity ('fi') requirements of eight popular audio units*

CD players, and in karaoke units, etc. Most such units do in fact give a medium-fi performance.

A low-fi system is one that gives a poor to moderate quality of sound reproduction, or responds well to only a restricted part of the audio frequency spectrum, but gives a performance that enables voice messages to be reproduced without significant loss of intelligibility. Most loud-hailers, public address systems, intercoms, and mobile phones give a low-fi performance.

## 'Harmonic' signal distortion

The prime aim of any audio system is to convey an acoustic audio signal from a source point to a destination point (the listener's ears) without undue loss of fidelity. Any loss of fidelity that occurs along the way can only be ascribed to changes in the signal's waveform, and such changes are called *distortion*. There are several basic forms of distortion, most of which can be classified as either an *harmonic* or a *frequency response* type of distortion.

If a perfect sinewave signal is applied to the input of an amplifier, the output signal will – if the amplifier is perfectly linear and gives exactly the same amount of amplification to every single part of the signal – also be a perfect sinewave, with no unwanted harmonics. If, on the other hand, the amplifier gives any form of non-linear amplification, the output signal will inevitably be distorted, and this distortion will manifest itself in the generation of a number of harmonic signals, at direct multiples of the original signal frequency.

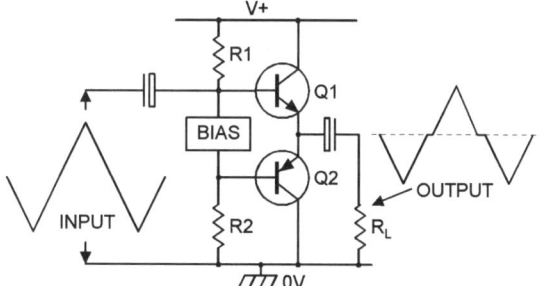

**Figure 1.6** *Example of cross-over distortion on a defectively biased complementary emitter follower output stage*

The three forms of harmonic distortion most often encountered in analogue audio systems are cross-over distortion, 'clipping' distortion, and simple non-linear distortion, and examples of these are shown in *Figures 1.6 to 1.8*, using triangle-shaped input signals for clarity.

*Figure 1.6* shows an example of cross-over distortion caused by a defectively biased complementary emitter follower output stage of the basic type used in most power amplifier ICs. In this type of circuit, R1 and R2 act as a potential divider that sets the bases of Q1 and Q2 at a mean half-supply voltage value, and the bias unit applies a voltage of (typically) about 1.2V between the two bases, so that each transistor is slightly biased on when no input signal is applied. Under this condition the mean charge in the load-driving electrolytic output capacitor tends to hold the emitters of both transistors close to the half-supply voltage value, and any positive-going input signal thus increases the drive to Q1 and reduces that to Q2, and any negative-going input signal causes the reverse of these actions, the net effect being that the output signal is a close but low-impedance copy of the original input waveform.

If, however, the bias network of this circuit develops a short, both transistors will normally be cut off, and Q1 will not turn on until the input swings 600mV positive, and Q2 will not turn on until the input swings 600mV negative, the net effect being that the circuit effectively removes the central 1.2V 'slice' of the output waveform, as shown in the diagram, as a consequence of this 'cross-over' defect. This type of distortion is highly undesirable, and generates many harmonic signals.

*Figure 1.7* shows an example of 'clipping' distortion caused by overdriving a simple common-emitter transistor amplifier. The output of this type of amplifier is an amplified and inverted version of the input signal, but if (as shown) the input signal is too large its larger positive parts will drive the transistor to saturation, causing the loss of the lower parts of the output signal, and its larger negative parts will drive the transistor to cut-off, causing

## 6  Audio basics

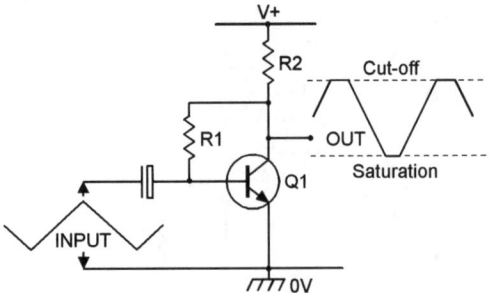

**Figure 1.7**  Example of 'clipping' distortion caused by overdriving an amplifier stage

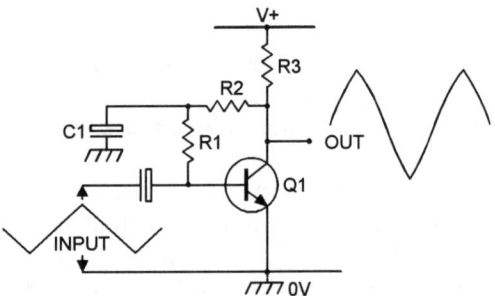

**Figure 1.8**  Example of simple non-linear distortion in a common-emitter amplifier

the loss of the upper parts of the output signal. 'Clipping' distortion is highly undesirable, and generates many harmonic signals.

*Figure 1.8* shows an example of simple non-linear distortion caused by operating a common-emitter transistor amplifier with insufficient signal negative feedback. Note in this circuit that biasing network R1–R2 is decoupled by C1 and thus provides dc negative feedback but not ac (signal) negative feedback, which is normally used to regulate the circuit's ac signal gain. Transistors are inherently non-linear devices, and their current gain varies with their collector current values. Consequently, the voltage gain of this circuit inevitably varies with the instantaneous signal amplitude, and the output is thus subject to non-linear distortion. This is a relatively trivial type of distortion, which normally generates only low-order harmonics. In most applications, the distortion is undesirable, but in some special applications it can be used to the advantage of the end-user.

### 'Frequency response' distortion

The other basic form of distortion is the *frequency response* type, in which the effective voltage gain or attenuation of a circuit or system element varies

Audio basics 7

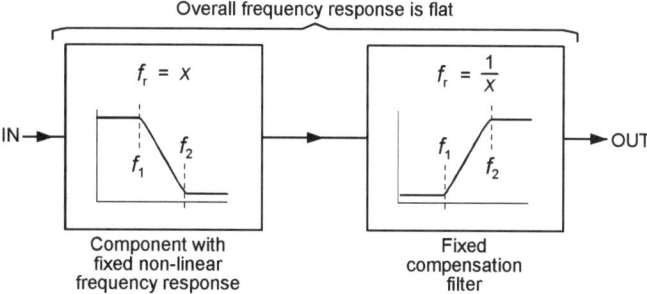

**Figure 1.9** *Frequency response non-linearity in one part of a circuit can be countered with a matching compensation filter, to give a flat overall frequency response*

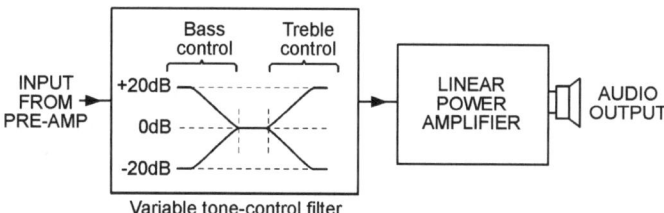

**Figure 1.10** *A variable tone-control filter can help compensate for defects in the user's acoustic frequency response*

with the applied frequency. The most obvious example of this kind of distortion is the human listener who sits at the money-paying end of the audio-system processing chain, and who (as shown in *Figures 1.2* and *1.3*) has an upper-frequency acoustic response that declines steeply with the passing of the years. Electromechanical devices such as microphones, loudspeakers and pickups also have a non-linear frequency response.

Modern active filter circuits can be designed to give virtually any desired frequency response, and can easily be used to counteract response defects that occur in any part of an audio system. *Figures 1.9* and *1.10* illustrate the basic principle. Thus, *Figure 1.9* shows how a component or device that has a fixed non-linear frequency response of '$x$' can be used in conjunction with a fixed compensation filter with a response of '$1/x$' (the exact inverse of '$x$') to give a flat overall frequency response.

*Figure 1.10* shows how a variable tone-control filter can be used to help compensate for frequency response defects or variations in the acoustic path that exists between the output stage of a power amplifier and the sound-perception mechanism of the audio listener, i.e. in the loudspeaker, the room acoustics, or in the listener's hearing. The filter shown in the diagram is a simple type in which bass and treble can each be boosted or cut by up to 20dB (i.e. by a 10:1 ratio), but in practice most modern tone-control filters also give similar control of the midband response. Thus, if the bass and midband controls are both set to maximum cut and the volume level is boosted by 20dB, the treble response can be effectively boosted by up to 40dB (i.e. by a 100:1 ratio).

## 'Noise' and dynamic range

If the input terminal of an audio power amplifier is shorted to ground, so that zero input signal is applied to the unit, and the volume control is then wound up to maximum, a distinct 'hissing' sound will be heard from the unit's loudspeaker. The signal responsible for this sound is generated by the equipment itself, and takes the form of a mixture of tones of randomly generated frequency and amplitude, and is known simply as 'noise'. If the r.m.s. amplitude of the noise signal is measured on an analogue meter, its r.m.s. value will be found to be reasonably constant even though its instantaneous value varies wildly from moment to moment. If an external variable-amplitude signal is now applied to the amplifier, it will produce no useful output until its r.m.s. amplitude significantly exceeds that of the noise signal, which effectively swamps all lesser signals. The unit's 'noise' signal value thus determines the minimum signal level that can be usefully handled by the equipment.

All electronic amplifiers and items such as microphones, tapes, discs, and CDs generate noise, and thus have finite 'minimum signal' handing limits. They also have intrinsic 'maximum signal' handling limits, beyond which any applied signal will become too distorted to have a practical value. In amplifiers, this limit is reached at the onset of signal clipping or some other form of severe distortion, and in tapes it occurs when the tape nears magnetic saturation.

An electronic item's maximum-to-minimum signal handling ratio or *absolute maximum dynamic operating range* is thus determined by the relative ratio of its 'maximum signal' to 'noise level' values, and is known as its signal-to-noise or 'S/N' ratio, and is defined by the formula:

S/N-ratio = maximum signal volts/noise volts.

Thus, if an amplifier can handle a maximum input of 1000mV and generates 1mV of noise, it has a 1000:1 or 60dB S/N-ratio.

Note at this point that, in audio systems, the input signal is usually a music/speech one that contains data in the form of complex audio patterns that can be subliminally interpreted by the listener's brain, and that the brain can still interpret these patterns even if a substantial portion of the original data is lost, i.e. if the signal is corrupted. Consequently, the practical value

| SIGNAL VOLTAGE | SIGNAL/NOISE VALUE | SIGNAL DATA LOST (%) | SIGNAL DATA RETAINED (%) |
|---|---|---|---|
| $V_{NOISE}$ | 0dB | 100% | 0% |
| $V_{NOISE}$ + 12% | +1dB | 89.3% | 10.7% |
| $V_{NOISE}$ + 26% | +2dB | 79.4% | 20.6% |
| $V_{NOISE}$ + 41% | +3dB | 70.9% | 29.1% |
| $V_{NOISE}$ + 58% | +4dB | 63.3% | 36.7% |
| $V_{NOISE}$ + 100% | +6dB | 50% | 50% |
| $V_{NOISE}$ + 151% | +8dB | 39.8% | 60.2% |
| $V_{NOISE}$ + 216% | +10dB | 31.6% | 68.4% |
| $V_{NOISE}$ + 462% | +15dB | 17.8% | 82.2% |
| $V_{NOISE}$ + 900% | +20dB | 10% | 90% |

**Figure 1.11** *The fidelity of a low-level signal depends on its magnitude relative to the system's 'noise' voltage*

of such a signal depends on its magnitude relative to that of the system's 'noise' signal, in the manner shown in *Figure 1.11*.

Thus, if the signal has the same amplitude as the noise signal (i.e. the signal/noise ratio equals 0dB) all of the signal data is lost, and the signal has zero practical value. The signal only achieves some slight practical value when its amplitude is double that of the noise (i.e. at +6dB), at which point 50% of its original data is retained, and begins to attain real value only when its amplitude rises to +10dB (about treble the noise value), at which point almost 70% of the original data is retained. The signal quality attains a reasonable 'hi-fi' standard only when its value rises to +20dB (ten times the noise value), at which point 90% of the signal data is retained. Consequently, in modern analogue audio systems, the following general empirical rules apply to the definition of a system's dynamic range:

(1) Absolute maximum dynamic range = S/N-ratio value.
(2) Absolute maximum *useful* dynamic range = S/N-ratio minus 6dB.
(3) Normal 'mid-fi' dynamic range = S/N-ratio minus 10dB.
(4) Normal 'hi-fi' dynamic range = S/N-ratio minus 20dB.

## Dynamic range manipulation

It is generally agreed that large orchestras are capable of generating sound levels – from the softest to the loudest values – over a 70dB (3162:1) dynamic

10  Audio basics

range. To reproduce this sound range with really good fidelity, a hi-fi pre-amplifier and power amplifier combination needs to have a S/N-ratio of 90dB or better, but in reality most modern hi-fi units fall some 10dB to 20dB short of this ideal. The biggest weakness in the *analogue* sound reproduction chain lays not, however, in the actual hi-fi units, but in the storage/transport media (tapes and phono discs) that are used to carry the music signals into the inputs of the hi-fi units.

Typically, modern cassette tapes and stereo phonograph discs have S/N-ratios of 55dB and 58dB respectively, and thus have *basic* hi-fi dynamic ranges of only 35dB and 38dB. Superficially, these performances seem too poor to be acceptable for hi-fi use, but in practice their *effective* dynamic ranges can be expanded by a further 10dB to 15dB by using dynamic range manipulation techniques, thus bringing their final replay performances up to the level of normal FM stereo broadcast signals. There are two basic types of dynamic range manipulation technique. They are known as the 'pre-emphasis' technique, and the 'compression' technique.

## Signal pre-emphasis

The amplitudes of most music and voice signals are dominated by bass frequencies, with the amplitudes of the higher 'treble' frequency signals – which are mainly harmonics of the bass signals – falling off at a 6dB/octave rate. When these signals are fed through an electronic system with a poor dynamic range, the higher-frequency signals are – because of their relatively low amplitudes – the first to be lost in the underlaying 'mush' of system noise, and the resulting audio output sounds flat and unpleasant. *Figure 1.12* shows how this problem can be partially overcome in tape or disc recording systems by feeding the input signals to the media via a treble-boosting pre-emphasis filter during the 'recording' stage, and then restoring the signals to their original form via a matching treble-cutting de-emphasis filter during the 'replay' stage.

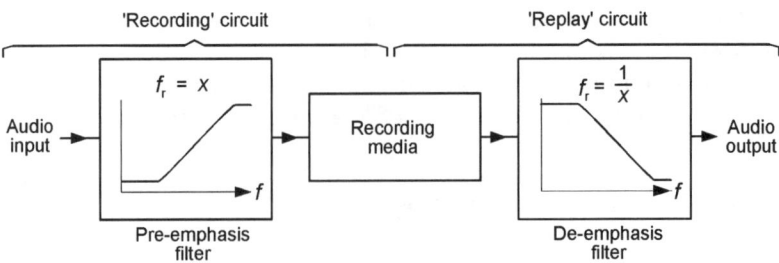

**Figure 1.12**  *Basic 'pre-emphasis' method of dynamic range manipulation*

Audio basics 11

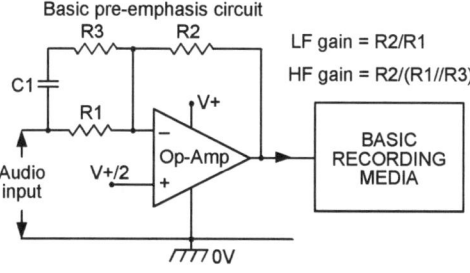

(a) Basic pre-emphasis 'record' circuit.

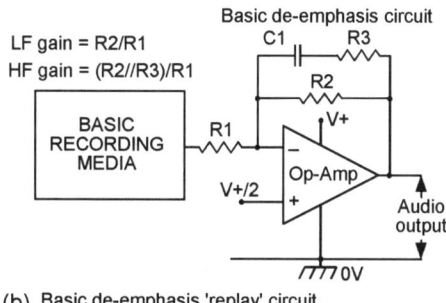

(b) Basic de-emphasis 'replay' circuit.

**Figure 1.13** *Basic pre-emphasis (treble-boost) and de-emphasis (treble-cut) filter circuits*

*Figure 1.13* shows the basic forms of the pre-emphasis and de-emphasis filters, which in this case have treble boost or cut slopes of 6dB/octave but have their maximum boost or cut limited to about 20dB by their resistance values and have their turnover frequency (usually about 800Hz) set via C1. Circuits of this type can improve the *effective* dynamic ranges of various media by about 12dB, and are often used in conjunction with other compensatory filters, as in the case of the RIAA phonograph system and with various tape/cassette playing systems.

## Signal compression

The 'signal compression' method of dynamic range manipulation operates in the basic way illustrated in *Figure 1.14*. Here, the initial input signal is applied to the input of a dynamic range compressor unit, which in this example has a 2:1 dynamic compression ratio and can convert an input signal with a 90dB dynamic range into an output with a 45dB dynamic range. This 'compressed' output is applied to the input of the 'poor-dynamic-range' transmission/recording medium, and when required is converted back into its original (90dB

12  Audio basics

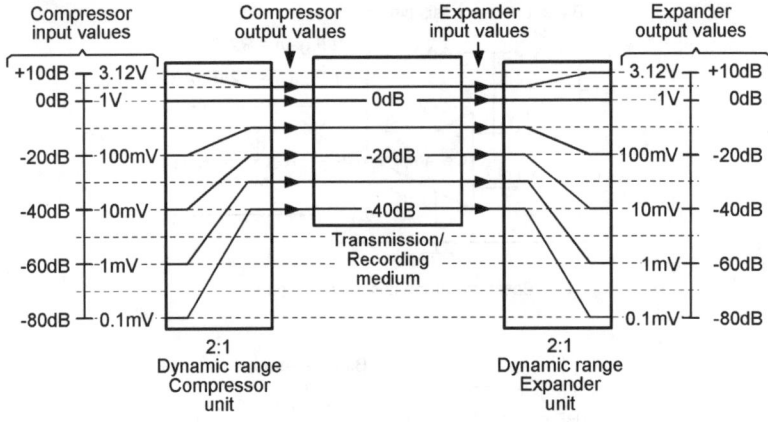

**Figure 1.14** *Diagram illustrating the basic principle of dynamic range companding, using a 2:1 companding ratio, over a 90dB operating range*

range) form by a matching dynamic range *expander* unit, which has characteristics that are the exact inverse of those of the compressor. This type of compressor–expander system is generally known as a 'compander' (or 'compandor') system, and was originally devised to improve the quality of various voice communication systems.

Note in *Figure 1.14* that the '2:1' compressor/expander ratio applies to the systems dynamic range in terms of dB, and not to its actual range in terms of input and output voltages. This point is made clear in the table of *Figure 1.15*, which shows that the compressor and expander work by giving highly non-linear variations in voltage gain to different input signals. The gain varies over a 175:1 range, from ×0.57 to ×100 in the compressor unit, and from ×1.75 to ×0.01 in the matching expander unit.

Practical compressor and expander circuits are both built around the basic dynamic range expander circuit that is shown in (a) descriptive and (b) symbolic forms in *Figure 1.16*. Here, the audio input signal is applied to the inputs of a current-controlled variable-gain cell (a high-grade operational transconductance amplifier, or OTA) and an electronic rectifier that converts the *mean* input signal voltage into a proportional dc output current, which controls the gain of the variable-gain cell. The action is such that if the signal input rises by 10dB (from, say, −40dB to −30dB) the gain also rises by 10dB, to give an overall increase in output voltage of 20dB, i.e. a 2:1 ratio of dynamic expansion. The gain-control attack and decay times are controlled by capacitor $C_T$.

The gain cell's output signal appears in the form of a current, rather than a voltage, but can be converted into a proportional voltage via a suitably wired op-amp. *Figure 1.17* shows ways of using the basic dynamic range expander to make practical voltage-in to voltage-out (a) expander or (b)

Audio basics  13

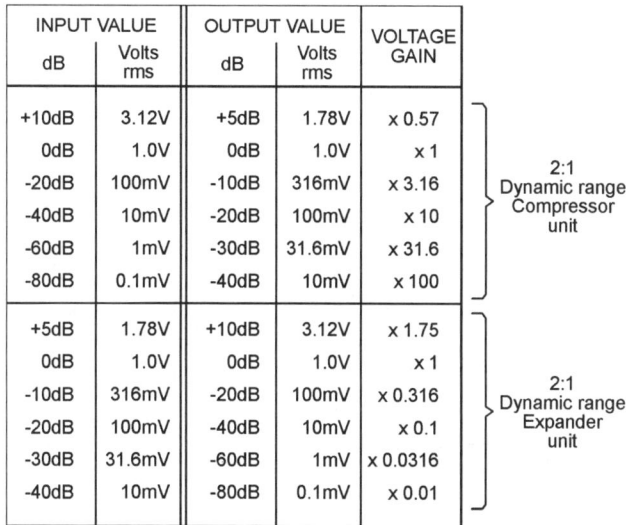

**Figure 1.15** *Table showing the voltage gain variations of the basic 2:1 dynamic range compressor/expander*

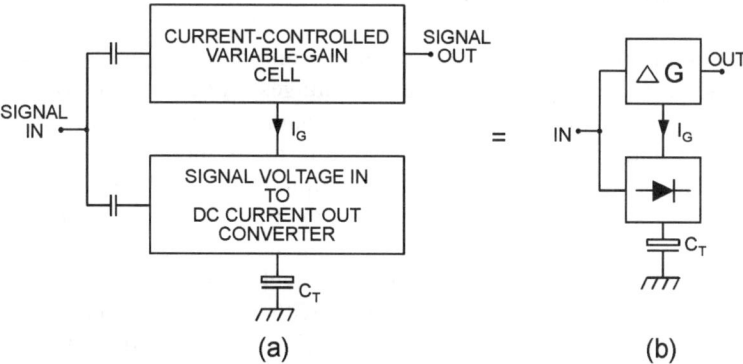

**Figure 1.16** *Diagram showing (a) descriptive and (b) symbolic versions of a basic dynamic range expander*

compressor units. In the expander circuit, the gain cell's output current is simply fed directly into the inverting input terminal of the op-amp, which gives direct current-to-voltage conversion. In the compressor, the audio input signal is fed into the op-amp's inverting input via R1, and the basic expander is wired in series with the op-amp's output-to-input negative feedback path, causing the overall circuit to act as a voltage-in to voltage-out dynamic range compresser with dynamic characteristics that are the exact inverse of those of the expander circuit.

14  Audio basics

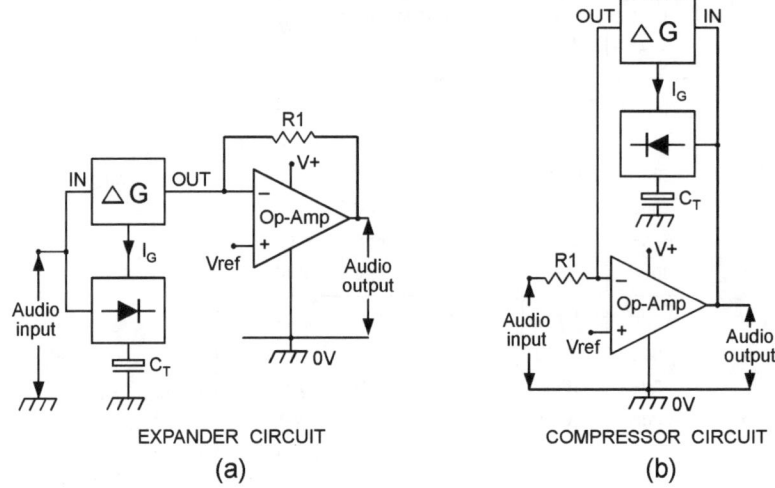

**Figure 1.17** *Ways of using the basic dynamic range expander to make practical voltage-in to voltage-out (a) expander or (b) compresser units*

### Hybrid dynamic range control

At first sight, pre-emphasis and compression methods of dynamic range control seem to offer huge practical advantages, but in reality both systems work by introducing various types of gain distortion and thus have weaknesses that limit their practical value in many medium-fi to hi-fi applications. The high-frequency gains of pre-emphasis filters should, for example, always be limited to about 20dB (in the manner shown in *Figure 1.13*) to avoid overdriving unusually large high-frequency signals that are generated as fundamental – rather than harmonic – waveforms.

Compander systems should be treated with special caution. They originated as strictly 'low-fi' voice processing units, as typified by the popular NE570 IC (see Chapter 3), and usually generate rather high levels of noise, THD, tracking distortion and output dc-tracking shift; these defects become magnified if the system's compression ratio is artificially raised above its basic 2:1 value. The system's greatest defects are caused by a problem known as 'breathing' or 'pumping', which occurs when a large transient input waveform is followed by a near-zero input signal, causing the expander gain to drop sharply on the arrival of the transient but (because of the system's AGC action) to rise sharply to its maximum value a few moments later, to produce the distinct hissing sound of system noise.

The best way to obtain really good dynamic range manipulation in hi-fi applications is to use a hybrid system that uses a subtle combination of pre-emphasis and/or filtered companding techniques, as in the cases of the dBx and Dolby magnetic recording/playback systems, which each give a large increase in useful dynamic range but without suffering 'pumping' problems.

## Digital audio

Audio systems are designed to convert acoustic input signals into remotely located acoustic output signals, and are basically analogue systems. In modern audio electronics there is, however, one area of the system in which *digital* electronic techniques offer definite advantages over analogue ones, and that is in the area of high-quality signal storage (as in CDs). This advantage occurs because digital signals have only two amplitude levels – either 'high' or 'low' – and thus (unlike analogue signals) cannot be corrupted by normal levels of system noise or non-linearity.

*Figure 1.18* shows the basic elements of a digital audio-signal 'storage' and 'replay' system. On the 'storage' side of the system, the audio input signal is subjected to normal signal processing (amplification and/or filtering, etc.) and is then applied to the input of an ADC (Analogue-to-Digital Converter) unit, which converts the analogue input signal into a digital equivalent, which is then superimposed on the system's storage medium (a CD or tape, etc.). On the 'replay' side of the system, the storage medium's digital signals are inspected by a DAC (Digital-to-Analogue Converter) unit, which converts them back into analogue form and passes them on to the outside world via another signal processing analogue circuit.

*Figure 1.19* illustrates some of the basic operating features of the ADC part of the system, as applicable to a normal 16-bit CD. Here, the system repeatedly takes high-speed samples of the audio input signal's instantaneous analogue amplitude, and converts each new sample's amplitude measurement into a multi-bit digital output word and passed it on to the recording media before carrying out a similar operation on the next sample. To be effective, the system's sampling frequency must be at least double that of the highest signal frequency of interest. Modern CD systems are designed to handle signal frequencies of up to 20kHz, and to attain this they use a

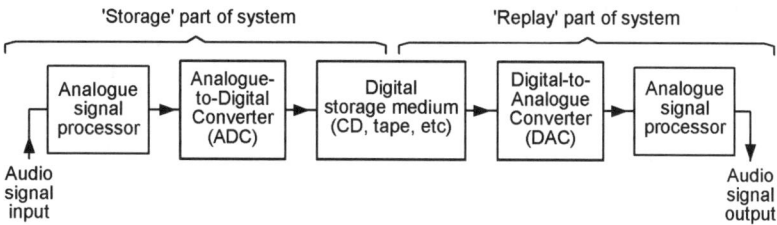

**Figure 1.18** *Diagram showing the basic elements of a digital audio-signal 'storage' and 'replay' system*

16  Audio basics

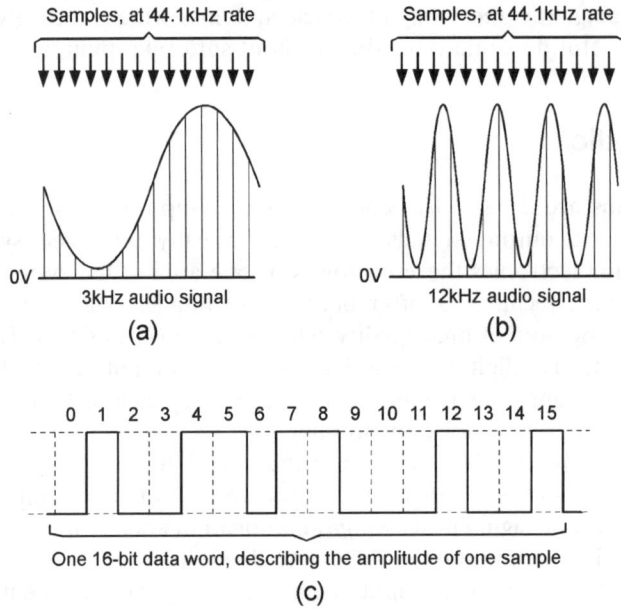

**Figure 1.19** *Diagram illustrating some basic features of analogue-to-digital conversion in 16-bit CD systems (see text for explanation)*

standard sampling frequency of 44.1kHz and thus execute 44.1 sampling operations during a 1kHz signal cycle, 14.7 samplings in a 3kHz cycle (see *Figure 1.19(a)*), and 3.67 in a 12kHz cycle (see *Figure 1.19(b)*).

The effective signal-to-noise ratio and dynamic range of an ADC unit's output is directly proportional to the unit's 'bit' size, and can be simply calculated from the equations:

(1) Signal-to-noise ratio = $6 \times n$dB.
(2) Useful dynamic range = $6 \times (n - 1)$dB,

where $n$ is the ADC's bit size. Thus, 16-bit ADCs have S/N-ratios of 96dB and have useful dynamic ranges of 90dB. The current generation of CDs are recorded via 16-bit ADC, which generate 16-bit outputs in the basic format shown in *Figure 1.19(c)* and can thus generate up to 65 536 different codes or level-measurement values. The next generation of CDs (for which players are already available) are scheduled to use 20-bit data recording, which offers S/N-ratios of 120dB and useful dynamic ranges of 114dB and can generate up to 1 048 576 different codes.

The above explanation of CD encoding is, of course, much simplified, and merely illustrates the basic principles. *Figure 1.20* shows a more realistic picture of the actual coding system used in 16-bit CDs. Here, each 16-bit data *word*

Audio basics 17

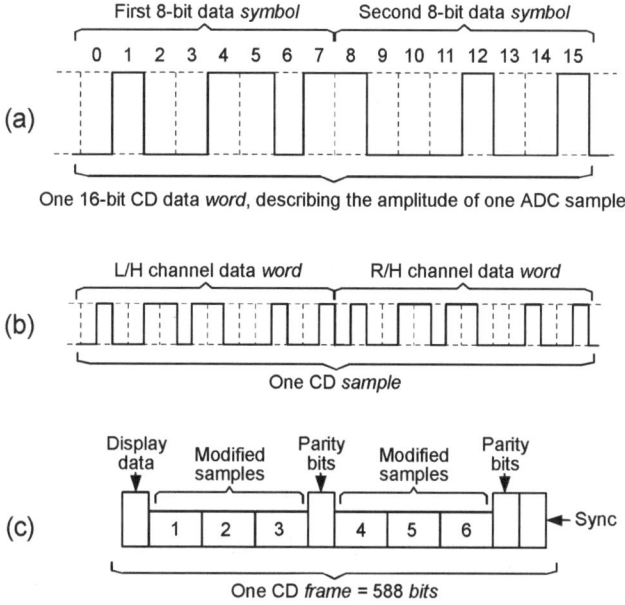

**Figure 1.20** *Diagram illustrating the formatting of a standard* frame *in a 16-bit CD; these frames are generated at a 7350 per second rate*

is made up of two 8-bit *symbols*, as shown in (a). All CDs give stereo outputs, so two *words* are generated at the same time, one for the L/F channel and one for the R/H channel, and are applied to the CD encoding system in series, as a 32-bit *sample*, as shown in (b). Each of these *samples* is modified and enlarged by the CD encoding system and is then incorporated in a CD *frame*. Each of these *frames* commenced with 8 bits of display data, followed by three of the modified *samples*, which are followed by a group of parity bits, plus three more *samples* and another group of parity bits, and ends with a synchronization code. The CD *frames* each contain 588 data bits, and are generated at a 7350 per second rate. The CD data is thus processed (at both the record and replay ends of the system) at a rate of 4.3218 megabits per second.

At the 'player' end of the CD system, the decoder circuitry copies and stores each frame of data as it becomes available, then converts its *samples* and *words* back into their original analogue forms (via a 16-bit or larger DAC) and passed them to the appropriate channels of the audio hi-fi unit – in the correct time sequence – before moving on to the next frame, and so on.

## The hi-fi unit

In its simplest form, a hi-fi unit may consist of little more than an input selector, a pre-amplifier, a power amplifier, and a pair of loudspeakers. Usually,

18  *Audio basics*

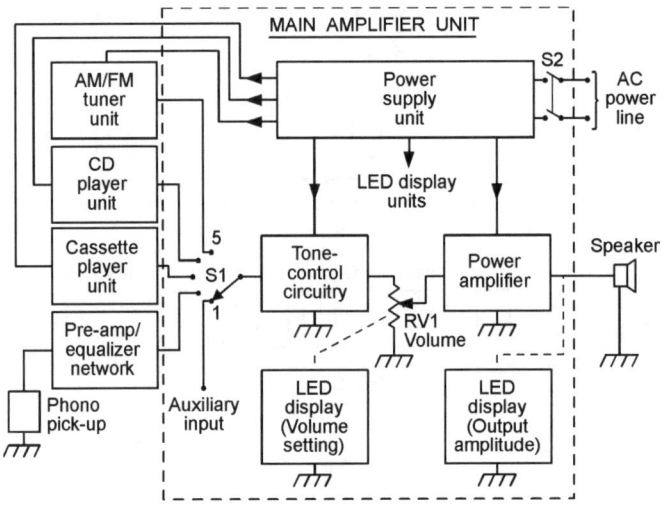

**Figure 1.21**  *Typical form of a modern hi-fi unit, showing only one of the stereo channels of the main amplifier unit*

however, the unit is fairly elaborate, and typically may take the form shown in *Figure 1.21*, which (in the main amplifier unit) shows just one channel of a stereo system. Thus, the main amplifier contains an input selector switch (S1), plus tone and volume control circuitry that feeds the input of the speaker-driving power amplifier. It also houses a power supply unit that can power the main amplifier and (usually) the other hi-fi units (tuner, CD player, etc.), plus one or two LED display units that give visual output indications of parameters such as the *volume control* setting and the instantaneous *output amplitude* values of the power amplifier.

In most cases, switch S1 can select inputs from an AM/FM tuner, a CD player, a cassette player unit, a phono pickup and pre-amplifier/equalizer unit, and from an 'Auxiliary' input terminal. This auxiliary input may be driven from a source such as an audio mixer unit, a TV sound-channel tuner, or a remote sound monitor such as a baby alarm, etc. Often, the main amplifier is designed for operation via a remote control unit, in which case the tone and volume control circuitry usually take the form of voltage-controlled units that are driven via the remote decoder, and S1 takes the form of a multi-way electronic switch that is driven via the decoder. If the hi-fi is a really elaborate one, it may also incorporate one or more audio delay lines, in the form of an ambience synthesizer or an echo-reverb unit.

The two most important basic items in the hi-fi system are its loudspeakers and its power amplifier. Inferior loudspeakers can make even the very best power amplifier sound bad, and an inadequate power amplifier can make even the very best of loudspeakers sound awful. Assuming, however, that a hi-fi unit's loudspeakers and power amplifier are both of excellent quality and are well matched, it will still be found that different listeners will each gain their own individual *subjective* impressions of the system's quality, and will probably tweak the tone controls until the system 'sounds right'. And 'sounding right' is, of course, the most important function of the entire hi-fi system.

# 2
# Op-amp audio processing circuits

An audio processor is a circuit that takes an audio input signal and generates an audio output that is directly related to that input but is modified in some way. The processor may, for example, simply invert the original signal and/or provide it with a fixed amount of voltage gain, as in the case of a linear amplifier, or it may give it an amount of gain that varies with signal frequency, as in the case of an active filter, or an amount of gain that varies with the signal's mean amplitude, as in the case of a 'constant-volume' or non-linear amplifier or an expander/compressor circuit. Practical examples of all these types of processor circuit are presented in this and the following chapter.

The present chapter deals exclusively with processing circuits based on operational-amplifier (op-amp) ICs and is divided into two major sections. The first of these deals with general-purpose audio-frequency processor circuits based on conventional op-amp ICs, and the second deals with circuits based on operational transconductance amplifier (OTA) types of op-amp, which can be used in a variety of voltage-controlled amplifier (VCA) circuits. The next chapter (Chapter 3) describes various types of 'dedicated' audio processing IC (such as voltage- or resistance-controlled amplifiers and switched-capacitor filter ICs) and their application circuits. Dedicated audio pre-amplifier and power amplifier ICs are not described in either of these chapters, but they are dealt with in great detail in later chapters of this volume.

## Op-amp basics

The best known and most versatile type of audio signal processing IC is the ordinary operational amplifier, which can be simply described as a high-gain

Op-amp audio processing circuits 21

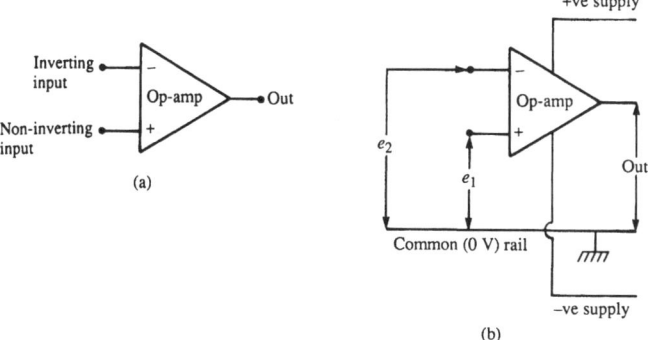

**Figure 2.1** (a) Symbol and (b) supply connections of a conventional op-amp

direct-coupled voltage amplifier block with a single output terminal but with both inverting and non-inverting high-impedance input terminals, thus enabling the device to function as either an inverting, non-inverting, or differential amplifier. *Figure 2.1(a)* shows the circuit symbol of the conventional op-amp.

Op-amps are very versatile devices. When coupled to suitable feedback networks they can be used to make precision ac and dc amplifiers, active filters, oscillators, voltage comparators, etc. They are normally powered from split supplies – as shown in *Figure 2.1(b)* – with positive (+ve), negative (–ve) and common (zero volt) supply rails, enabling the op-amp output to swing either side of the zero volts value and to be set at zero volts when the differential input voltage is zero. They can, however, also be powered from single-ended supplies, if required.

The output signal voltage of an op-amp is proportional to the differential signal voltage between its two input terminals and, at low audio frequencies, is given by

$$e_{out} = A_0(e1 - e2),$$

where $A_0$ is the op-amp's low frequency open-loop voltage gain (typically 100dB, or ×100 000), $e1$ is the non-inverting terminal input signal voltage, and $e2$ is the inverting terminal input signal voltage.

Thus, an op-amp can be used as a high-gain inverting ac amplifier by grounding its non-inverting terminal and feeding the input signal to its inverting pin via C1 and R1, as in *Figure 2.2(a)*, or as a non-inverting ac amplifier by reversing the two input connections as in *Figure 2.2(b)*, or as a differential amplifier by feeding the two input signals to the op-amp as shown in *Figure 2.2(c)*. Note in the latter case that if both input signals are identical the op-amp should, ideally, give zero output signal.

22  Op-amp audio processing circuits

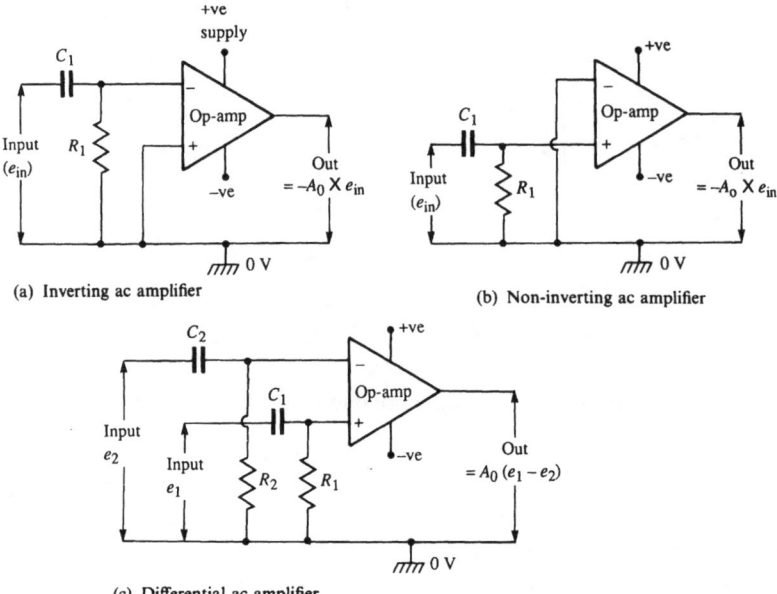

**Figure 2.2** Methods of using the op-amp as a high-gain open-loop ac amplifier

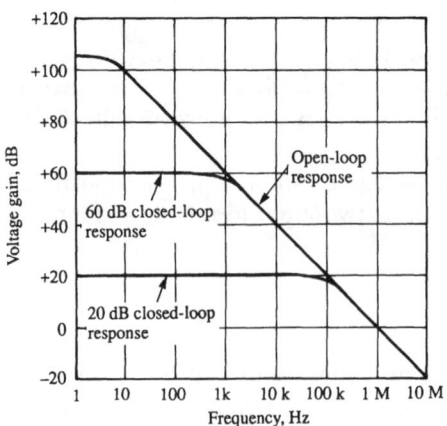

**Figure 2.3** Typical frequency response curve of the 741 op-amp

The voltage gains of the *Figure 2.2* circuits depend on the open-loop gains of individual op-amps and on input signal frequencies. *Figure 2.3*, for example, shows the typical frequency response graph of the well known 'type 741' op-amp; its voltage gain is greater than 100dB at frequencies below

Op-amp audio processing circuits   23

10Hz, but falls off at a 6dB/octave (= 20dB/decade) rate at frequencies above 10Hz, reaching unity (0dB) at an $f_T$ 'unity gain transition' frequency of 1MHz. This graph is typical of most op-amps, although individual types may offer different $A_0$ and $f_T$ values.

## Closed-loop amplifiers

The best way of using an op-amp as an ac amplifier is to wire it in the closed-loop mode, with negative feedback applied from output to input as shown in the circuits of *Figure 2.4*, so that the overall gain is determined by the external feedback components values, irrespective of the individual op-amp characteristics (provided that the open-loop gain, $A_0$, is large relative to the closed-loop gain, $A$). Note from *Figure 2.3* that the signal bandwidth of such circuits equals the IC's $f_T$ value divided by the circuit's closed-loop '$A$' value. Thus, the 741 gives a 100kHz bandwidth when the gain is set at ×10 (= 20dB), or 1kHz when the gain is set at ×1000 (= 60dB).

*Figure 2.4(a)* shows the op-amp wired as a fixed-gain inverting ac amplifier. Here, the circuit's voltage gain ($A$) is determined by the R1 and R2

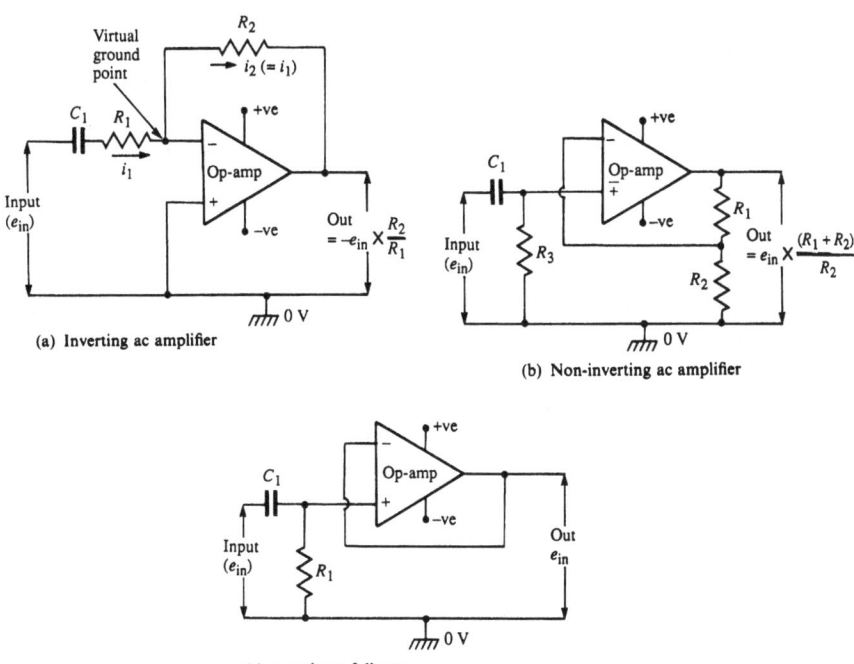

**Figure 2.4** *Closed-loop ac amplifier circuits*

ratios and equals R2/R1, and its input impedance equals the R1 value; the circuit can thus easily be designed to give any desired values of gain and input impedance. R1 and R2 have no effect on the voltage gain of the actual op-amp, so the signal voltage appearing at its output is $A_0$ times greater than that appearing on its input terminal; consequently, the signal current induced in R2 is $A_0$ times greater than that caused by the input terminal signal alone, and this terminal thus acts as though it has an impedance of R2/$A_0$ connected between the terminal and ground; the terminal thus acts like a low-impedance 'virtual ground' point.

*Figure 2.4(b)* shows how to connect the op-amp as a fixed-gain non-inverting ac amplifier. In this case the voltage gain equals (R1+R2)/R2. The input impedance, looking into the op-amp's input terminal, equals $(A_0/A)Zin$, where $Zin$ is the open-loop input impedance of the op-amp; this impedance is shunted by R3, however, so the input impedance of the actual circuit is less than the R3 value.

The above circuit can be made to function as a precision ac voltage follower by wiring it as a unity-gain non-inverting amplifier, as shown in *Figure 2.4(c)*, where the op-amp operates with 100% negative feedback. The op-amp input impedance is very high in this circuit (roughly $A_0 \times Zin$), but is shunted by R1, which thus determined the circuit's input impedance value.

## Practical op-amps

Practical op-amps are available in a variety of types of IC construction (bipolar, MOSFET, JFET, etc.), and in a variety of types of packaging styles (plastic DIL, metal-can TO5, etc.). Some of these packages house two or four op-amps, all sharing common supply line connections. *Figure 2.5* gives the parameter and outline details of eight popular 'single' op-amp types, all of which use 8-pin DIL (dual in-line) packaging.

The 741 and NE531 are bipolar types. The 741 is a very popular general-purpose type featuring internal frequency compensation and full overload protection on inputs and output. The NE531 is a high-performance type with a very high 'output slew rate' capability; an external compensation capacitor (100pF) wired between pins 6 and 8 is needed for stability, but can be reduced to a very low value (1.8pF) to give a very wide bandwidth at high gain.

The CA3130 and CA3140 are MOSFET-input op-amps that can operate from single or dual power supplies, can sense inputs down to the negative supply rail value, have very high input impedances (1.5 million megohms), and have outputs that can be strobed. The CA3130 has a CMOS output stage; an external compensation capacitor (typically 47pF) between pins 1 and 8 permits adjustment of bandwidth characteristics. The CA3140 has a bipolar output stage and is internally compensated.

*Op-amp audio processing circuits* 25

| Parameter | Bipolar op-amps | | MOSFET op-amps | | JFET op-amps | | | |
|---|---|---|---|---|---|---|---|---|
| | 741 | NE531 | CA3130E | CA3140E | LF351 | LF411 | LF441 | LF13741 |
| Supply voltage range | ±3 V to ±18 V | ±5 V to ± 22V | ±2V5 to ±8 V OR 5 V to 16 V | ±2 V to ±18 V OR 4 V to 36 V | ←—— ±5 V to ±18 V ——→ | | | |
| Supply current | 1.7 mA | 5.5 mA | 1.8 mA | 3.6 mA | 800 µA | 1.8 mA | 150 µA | 2 mA |
| Input offset voltage | 1 mV | 2 mV | 8 mV | 5 mV | 5 mV | 0.8 mV | 1 mV | 5 mV |
| Input bias current | 200 nA | 400 nA | 5 pA | 10 pA | 50 pA | 50 pA | 10 pA | 50 pA |
| Input resistance | 1M0 | 20M | 1.5 TΩ | 1.5 TΩ | 1TΩ | 1TΩ | 1TΩ | 0.5 TΩ |
| Voltage gain, $A_o$ | 106 dB | 96 dB | 110 dB | 100 dB | 88 dB | 106 dB | 100 dB | 100 dB |
| CMRR | 90 dB | 100 dB | 90 dB | 90 dB | 100 dB | 100 dB | 95 dB | 90 dB |
| $f_T$ | 1 MHz | 1 MHz | 15 MHz | 4.5 MHz | 4 MHz | 4 MHz | 1 MHz | 1 MHz |
| Slew rate | 0.5 V/µS | 35 V/µS | 10 V/µS | 9 V/µS | 13 V/µS | 15 V/µS | 1 V/µS | 0.5 V/µS |
| 8-pin DIL outline | b | a | c | c | b | b | b | b |

**Figure 2.5** *Parameter and outline details of eight popular 'single' op-amp types*

The LF351, 411, 441 and 13741 are JFET type op-amps with very high input impedances. The LF351 and 411 are high-performance types, while the LF441 and 13741 are general-purpose types that can be used as direct replacements for the popular 741. Note that the LF441 quiescent current consumption is less than one tenth of that of the 741.

## Linear amplifier circuits

*Figures 2.6* to *2.12* show a variety of ways of using op-amps to make practical linear amplifier circuits. Note that although type 741 op-amps are specified in the diagrams, any of the op-amp types listed in *Figure 2.5* can in fact be used in these circuits.

*Figure 2.6* shows the op-amp connected as an inverting ac amplifier with a ×10 overall voltage gain. Note that the pin-3 non-inverting input terminal is tied to ground via R3, which has the same value as R2 in order to preserve the dc balance of the op-amp.

*Figure 2.7* shows the op-amp wired as a non-inverting ac amplifier with an overall voltage gain of ×10 (= (R1+R2)/R2). Note that R1 and R2 are isolated from ground via C2; at normal operating frequencies C2 has negligible ac

26  Op-amp audio processing circuits

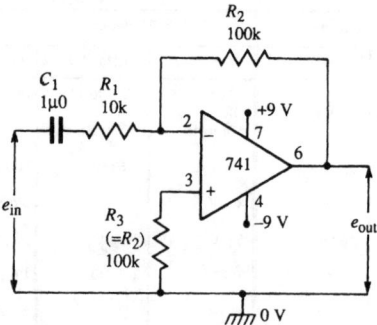

**Figure 2.6**  *Inverting ac amplifier with ×10 gain*

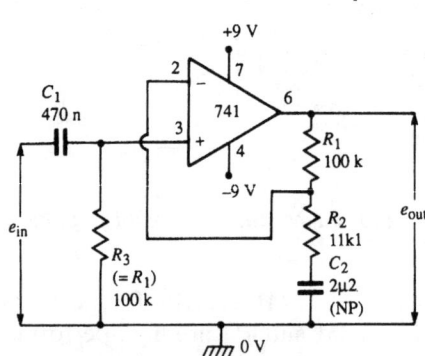

**Figure 2.7**  *Non-inverting ac amplifier with ×10 gain*

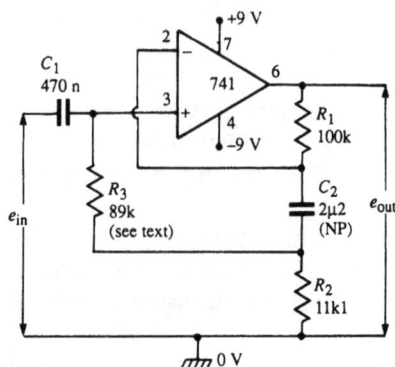

**Figure 2.8**  *Non-inverting ×10 ac amplifier with 50 megohm input impedance*

impedance, so ac voltage gain is determined by the R1 and R2 ratios, but the op-amp's inverting terminal is subjected to 100% dc negative feedback via R1, so the circuit has excellent dc stability. For optimum biasing, R3 has the same value as R1. The op-amp's input impedance, looking directly into pin-3, is several hundred megohms, but is shunted by R3, which reduces the circuit's input impedance to 100k.

*Figure 2.8* shows the above circuit modified to give a 50 megohm input impedance. Here, the C2 position is changed and the low end of R3 is taken to the C2–R2 junction rather than directly to ground. The ac feedback signal appearing on this junction is virtually identical to the pin-3 input signal, so near-identical signal voltages appear on both ends of R3, which thus passes negligible signal current. The apparent impedance of R3 is thus raised to near-infinity by this 'bootstrap' feedback action. In practice, the circuit's input impedance is limited to about 50 megohms by PCB leakage impedances. For optimum biasing, the sum of the R2 and R3 values should ideally equal R1, but in practice the R3 value can differ from the ideal by up to 30%, enabling an actual R3 value of 100k to be used if desired.

## Voltage follower circuits

An op-amp non-inverting ac amplifier will act as a precision ac voltage follower if wired to give unity voltage gain, and *Figure 2.9* shows some of the design possibilities of such a circuit, which has 100% negative feedback applied from output to input via R2. Ideally, R1 (which determines the circuit's input impedance) and R2 should have equal values, but in practice the R2 value can be varied from zero to 100k without greatly upsetting circuit accuracy. If a low $f_T$ op-amp (such as the 741) is used, the R2 value can

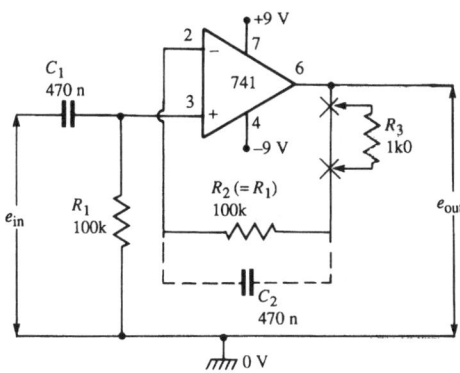

**Figure 2.9** *Ac voltage follower with 100k input impedance*

28  Op-amp audio processing circuits

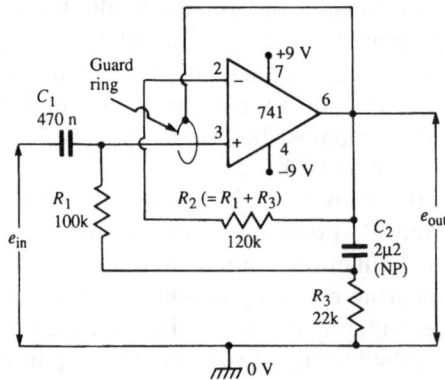

**Figure 2.10**  *Ac voltage follower with 50 megohm input impedance without the guard ring, or 500 megohm with the guard ring*

usually be reduced to zero; note, however, that 'high $f_T$' op-amps tend towards instability when used in the unity-gain mode, and in such cases stability can be assured by giving R2 a value of 1k0 or by replaced it with a 1k0 and 100k resistor wired in series (as shown in *Figure 2.9*), with a 470n capacitor wired across the 100k resistor to reduce its ac impedance.

If a very high input impedance is wanted from an ac follower it can be obtained by using the circuit of *Figure 2.10*, in which R1 is 'bootstrapped' from the op-amp output via C2, so that the R1 impedance is increased to near-infinity. In practice, this circuit gives an input impedance of about 50 megohms from a 741 op-amp, this limit being set by the leakage impedances of the op-amp's IC socket and the PCB.

If even greater input impedances are needed, the PCB area surrounding the op-amp input pin must be given a printed 'guard ring' that is

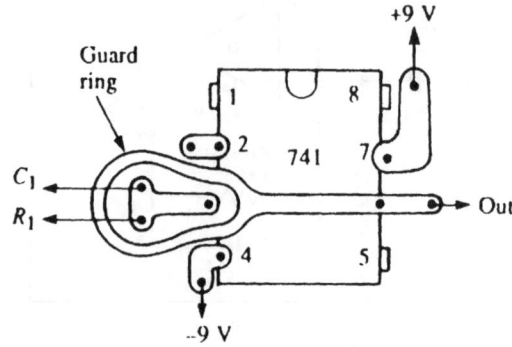

**Figure 2.11**  *Guard ring etched on a PCB and viewed through the top of the board*

*Op-amp audio processing circuits* 29

driven from the op-amp output, as shown, so that the leakage impedances of the PCB, etc., are also bootstrapped and raised to near-infinite values. In this case the *Figure 2.10* circuit will give an input impedance of about 500 megohms when using a 741 op-amp, or even higher if a FET-input op-amp is used. *Figure 2.11* shows an example of a guard ring etched on a PCB.

## An audio mixer

When describing the basic inverting amplifier circuit of *Figure 2.4(a)* it was pointed out that its voltage gain equals R2/R1; the signal currents flowing in R1 and R2 are thus always exactly equal (but are opposite in phase), irrespective of the R1 and R2 values. Thus, if this circuit is modified as shown in *Figure 2.12*, where four identical input networks are wired in parallel, the feedback signal current flowing in R6 will inevitable equal the sum of the input signal currents flowing in resistors R1 to R4, and the circuit's output signal voltage is thus proportional to the sum of the audio input signal voltages. When (as in the diagram) the input and feedback resistors have equal values, this circuit gives unity voltage gain between each input and the output.

It can thus be seen that this circuit actually functions as a unity-gain 4-input (or 4-channel) audio mixer that gives an output equal to the sum of the four input signal voltages. If desired, this simple circuit can be converted into a practical audio mixer by feeding each input signal to its input network via a 10k 'volume control' pot. If desired, the circuit can be made to give a voltage gain greater than unity by increasing the R6 value; the number of available input channels can be increased (or reduced) by adding (or deleting) one new C1–R1 network for each new channel.

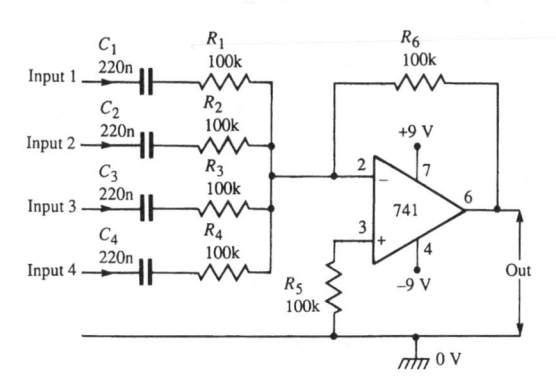

**Figure 2.12** *4-input audio mixer*

## Active filters

Filter circuits are used to reject unwanted frequencies and pass only those wanted by the designer. A simple R–C low-pass filter (*Figure 2.13(a)*) passes low-frequency signals, but rejects high-frequency ones. The output falls by 3dB at a 'break' or 'cross-over' frequency ($f_C$) of $1/(2\pi RC)$, and then falls at a rate of 6dB/octave (= 20dB/decade) as the frequency is increased (see *Figure 2.13(b)*). Thus, a 1kHz filter gives about 12dB of rejection to a 4kHz signal, and 20dB to a 10kHz one.

A simple R–C high-pass filter (*Figure 2.13(c)*) passes high-frequency signals but rejects low-frequency ones. The output is 3dB down at a break frequency of $1/(2\pi RC)$, and falls at a 6dB/octave rate as the frequency is decreased below this value (see *Figure 2.13(d)*). Thus, a 1kHz filter gives 12dB of rejection to a 250Hz signal, and 20dB to 100 Hz.

Each of the above two filter circuits uses a single R–C stage, and is known as a '1st-order' filter. If a number ($n$) of these filter stages could be simply cascaded, the resulting circuit would be known as an '$n$th order' filter and would have an output slope, beyond $f_C$, of ($n \times$ 6dB)/octave. Thus, a 4th-order 1kHz low-pass filter would have a slope of 24dB/octave, and would give 48dB of rejection to a 4kHz signal, and 80dB to a 10kHz signal.

Unfortunately, simple R–C filters cannot be directly cascaded, since they would then interact and give poor results; they can, however, be effectively cascaded by incorporating them into the feedback networks of

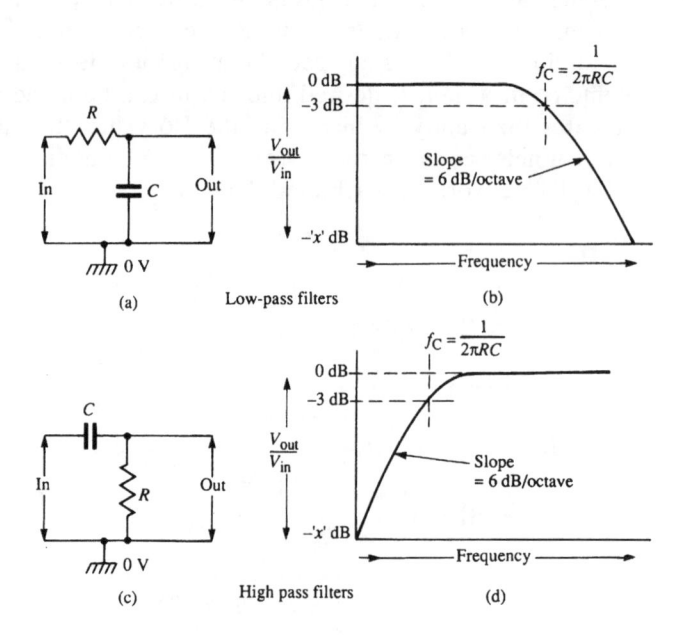

**Figure 2.13** *Circuits and response curves of simple 1st-order R–C filters*

Op-amp audio processing circuits 31

suitable op-amp circuits. Such circuits are known as 'active filters', and *Figures 2.14* to *2.20* show practical examples of some of them.

## Active filter circuits

*Figure 2.14* shows the practical circuit and formula of a maximally-flat (Butterworth) unity-gain 2nd-order low-pass filter with a 10kHz break frequency. To alter its break frequency, change either the R or the C value in proportion to the frequency ratio relative to *Figure 2.14*; reduce the values by this ratio to increase the frequency, or increase them to reduce the frequency. Thus, for 4kHz operation, increase the R values by a factor of 10kHz/4kHz, or 2.5 times.

A minor snag with the *Figure 2.14* circuit is that one of it's 'C' values should ideally be precisely twice the value of the other, and this can result in some rather odd component values. *Figure 2.15* shows an alternative 2nd-order

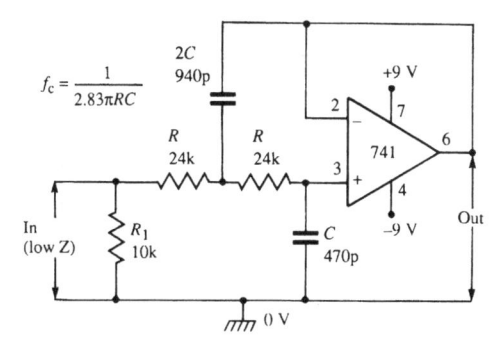

**Figure 2.14** *Unity-gain 2nd-order 10kHz low-pass active filter*

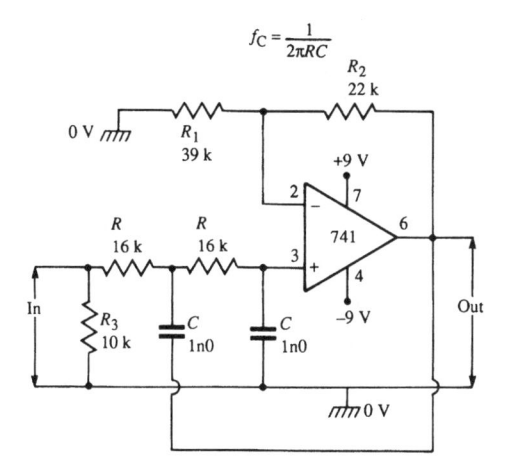

**Figure 2.15** *'Equal components' version of 2nd-order 10kHz low-pass active filter*

32  Op-amp audio processing circuits

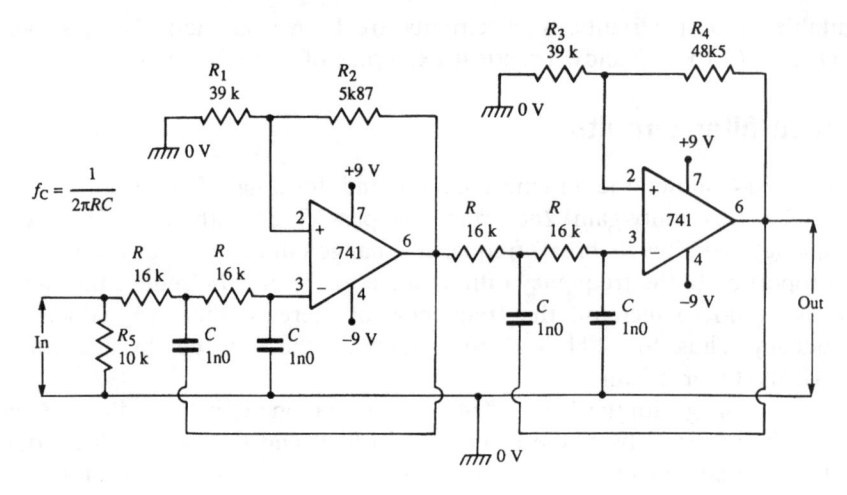

**Figure 2.16**  *4th-order 10kHz low-pass filter*

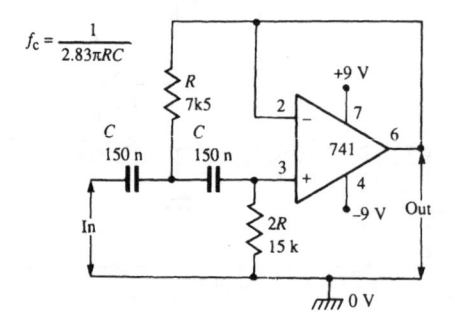

**Figure 2.17**  *Unity-gain 2nd-order 100Hz high-pass filter*

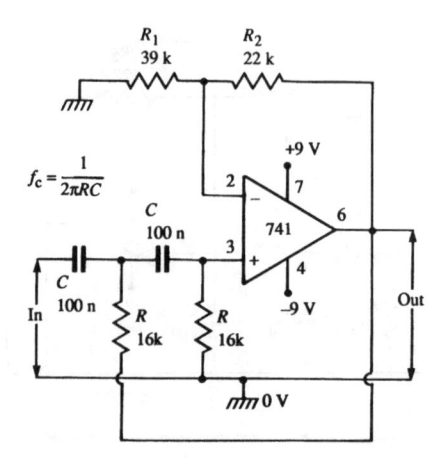

**Figure 2.18**  *'Equal components' version of 2nd-order 100Hz high-pass filter*

Op-amp audio processing circuits 33

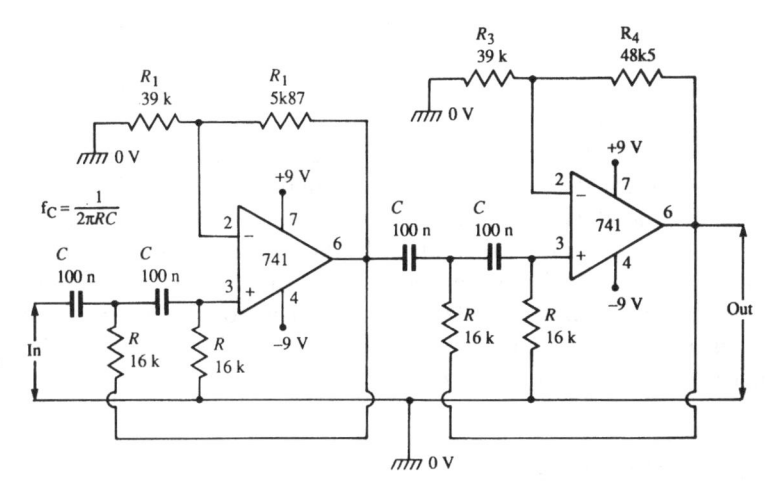

**Figure 2.19** *4th-order 100Hz high-pass filter*

10kHz low-pass filter circuit that overcomes this snag and uses equal component values. Note here that the op-amp is designed to give a voltage gain of 4.1dB via R1 and R2, which must have the values shown.

*Figure 2.16* shows how two of these 'equal component' filters can be cascaded to make a 4th-order low-pass filter with a slope of 24dB/octave. In this case gain-determining resistors R1/R2 have a ratio of 6.644, and R3/R4 have a ratio of 0.805, giving an overall voltage gain of 8.3dB. The odd values of R2 and R4 can be made by series-connecting standard 5% resistors.

*Figures 2.17* and *2.18* show unity-gain and 'equal component' versions respectively, of 2nd-order 100Hz high-pass filters, and *Figure 2.19* shows a

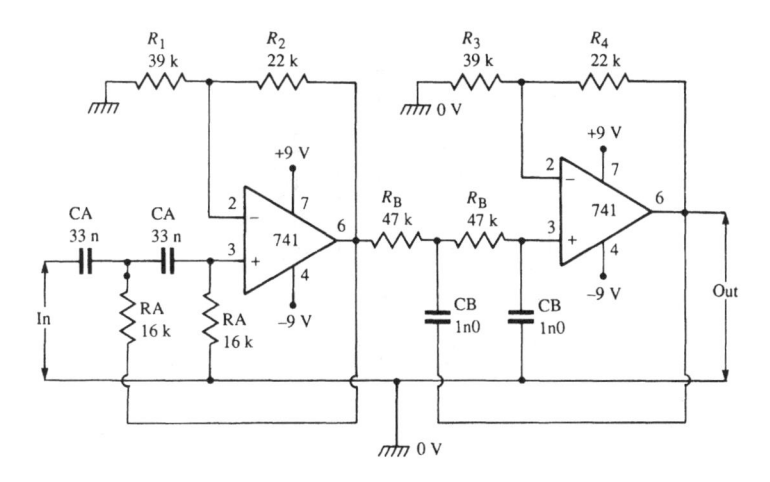

**Figure 2.20** *300Hz to 3.4kHz speech filter with 2nd-order response*

## 34  Op-amp audio processing circuits

4th-order high-pass filter. The operating frequencies of these circuits, and those of *Figures* 2.15 and 2.16, can be altered in exactly the same way as in *Figure* 2.14, i.e. by increasing the R or C values to reduce the break frequency, or vice versa.

Finally, *Figure* 2.20 shows how the *Figure* 2.18 high-pass and *Figure* 2.15 low-pass filters can be wired in series to make (with suitable component value changes) a 300Hz to 3.4kHz speech filter that gives 12dB/octave rejection to all signals outside of this range. In the case of the high-pass filter, the 'C' values of *Figure* 2.18 are reduced by a factor of three, to raise the break frequency from 100Hz to 300Hz, and in the case of the low-pass filter the 'R' values of *Figure* 2.15 are increased by a factor of 2.94, to reduce the break frequency from 10kHz to 3.4kHz.

## Variable active filters

The most useful type of active filter is that in which the cross-over frequency is fully and easily variable over a fairly wide range, and *Figures* 2.21 to 2.23 show three practical examples of 2nd-order versions of such circuits.

The *Figure* 2.21 circuit is a simple development of the high-pass filter of *Figure* 2.17, but has its cross-over frequency fully variable from 23.5Hz to 700Hz via RV1. The reader should note that this circuit can be used as a high-quality turntable disc (record) 'rumble' filter; 'fixed' versions of such filters usually have a 50Hz cross-over frequency.

The *Figure* 2.22 circuit is a development of the high-pass filter of *Figure* 2.14, but has its cross-over frequency fully variable from 2.2kHz to 24kHz via RV1. Note that this circuit can be used as a high quality 'scratch' filter; 'fixed' versions of such filters usually have a 10kHz cross-over frequency.

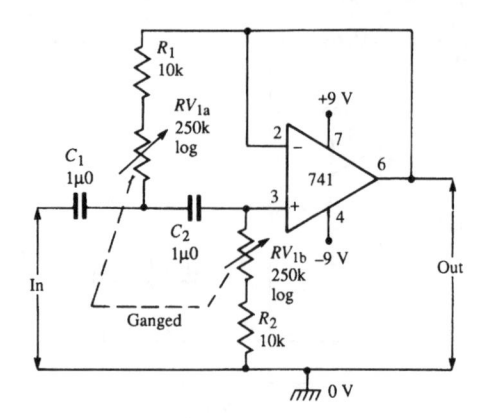

**Figure 2.21**  *Variable high-pass filter, covering 23.5Hz to 700Hz*

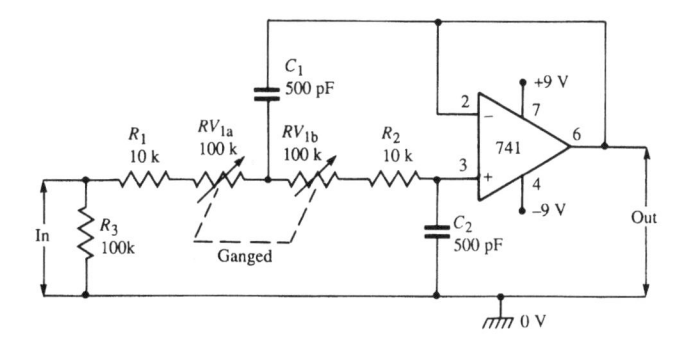

**Figure 2.22** *Variable low-pass filter, covering 2.2kHz to 24kHz*

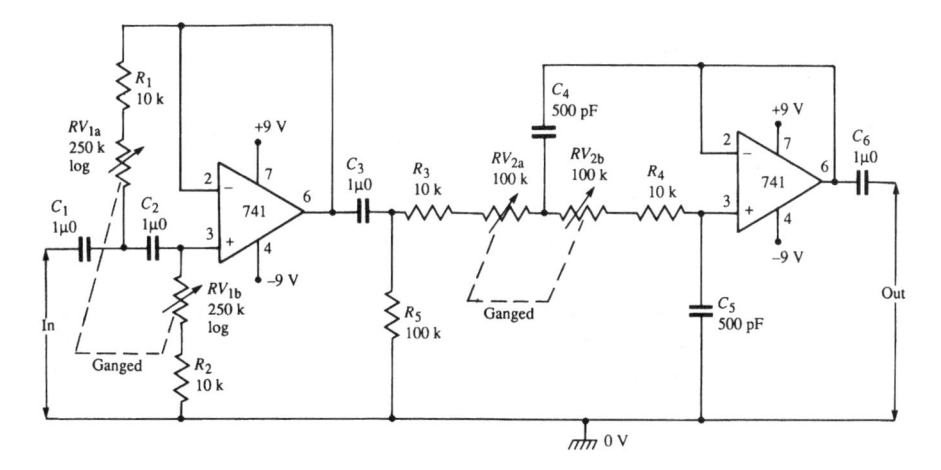

**Figure 2.23** *Variable high-pass/low-pass or rumble/scratch/speech filter*

*Figure 2.23* shows how the above two filter circuits can be combined to make a truly versatile variable high-pass/low-pass or rumble/scratch/speech filter. The high-pass cross-over frequency is fully variable from 23.5Hz to 700Hz via RV1, and the low-pass frequency is fully variable from 2.2kHz to 24kHz via RV2.

## Tone-control networks

The most widely used types of variable filter circuit are those used in audio tone-control applications. These allow the user to alter a system's frequency response to either suit his/her individual needs/moods, or to compensate for anomalies in room acoustics. Before looking at practical examples of such

36  Op-amp audio processing circuits

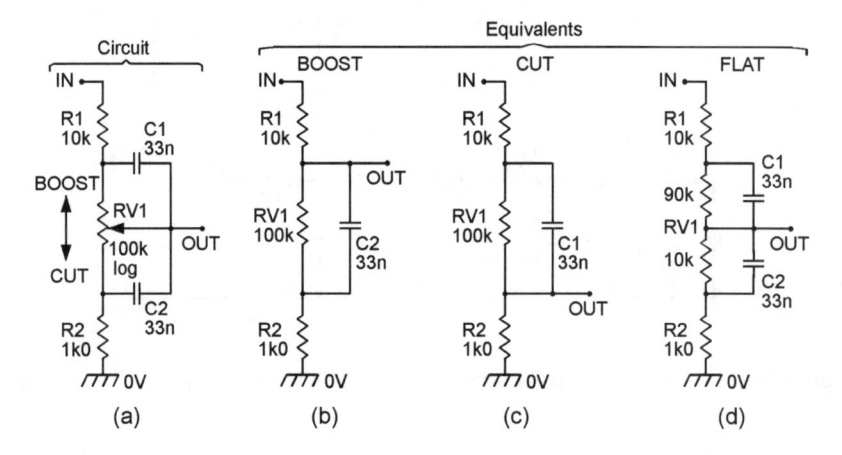

**Figure 2.24**  *Circuit and equivalents of BASS tone-control network*

circuits, it is best to look briefly at some basic tone-control networks, as follows.

*Figure 2.24(a)* shows the typical circuit of a passive bass tone-control network (which can be used to effectively boost or cut the low-frequency parts of the audio spectrum), and *Figures 2.24(b)* to *2.24(d)* show the equivalent of this circuit when RV1 is set to the maximum BOOST, maximum CUT, and FLAT positions, respectively. C1 and C2 are effectively open circuit when the frequency is at its lowest bass value, so it can be seen from *Figure 2.24(b)* that the BOOST circuit is equivalent to a 10k-over-101k potential divider, and gives only slight attenuation to bass signals.

The *Figure 2.24(c)* CUT circuit, on the other hand, is equal to a 110k-over-1k0 divider, and gives roughly 40dB of bass signal attenuation. Finally, when RV1 is set to the FLAT position shown in *Figure 2.24(d)* (with 90k of RV1 above the slider, and 10k below it), the circuit is equal to a 100k-over-11k divider, and gives about 20dB of attenuation at all frequencies. Thus, the circuit gives a maximum of about 20dB of bass boost or cut relative to the FLAT signals.

*Figure 2.25* shows the typical circuit of a passive treble tone-control network (which can be used to effectively boost or cut the high-frequency parts of the audio spectrum), together with its equivalent circuits under the maximum BOOST, maximum CUT, and FLAT operating conditions. This circuit gives about 20dB of signal attenuation when RV1 is in the FLAT position, and gives maximum treble boost or cut values of 20dB relative to the FLAT performance.

*Figure 2.26* shows how the *Figure 2.24(a)* and *2.25(a)* circuits can be combined to make a complete passive Bass and Treble tone-control network;

Op-amp audio processing circuits 37

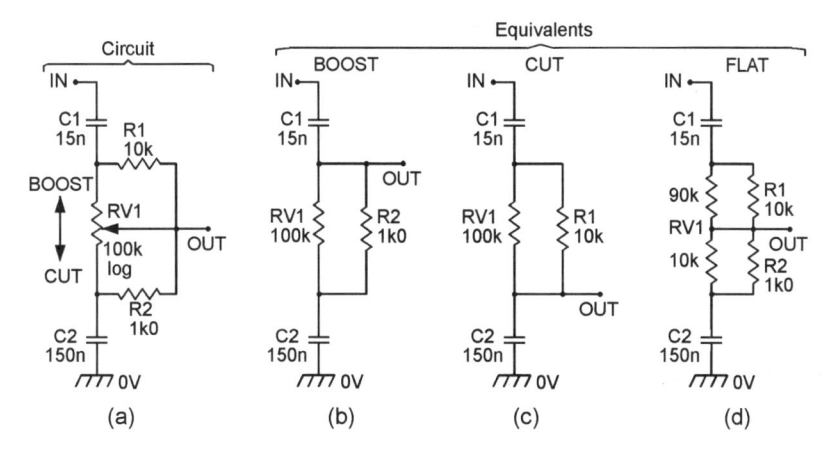

**Figure 2.25** Circuit and equivalents of TREBLE tone-control network

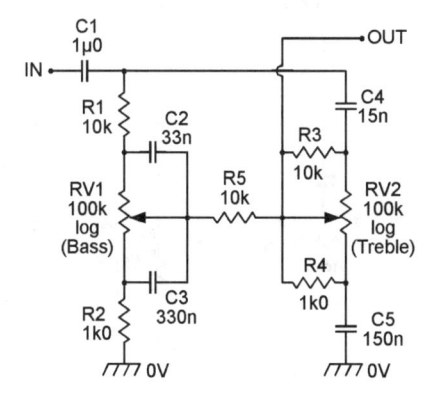

**Figure 2.26** Passive BASS and TREBLE tone-control network

10k resistor R5 helps minimize unwanted interaction between the two circuit sections. This tone-control circuit can be interposed between an amplifier's volume control and the input of its power amplifier stage.

## Active tone controls

An active tone-control circuit can easily be made by wiring a passive tone-control network into the negative feedback loop of an op-amp linear amplifier, so that the system gives an overall signal gain (rather than attenuation) when its controls are in the FLAT position. Such networks can take the form of simplified versions of the basic *Figure 2.26* circuit, but more often are based on the alternative passive tone-control circuit shown in *Figure 2.27*,

## 38 Op-amp audio processing circuits

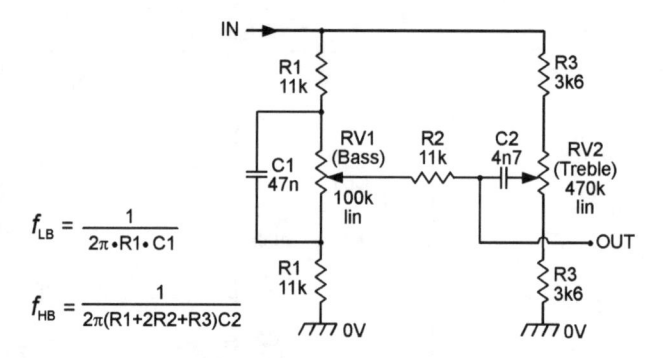

$$f_{LB} = \frac{1}{2\pi \cdot R1 \cdot C1}$$

$$f_{HB} = \frac{1}{2\pi(R1+2R2+R3)C2}$$

**Figure 2.27** *Alternative tone-control circuit*

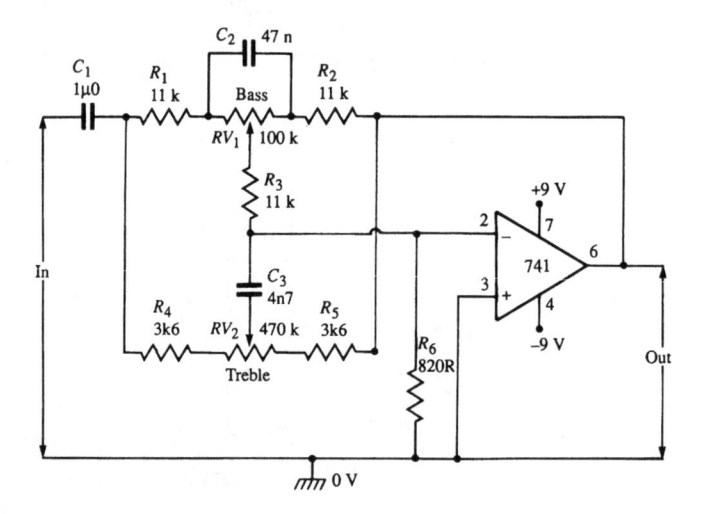

**Figure 2.28** *Active tone-control circuit*

which gives a similar performance but uses fewer components and uses linear control potentiometers (pots).

Looking at *Figure 2.27*, it can be seen that at very low frequencies (when the two capacitors act like open circuits) the output signal amplitudes are controlled entirely by RV1 (since RV2 is isolated from the output via C2), but that at high frequencies (when the two capacitors act like short circuits) the output signal amplitudes are controlled entirely by RV2 (since RV1 is shorted out via C1). The low-frequency (BASS) break point of the circuit is determined by the R1–C1 values, and the high-frequency (TREBLE) break point is determined by C2 and the values of R1 to R3.

*Figure 2.28* shows how to use the above network to make a practical active tone-control circuit that can give up to 20dB of boost or cut to bass or treble signals. This is an excellent high-quality design.

Op-amp audio processing circuits  39

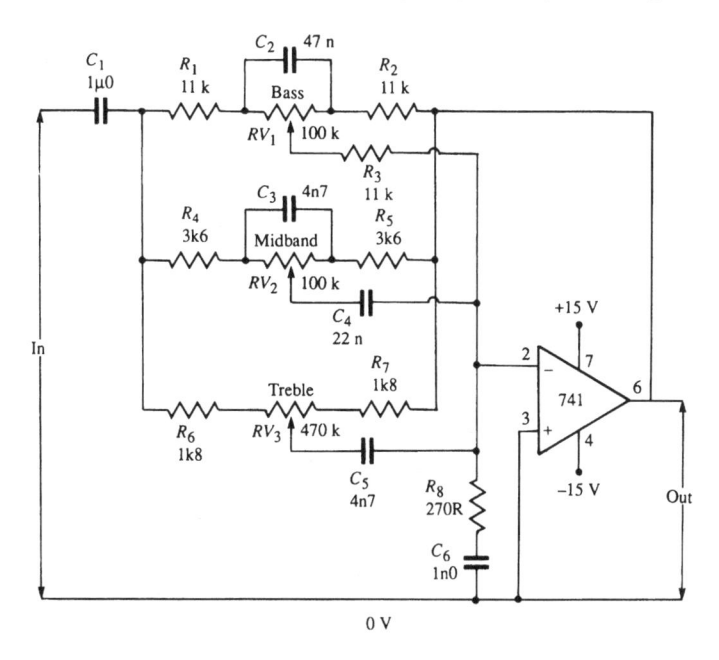

**Figure 2.29** *Three-band (BASS, MIDBAND, TREBLE) active tone-control circuit*

An even more useful circuit is shown in *Figure 2.29*. This design is similar to the above, but has an additional filter control network that is centred on the 1kHz 'midband' part of the spectrum, thus enabling this part of the audio band to also be boosted or cut by up to 20dB.

## Graphic equalizers

The ultimate (most sophisticated) type of tone-control system is the so-called graphic equalizer. This consists of a number of parallel-connected overlapping narrow-band variable-response filters that cover the entire audio spectrum, thus enabling an amplifier system's spectral response to be precisely adjusted to suit individual needs. Usually, the filter centre frequencies are spaced at one octave intervals, and such systems are thus also known as octave equalizers.

*Figure 2.30* shows the basic circuit of a typical octave (graphic) equalizer section. This circuit is in fact very similar to that of the *Figure 2.28* active tone control, except that the C2–R2 'treble control' network is fixed, rather than variable, and the bass and treble break frequencies are fairly closely spaced, so that the two response curves overlap. The net effect of this is that the *Figure 2.30* circuit acts as a narrow-band filter which has a centre-frequency response that is fully adjustable between +12dB (full boost) and –12dB (full cut) via RV1.

40  Op-amp audio processing circuits

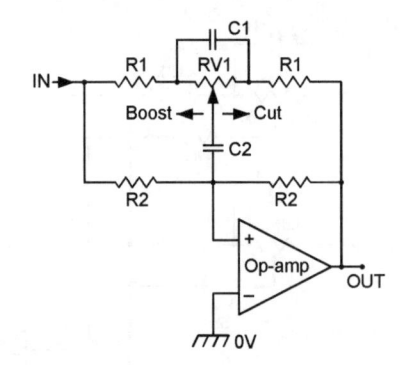

**Figure 2.30**  *Typical octave (graphic) equalizer section*

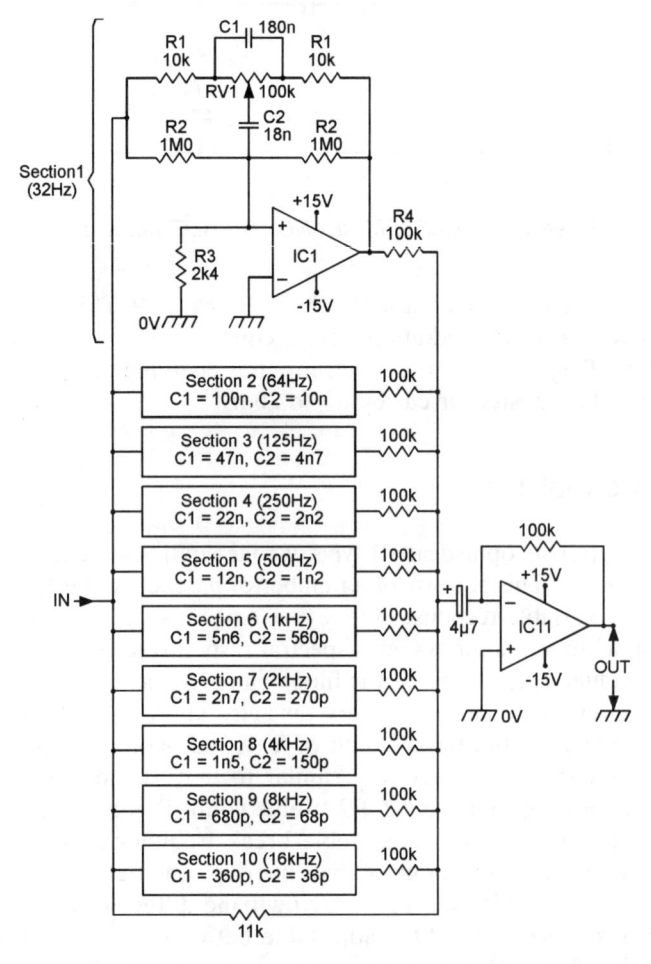

**Figure 2.31**  *Ten-band octave (graphic) equalizer circuit*

*Figure 2.31* shows how ten of the above circuits can be interconnected to make a practical high-quality ten-band graphic equalizer; the ten equaliser sections are wired in parallel, and their outputs are added together in the IC11 output stage. The ICs used here can be type-741 'single' or 'quad' op-amps. Two complete *Figure 2.31* circuits are needed in a normal stereo amplifier system.

## RIAA equalization

Three types of phonograph record (disc) pickup are in general use, these being the ceramic, crystal, and magnetic types. The first two of these are fairly cheap types which give large amplitude outputs, have a reasonably linear frequency response, and are mainly used in low-fi to mid-fi equipment. The third (magnetic) type, on the other hand, gives a low amplitude output and has a non-linear frequency response, but is widely used on good quality hi-fi equipment.

If the reader were to take a test disc (phonograph record) on which a constant amplitude 20Hz to 20kHz three-decade span of sinewave tone signals had been recorded with perfect linearity (constant signal amplitude in the recording groves), and were to then play that disc through a good quality magnetic pickup, he/she would find that it generates a non-linear frequency response that rises at a rate of 6dB per octave (= 20dB per decade). Thus, output signals would be very weak at 20Hz, but would be one thousand times greater (= +60dB) at 20kHz. This non-linear frequency response is inherent in all magnetic pickups, since their output voltage is directly proportional to the rate of movement of the pickup needle, which in turn is proportional to the recording frequency.

In practice, disc recording equipment does not give an exactly linear frequency response. To help enhance the effective dynamic range and signal-to-noise ratio performance of discs, frequencies below 50Hz and those in the 500Hz to 2.12kHz midband range are recorded in a non-linear fashion that is precisely defined by Record Industry Association of America (RIAA) standards. This non-linearity is such that it causes a midband drop of 12dB when played through linear-response ceramic or crystal pickups, but this modest decrease is too small to be objectionable in most lo-fi to mid-fi playback equipment.

When a practical RIAA test frequency disc is played through a magnetic pickup, the pickup produces the frequency response curve shown in *Figure 2.32*. Here, the dashed line shows the idealized shape of this curve, which is flat up to 50Hz, then rises at a 6dB/octave rate to 500Hz, then is flat to 2120Hz, and then rises at a 6dB/octave rate beyond that. The solid line shows the practical shape of the curve.

The really important thing that the reader should note from all this is that when a disc is played through a magnetic pickup in a good quality hi-fi

42  Op-amp audio processing circuits

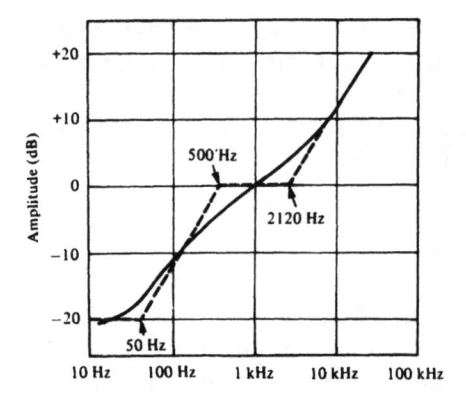

**Figure 2.32** *Typical phono disc playback frequency response curve*

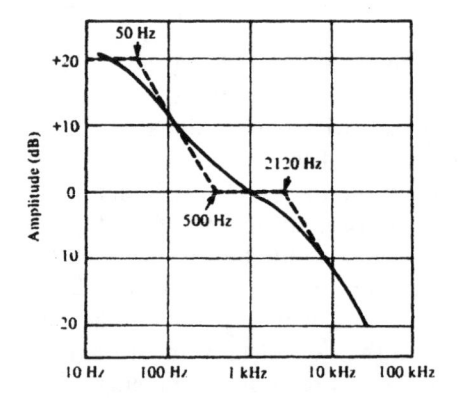

**Figure 2.33** *RIAA playback equalization curve*

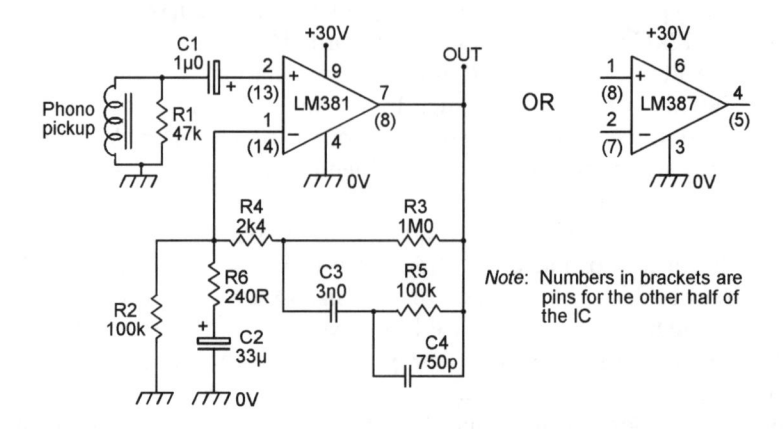

**Figure 2.34** *LM381 (or LM387) low-noise phono pre-amp (RIAA)*

system, the output of the pickup must be passed to the power amplifier circuitry via a pre-amplifier that has a frequency equalization curve that is the exact inverse of that shown in *Figure 2.32*, so that a linear overall record-to-replay response is obtained. *Figure 2.33* shows the actual form of the necessary RIAA equalization curve, and *Figure 2.34* shows a practical example of a modern low-noise phono pre-amplifier with integral RIAA magnetic-pickup equalization.

## RIAA phono pre-amp

Magnetic pickups are low-sensitivity devices, and give typical midband outputs of only a few millivolts. Consequently, their outputs must be passed to main amplifiers via dedicated low-noise pre-amplifier ICs (rather than via simple op-amps), and the *Figure 2.34* circuit is thus designed around LM381 or LM387 ICs of this type. These are 'dual' ICs, and in the diagram the pin numbers of one half of the IC are shown in plain numbers, and the other half are shown in brackets. Two of these pre-amp circuits are needed in a stereo audio system, and these can thus be obtained from one 'dual' IC.

The operating theory of the *Figure 2.34* circuit is fairly simple. The IC is wired as a non-inverting amplifier, with negative feedback applied from the output to the inverting input pin; potential divider (R3+R4) – R2 sets the circuit's dc biasing, and the R5–C3–C4–R4 and R6–C2 networks determine the ac signal gain. At the 1kHz midband frequency, C2 and C3 have low impedances and C4 has a high impedance, so the ac gain is determined mainly by R5/R6, and equals ×400. At lower frequencies the impedance of C3 starts to become significant and causes the ac gain to increase until eventually, at very low frequencies, it is limited to ×4000 by the R3/R6 ratio. At high frequencies, on the other hand, the impedance of C4 falls to significant levels and shunts R5, thus causing the ac gain to decrease until eventually, at very high frequencies, it is limited to ×10 by the R4/R6 ratio. The pickup signals are ac coupled to the IC via C1, and the circuit can be used with any type of magnetic pickup unit.

## Non-linear amplifiers

An op-amp circuit can be made to act as a non-linear amplifier by simply incorporating a non-linear device in its negative feedback network, as shown in *Figure 2.35*, where the feedback elements comprise a pair of silicon diodes connected back-to-back. When very small signals are applied to this circuit the diodes act like very high resistances, so the circuit gives high voltage gain, but when large signals are applied the diodes act like low resistances, so the

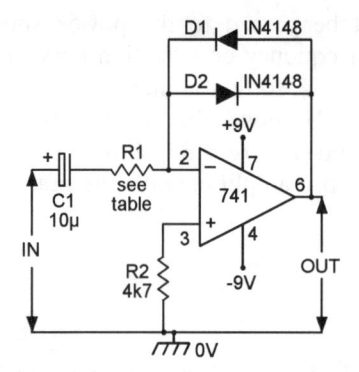

## 44  Op-amp audio processing circuits

| $V_{in}$ (rms) | R1 = 1k0 |  | R1 = 10k |  |
|---|---|---|---|---|
|  | $V_{out}$ (rms) | $V_{gain}$ | $V_{out}$ (rms) | $V_{gain}$ |
| 1mV | 110mV | x110 | 21mV | x21 |
| 10mV | 330mV | x33 | 170mV | x17 |
| 100mV | 450mV | x4.5 | 360mV | x3.6 |
| 1V | 560mV | x0.56 | 470mV | x0.47 |
| 10V | 600mV | x0.07 | 560mV | x0.056 |

**Figure 2.35**  *Circuit and performance table of non-linear (semi-log) amplifier*

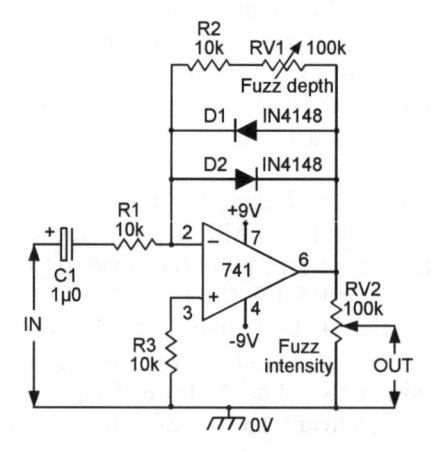

**Figure 2.36**  *'Fuzz' circuit*

circuit gives low gain. The gain in fact varies in a semi-logarithmic fashion, and circuit sensitivity can be varied by altering the R1 value; the table shows actual performance details. Note that a 1000:1 change in input signal amplitude can cause as little as a 2:1 change in output level, so, by taking its output to a simple ac millivoltmeter, this circuit can be used as a single-range bridge-balance detector or signal strength indicator.

When this circuit is fed with a sinewave input its two diodes clip the output to about 1.4V peak-to-peak, giving an almost square output that is rich in odd harmonics, and which, when fed through an amplifier, sounds like a clarinet; in the music world this is known as a 'fuzz' effect. *Figure 2.36* shows how the circuit can be modified to make a practical 'fuzz' generator; RV1 controls the level at which fuzz clipping begins, and RV2 controls the circuit's output level or fuzz intensity.

## Constant-volume amplifier

The non-linear amplifier of *Figure 2.35* gives a near-constant-amplitude output signal over a wide range of input levels, but does so by causing heavy amplitude distortion of the signal. *Figure 2.37* shows an alternative type of 'constant-volume' amplifier in which amplitude control is obtained without generating signal distortion; this is achieved by using a voltage-controlled linear element (rather than a non-linear element) in its negative feedback loop.

Here, the op-amp is wired as an ac amplifier with its gain controlled by the R2/R1 ratio and by the ac potential divider formed by R4 and the drain impedance of field-effect transistor Q1, which is used as a voltage-controlled resistor with its control voltage derived from the op-amp output via the D1–R5–R6–C3 network, which generates a voltage proportional to the mean value (integrated over hundreds of milliseconds) of the output signal. With zero bias applied to Q1 gate the FET acts like a low resistance (a few hundred ohms), but with a large negative bias voltage applied it acts like a high resistance (a few megohms).

Thus, when a very small input signal is applied, the op-amp's output also tends to be small, so D1–R5–C3 feed near-zero negative bias to the gate of the FET, which thus acts like a resistance of only a few hundred ohms. Under this condition the R4–Q1 divider causes very little negative feedback to be applied to the op-amp, which thus gives a high voltage gain. When, however, a large input signal is applied to the op-amp, the op-amp's output tends to be large, so a large negative bias is developed on the gate of the FET, which thus acts like a very high resistance and causes heavy negative feedback to be applied to the op-amp, which thus gives very low voltage gain. The net effect of this action is that the mean level of the output tends to self-regulate at 1.5V to 2.85V over a 50:1 (500mV to 10mV) range of input signal levels, and does so without generating significant distortion. The R1 value determines the

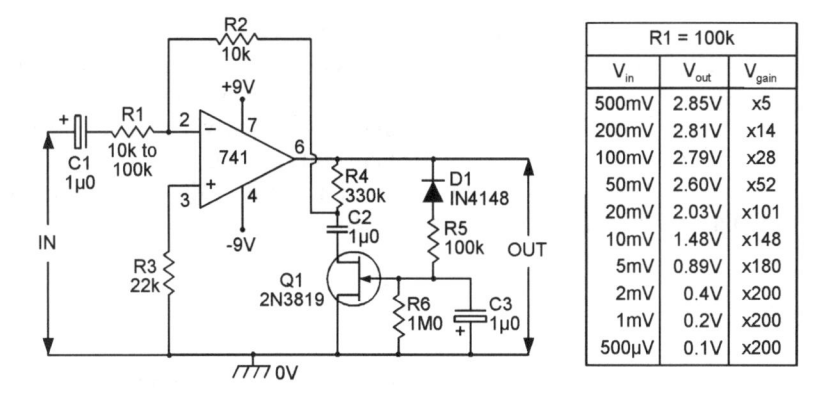

| R1 = 100k | | |
|---|---|---|
| $V_{in}$ | $V_{out}$ | $V_{gain}$ |
| 500mV | 2.85V | x5 |
| 200mV | 2.81V | x14 |
| 100mV | 2.79V | x28 |
| 50mV | 2.60V | x52 |
| 20mV | 2.03V | x101 |
| 10mV | 1.48V | x148 |
| 5mV | 0.89V | x180 |
| 2mV | 0.4V | x200 |
| 1mV | 0.2V | x200 |
| 500µV | 0.1V | x200 |

**Figure 2.37** *Circuit and performance table of constant-volume amplifier*

46  Op-amp audio processing circuits

sensitivity of the circuit, and is selected to suit the maximum input signal amplitude that the circuit is expected to handle, on the basis of 200k per r.m.s. volt of input signal; thus, for a maximum input of 50V R1 has a value of 10M, and for 50mV it has a value of 10k. C3 determines the VCA time constant, and its value can be altered to suit individual needs

## OTA devices

All but one of the circuits described so far in this chapter are designed around conventional voltage-in–voltage-out op-amps. There is another type of op-amp, however, that can also be used in many audio processing applications, but uses a voltage-in–current-out (transconductance) form of operation in which the gain is externally variable via one control terminal; such devices are known as operational transconductance amplifiers, or OTAs.

*Figure 2.38* shows the standard circuit symbol and basic operating formula of the OTA. The device has conventional differential voltage input terminals, which accept inputs $e1$ and $e2$, and gives an output *current* that equals the difference between these signals multiplied by the OTA's transconductance or '$gm$' value, which in turn typically equals 20 times the external $I_{bias}$ bias current value. Thus, the gain can be controlled by the bias current which, in practice, can easily be varied over a 10 000:1 range.

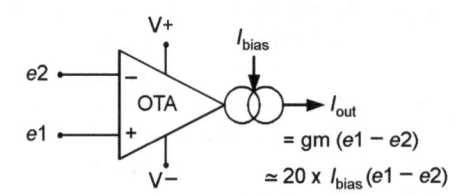

$$I_{out} = gm\,(e1 - e2)$$
$$\approx 20 \times I_{bias}\,(e1 - e2)$$

**Figure 2.38** Symbol and basic formulae of a conventional OTA

OTAs are reasonably versatile devices. They can be made to act like voltage-in–voltage-out op-amps by simply feeding their output current into a load resistor, thus converting the output current into a voltage, and their gain can be voltage-controlled by applying the voltage via a series resistor, thus converting the voltage into a control current. By using these techniques, the OTA can be used as a voltage-controlled amplifier (VCA), as an amplitude modulator, or as a ring modulator or four-quadrant multiplier.

## Practical OTAs

The two best known versions of the OTA are the CA3080 and the LM13600. The CA3080 is a simple first-generation device that can accept bias currents

Op-amp audio processing circuits 47

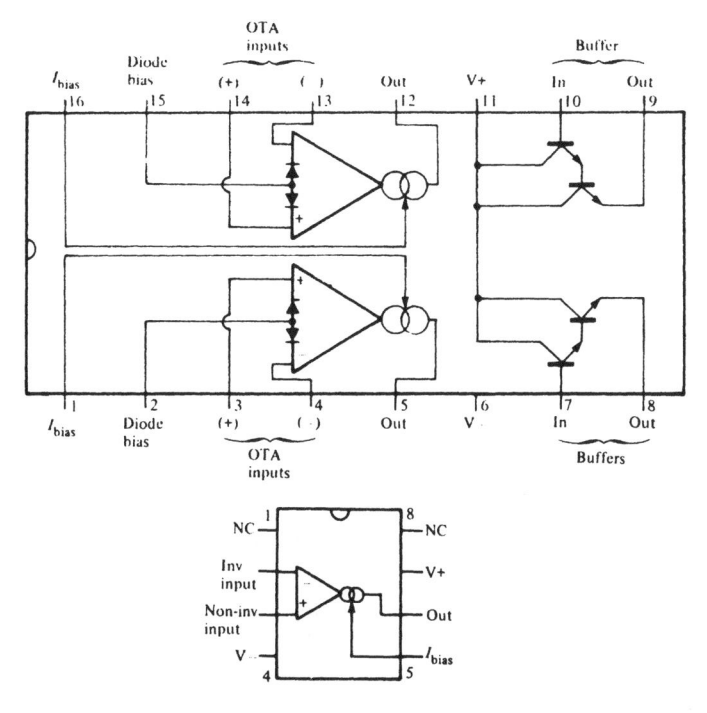

**Figure 2.39** *Outline and pin notations of the CA3080 (lower) and LM13600 (upper) OTAs*

in the range 100nA to 1mA and can operate from split power supplies in the 2V to 15V range; the device is housed in an 8-pin DIL package with the outline and pin notation shown in *Figure 2.39*. A minor defect of this IC is that it generates a significant amount of signal distortion.

The LM13600 is an improved second-generation version of the OTA, and incorporates linearizing input diodes that greatly reduce signal distortion, plus a coupled output buffer stage that can be used to give a low-impedance output. The LM13600 is actually a dual OTA, and *Figure 2.39* also shows the circuit symbol, IC outline, and pin notations of this device, which is housed in a 16-pin DIL package.

## Basic OTA circuits

The CA3080 and LM13600 OTAs are very easy ICs to use, and in this section they are shown used in basic fixed-gain ac amplifier applications. Looking first at the CA3080, its pin-5 $I_{bias}$ terminal is actually connected to the pin-4 negative supply rail via an internal base–emitter junction, so the biased voltage of pin-5 is about 600mV above the pin-4 voltage. $I_{bias}$ can thus be

48   Op-amp audio processing circuits

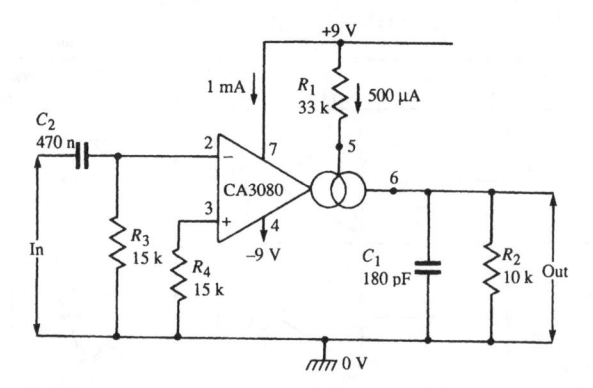

**Figure 2.40**   Ac-coupled 40dB inverting amplifier

obtained by simply connecting pin-5 to either the common rail or the positive supply rail via a current-limiting resistor of suitable value.

*Figure 2.40* shows a simple but instructive way of using the CA3080 as an ac-coupled inverting amplifier with a voltage gain of about 40dB. The circuit is operated from split 9V supplies, so 17.4V are generated across bias resistor R3, which thus feeds roughly 500µA into pin-5 and thus causes the IC to consume 1mA (20 times $I_{bias}$) from its supply rails.

At a bias current of 500µA the *gm* (forward transconductance) of the CA3080 is about 10mmho (a mho is a standard unit of conductance). The output of the *Figure 2.40* circuit is loaded by a 10k resistor (R4), and thus gives an overall voltage gain of 10mmho × 10k = ×100, or 40dB. The peak current that can flow into this 10k load is 500µA (= $I_{bias}$), so the peak available output voltage is plus or minus 5V. The output is also loaded by 180pF capacitor C2, which limits the output slew rate to about 2.8V/µs.

Each of the two OTA devices housed in the LM13600 package can, if desired, be used in exactly the same way as the CA3080 shown in *Figure 2.40*. One of the main features of the LM13600, however, is that it incorporates linearizing diodes that help reduce signal distortion, so in practice each half of the device is best used as an inverting ac amplifier by wiring it as shown in *Figure 2.41*, where the two input diodes are biased, via R3, with current $I_D$, which flows to ground via R2 and R4.

Note in this circuit that the input signal is applied to the non-inverting input terminal via R1, and that R1 and R2 form a voltage divider that attenuates the input signal, and that the circuit consequently gives an overall voltage gain of slightly less than unity. The gain of this circuit is in fact proportional to the value of $I_{bias}/I_D$, and can thus be varied by altering the value of either $I_{bias}$ or $I_D$.

The above circuit can be made to act as a non-inverting amplifier by modifying the input circuitry as shown in *Figure 2.42*, which also shows how

Op-amp audio processing circuits 49

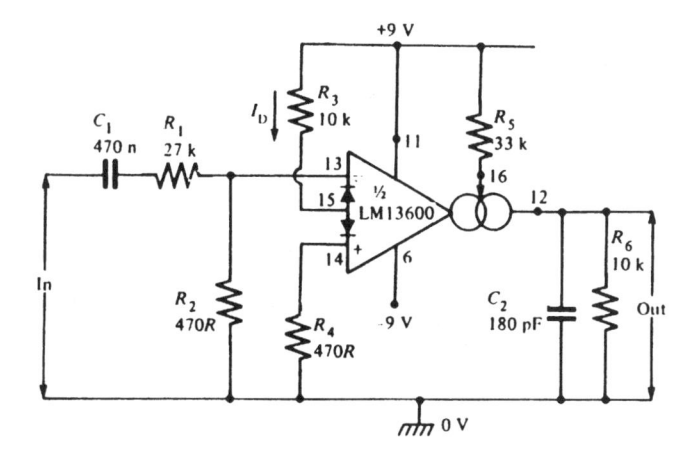

**Figure 2.41** *Inverting ac amplifier with near-unity overall voltage gain*

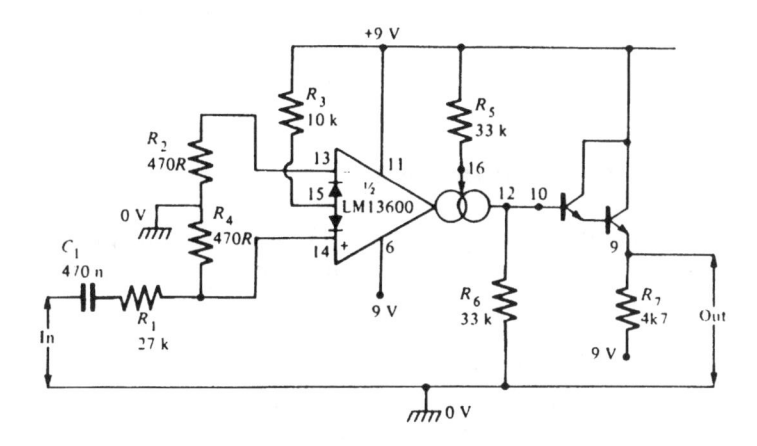

**Figure 2.42** *Non-inverting ac amplifier with buffered output*

it can be made to give a low-impedance output by feeding the OTA output to the outside world via one of the IC's internal buffer amplifiers; this modification enables the R6 value to be increased to 33k, with a consequent increase in overall voltage gain.

The graph of *Figure 2.43* shows the typical signal distortion figures obtained from the LM13600 when used with and without the internal linearizing diodes. With an input terminal signal of 20mV peak-to-peak the device generates less than 0.02% distortion with the diodes, but about 0.3% without them; these figures rise to 0.035% and 1.5%, respectively, at 40mV input.

50  Op-amp audio processing circuits

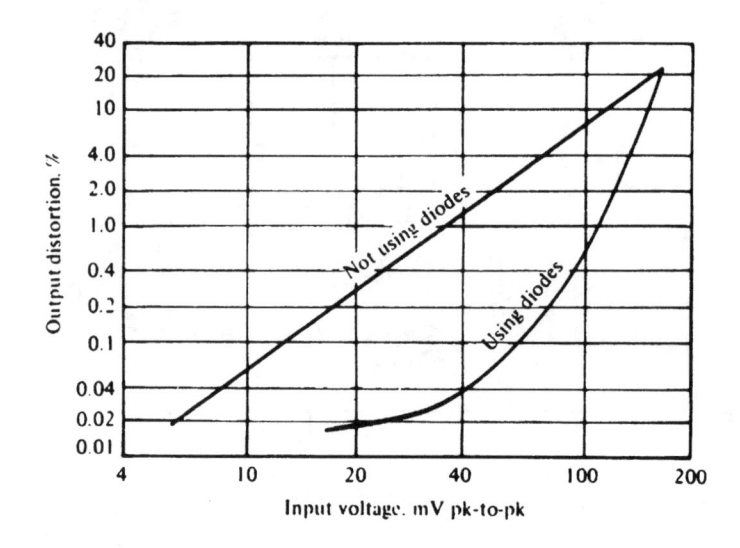

**Figure 2.43**  *Typical distortion levels of the LM13600 OTA, with and without the use of the linearizing diodes*

## CA3080 variable-gain circuits

*Figure 2.44* shows how the basic *Figure 2.40* inverting amplifier circuit can be modified so that its gain is variable from ×5 to ×100 via RV2, which enables $I_{bias}$ to be varied from 12.4µA to 527µA. In this type of application the input bias levels of the IC must be balanced so that the output dc level does not shift as the gain is varied, and this is achieved via offset-nulling preset pot RV1. To set up the circuit, set RV2 to its minimum (maximum gain) value, and then trim RV1 to give zero dc output.

The *Figure 2.44* circuit can be converted into a voltage-controlled amplifier by removing R4 and RV2 and connecting the $V_{IN}$ voltage-control input to pin-5 of the CA3080 via a 33k series resistor. In this case the circuit gives a gain of ×100 when $V_{IN}$ equals the positive supply rail value, and near-zero gain when $V_{IN}$ is 600mV above the negative supply rail value. Thus, to give the full range of gain control, $V_{IN}$ must be referenced to the negative supply rail.

*Figure 2.45* shows a more useful type of VCA, in which $V_{IN}$ is referenced to the common (zero) supply rail. Here, Q1 and the 741 op-amp form a linear voltage-to-current converter (with a 100µA/V conversion rate) which responds to positive $V_{IN}$ values only. Thus, when $V_{IN}$ equals zero or less, the VCA gives near-zero gain, but when $V_{IN}$ equals 5V it gives a basic gain of ×100. Note that Rx is shown wired in series with the signal input line; Rx

Op-amp audio processing circuits  51

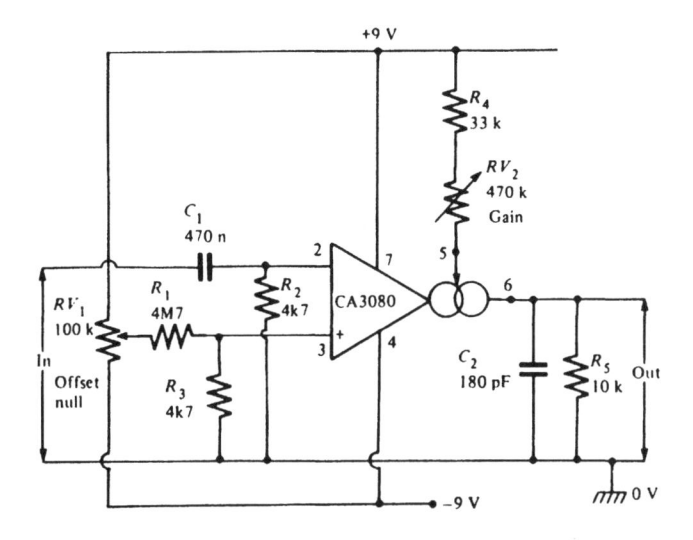

**Figure 2.44** Variable gain (×5 to ×100) ac amplifier

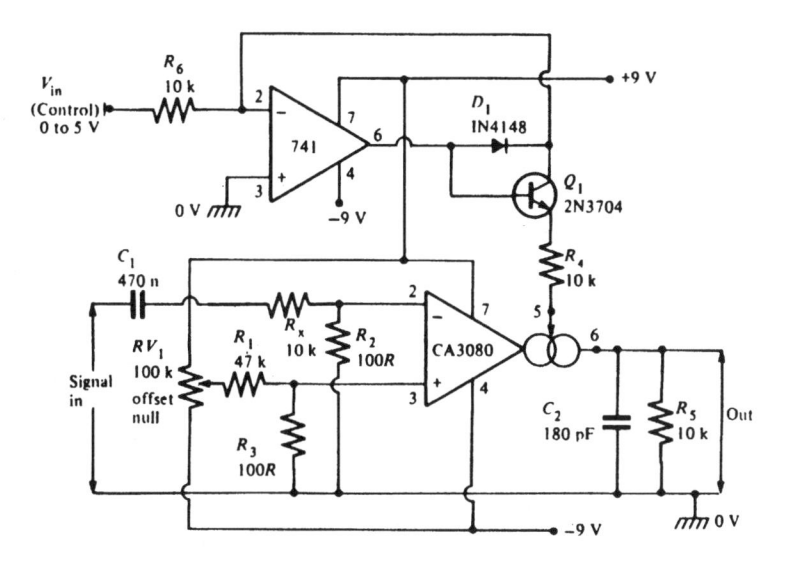

**Figure 2.45** Voltage-controlled amplifier (VCA)

and R2 actually form a potential divider that reduces the pin-2 input signal amplitude (and thus the overall voltage gain) of the circuit; the Rx value should be chosen to limit the pin-2 signal voltage to a maximum of 20mV pk-to-pk, to minimize signal distortion.

## LM13600 variable-gain circuits

The LM13600 OTA can be used (with or without its linearizing diodes) in any of the basic variable-gain amplifier configurations described above. *Figure 2.46*, for example, shows it used with linearizing diodes in the VCA

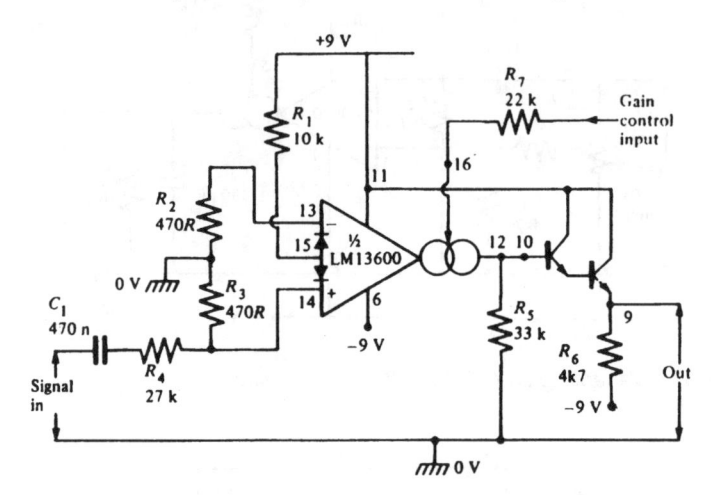

**Figure 2.46** *Voltage-controlled amplifier*

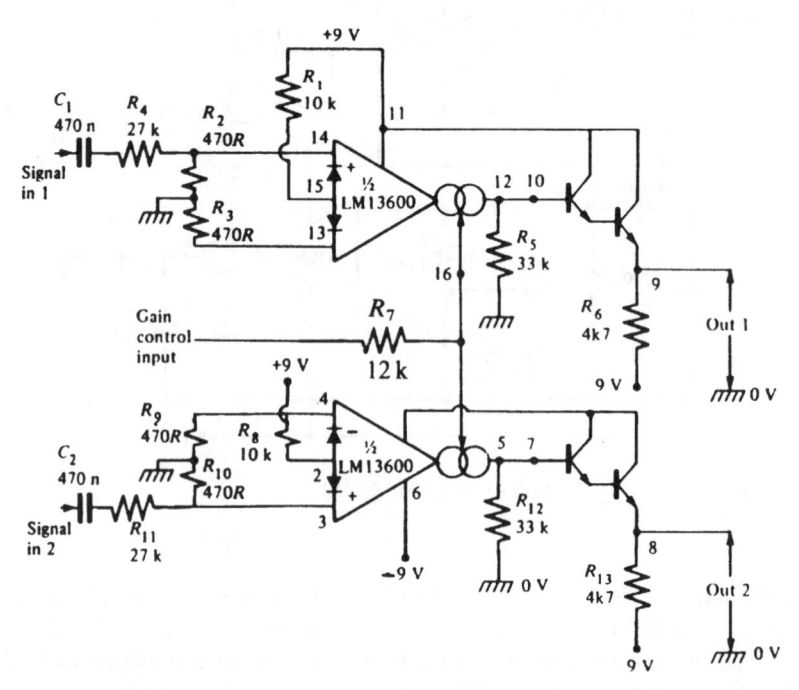

**Figure 2.47** *Stereo VCA*

Op-amp audio processing circuits 53

configuration in which the gain-control voltage is referenced to the negative supply rail; this circuit gives a gain of ×1.5 when $V_{IN}$ equals the positive supply rail value, and gives an attenuation of 80dB when $V_{IN}$ equals the negative supply rail value.

*Figure 2.47* shows how two of the above circuits can be joined together to make a stereo VCA unit that is controlled via a single input voltage, which may be derived from a 'volume control' pot wired between the two supply rails, in which case a 10μF capacitor can be wired across the lower half of the pot so that the circuit acts as a 'noiseless' volume control system.

## Amplitude modulation

A VCA circuit can be used as an amplitude modulator (AM) circuit by feeding a carrier signal to its input terminal and using a modulating signal to control the output amplitude via the gain-control input terminal. *Figure 2.48* shows a CA3080 used in a dedicated version of such a circuit, and *Figure 2.49* shows an LM13600 version of the same basic design.

The *Figure 2.48* circuit acts as an inverting amplifier; its dc gain is set via R4 and R6, but its ac gain is variable via signals applied to C2. Input bias resistors R1 and R2 have low values to minimize the IC's noise levels and enhance stability; offset biasing is applied via R3–RV1. The carrier input signal is applied to pin-2 via potential divider Rx–R1; when Rx has the value shown, the circuit gives near-unity overall voltage gain with zero modulation input; the gain doubles when the modulation terminal swings to +9V, and falls to near-zero (actually –80dB) when the terminal swings to –9V. The *Figure 2.49* circuit acts in a similar way.

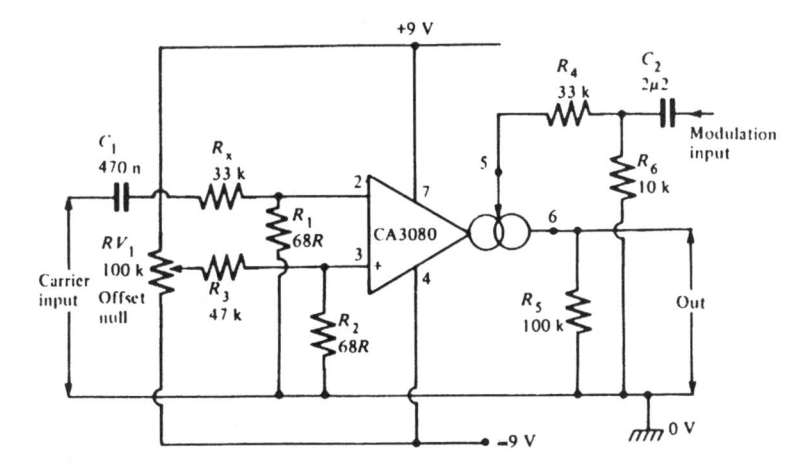

**Figure 2.48** *Amplitude modulator*

54   Op-amp audio processing circuits

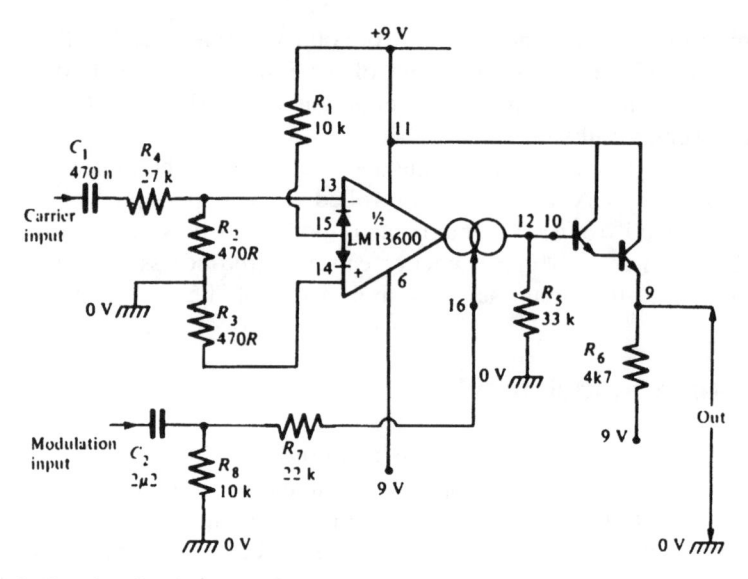

**Figure 2.49**  *Amplitude modulator*

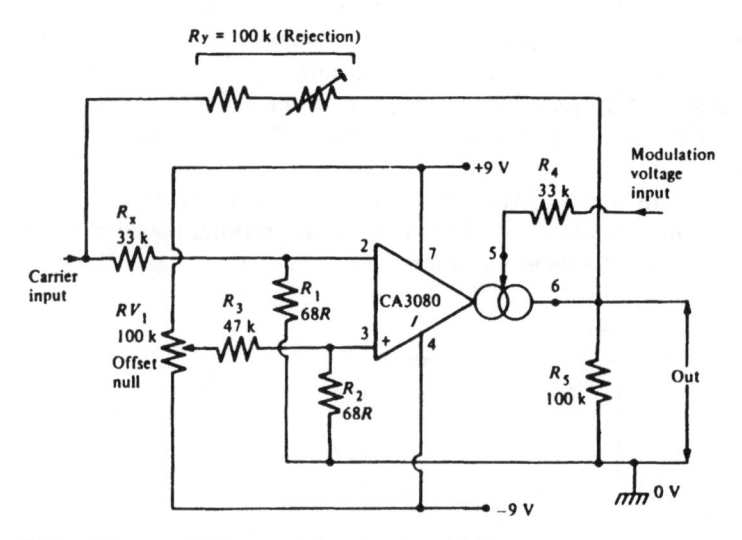

**Figure 2.50**  *Ring modulator or 4-quadrant multiplier*

Note in the above two circuits that the instantaneous polarity of the output signals is determined entirely by the instantaneous polarity of the carrier input signal, which has two possible states (positive or negative), and is independent of the modulation signal, which has only one possible state (positive). This type of circuit is thus known as a 2-quadrant multiplier. There is another type of modulator circuit, known as a ring modulator or 4-quadrant multiplier, in which the output signal polarity depends on the polarities of both the input signal and the modulation voltage. *Figure 2.50* shows a CA3080-based version of such a circuit.

## Ring modulators

The *Figure 2.50* circuit is similar to that of *Figure 2.48*, except that Ry is wired between input and output and is adjusted so that, when the modulator input is tied to the zero-volts rail, the input-derived signal currents feeding into R5 via Ry are exactly balanced by the inverted signal currents feeding into R5 from the OTA output, so zero output is generated across R5. If the modulation input then goes +ve, the OTA output current exceeds that of the Ry network, and an inverted gain-controlled output is obtained, but if the modulation input goes −ve the Ry output current exceeds that of the OTA, and a non-inverted gain-controlled output is obtained. Thus, both the phase and the amplitude of the output signal of this 4-quadrant multiplier are controlled by the modulation signal. The circuit can be used as a ring modulator by feeding independent ac signals to the two inputs, or as a frequency doubler by feeding identical sinewave signals to the two inputs.

With the Rx and Ry values shown this circuit gives a voltage gain of ×0.5 when the modulation terminal is tied to the +ve or −ve supply rail; the gain doubles if the values of Rx and Ry are halved.

*Figure 2.51* shows how one half of an LM13600 can be used as a ring-modulator or 4-quadrant multiplier. This circuit is similar to *Figure 2.49*

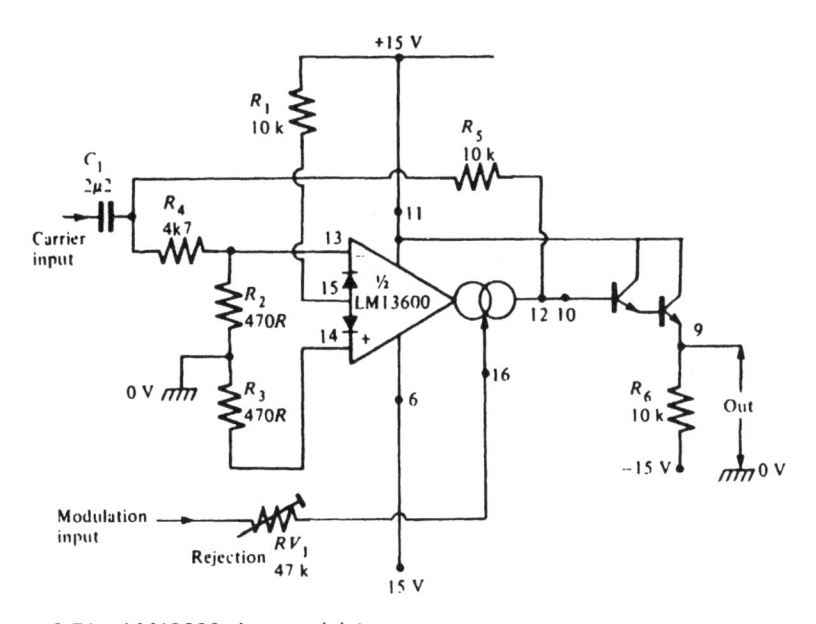

**Figure 2.51** *LM13600 ring modulator*

56  Op-amp audio processing circuits

except that R5 is wired between the input signal and the OTA output, and $I_{bias}$ is presettable via RV1. The basic circuit action is such that the carrier input feeds a signal current into one side of R5, and the OTA output feeds an inverted signal current into the other side of R5, so these two currents tend to self-cancel. In use, the OTA gain is pre-set via RV1 so that the two currents are exactly balanced when the modulation input is tied to the common zero volts line, and under this condition the circuit gives zero carrier output. Consequently, when the modulation input moves positive, the OTA gain increases and its output current to R5 exceeds that caused by the direct input signal, so an inverted output carrier signal is generated. Conversely, when the modulation input moves negative the OTA gain decreases and the direct signal current of R5 exceeds that generated by the OTA output, so a non-inverted output signal is generated.

## An AGC amplifier

The gain of the LM13600 OTA can be varied by altering either its $I_{bias}$ or $I_D$ current. *Figure 2.52* shows how $I_D$ variation can be used to make an AGC (automatic gain control) amplifier in which a 100:1 change in input signal amplitude causes only a 5:1 change in output amplitude.

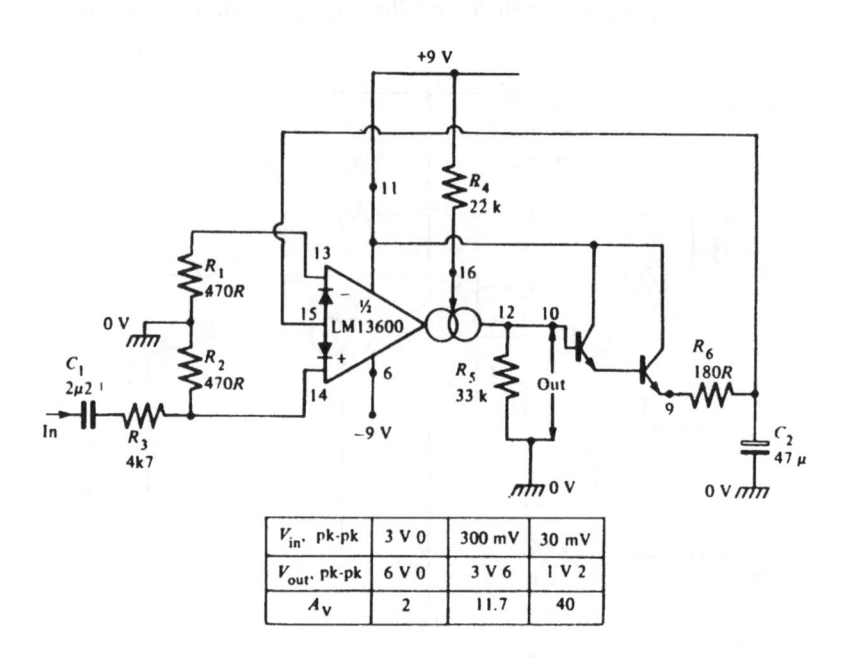

| $V_{in}$. pk-pk | 3 V 0 | 300 mV | 30 mV |
|---|---|---|---|
| $V_{out}$. pk-pk | 6 V 0 | 3 V 6 | 1 V 2 |
| $A_V$ | 2 | 11.7 | 40 |

**Figure 2.52**  *Circuit and performance table of an AGC amplifier*

In this circuit, $I_{bias}$ is fixed by R4, and the output signal is taken directly from the OTA via R5. The output buffer and R6–C2 are used to rectify and smooth the OTA output and to then apply an $I_D$ current to the OTA's linearizing diodes. No $I_D$ current is generated until the OTA output exceeds the 1.8V peak (equals three base–emitter volt drops) needed to turn on the Darlington buffer and the linearizing diodes, but any increase in $I_D$ then reduces the OTA gain and, by negative feedback action, tends to hold $V_{out}$ constant at that level.

The basic gain of this amplifier, with zero $I_D$, is ×40. Thus, with an input signal of 30mV pk–pk, the OTA output of 1V2 pk–pk is not enough to generate an $I_D$ current, and the OTA operates at full gain. At 300mV input, however, the OTA output is enough to generate significant $I_D$ current, and the circuit's negative feedback automatically reduces the output level to 3V6 pk–pk, giving an overall gain of ×11.7. With an input of 3V, the gain falls to ×2, giving an output of 6V pk–pk. The circuit thus gives 20:1 signal compression over this range.

## The LM13700

Finally, to conclude this chapter, note that the LM13700 is a dual OTA IC that is identical to the LM13600 except for relatively minor differences in its output buffer stages, but is sometimes more readily available than the LM13600 and can be used as a pin-for-pin replacement for it in all LM13600 circuits shown in this chapter.

# 3
# Dedicated audio processing IC circuits

The preceding chapter showed how standard op-amp and operational transconductance amplifier (OTA) ICs can be used in a variety of useful audio signal processing applications. The present chapter continues the 'signal processing' theme by showing ways of using various dedicated audio signal processing ICs, including the MC3340P electronic attenuator IC, the NE570/571 dual 'compander' ICs, several multi-way 'analogue switching' and voltage- or digitally-controlled 'gain' and 'tone-control' ICs, and the ever-popular MF10C universal dual switched capacitor filter IC.

## The MC3340P

The Motorola MC3340P is an old but very popular dedicated 'electronic attenuator' IC. *Figure 3.1* shows the outline, pin notation and basic details of the device, which is housed in an 8-pin DIL package; only six of these pins perform useful functions, and two of these are used for power supply connections. Of the remainder, pins 1 and 7 provide input and output signal connections, pin 6 controls roll-off of the device's frequency response, and pin 2 is the device's gain-control terminal.

The MC3340P is really a simple operational transconductance amplifier (OTA) of the type described in Chapter 2, but is configured as a voltage-controlled amplifier. Its basic action is such that it acts as a linear voltage amplifier with 13dB of signal gain when its pin-2 CONTROL terminal is tied to ground via a 4k0 resistance or is connected to a dc potential of 3.5V. This gain decreases if the control resistance/voltage is increased above these values, falling by 90dB (to –77dB) when the values are increased to 32k or 6V. The device's attenuation (or gain) can thus be controlled over a wide range via either a resistance or a voltage.

*Figure 3.2* shows a practical example of a voltage-controlled MC3340P electronic attenuator, together with its performance graph, and *Figure 3.3*

Dedicated audio processing IC circuits 59

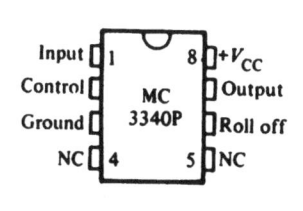

| Parameter | Min. | Typ. | Max. |
|---|---|---|---|
| Supply volts | +9 V | | +18 V |
| Control pin sink current | | | 2 mA |
| Input voltage, rms | | | 0.5 V |
| Voltage gain | | 13 dB | |
| Attenuation range | | 90 dB | |
| Total harmonic distortion | | 0.6% | |

**Figure 3.1** Outline and main characteristics of the MC3340P IC

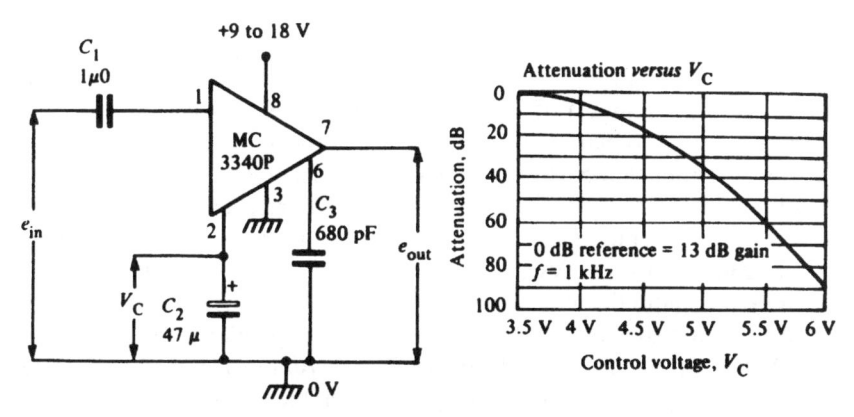

**Figure 3.2** Circuit and performance graph of a voltage-controlled electronic attenuator

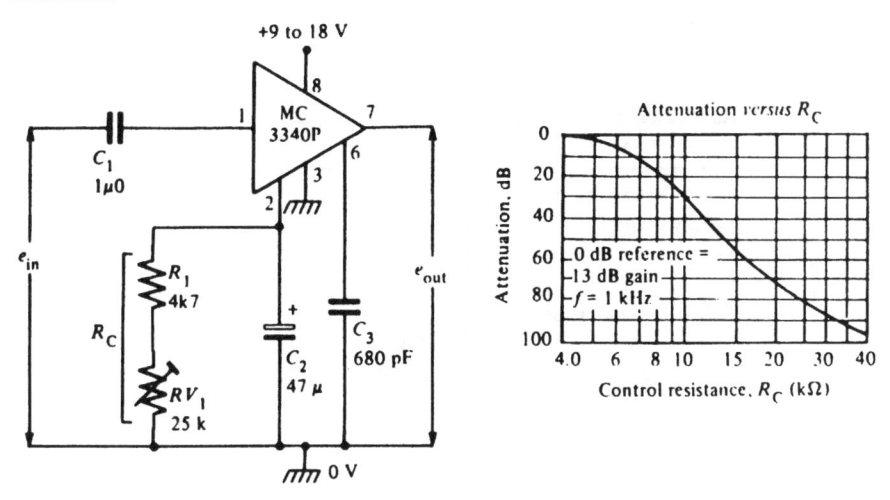

**Figure 3.3** Circuit and performance graph of a resistance-controlled electronic attenuator

shows a resistance-controlled version of the device. In each of these circuits, C2 is wired to the control terminal to eliminate control noise and transients, thus giving a 'noiseless' form of gain control and enabling the control resistance/voltage to be remotely located, and 680pF capacitor C3 is wired to pin 6 of the IC and limits the upper-frequency response of the circuit to the high audio range. Without C3, the response extends to several MHz, but the circuit tends to be unstable. Note that this IC gives only slight signal distortion at low attenuation levels, but that the distortion rises to about 3% at maximum attenuation values.

## The NE570/571 IC

The Signetics NE570 is known as a dual 'compandor' (compressor-expander) IC but is really a rather sophisticated dual VCA (voltage-controlled amplifier). Each half (channel) of the IC contains an identical circuit, comprising a current-controlled variable gain cell (actually a high-quality OTA), an electronic rectifier that converts an ac input signal voltage into an OTA gain-control current, an op-amp, a precision 1.8V reference, and a resistor network. These elements can be externally configured so that each channel acts as either a normal VCA, as a constant-volume or VOGAD (voice-operated-gain audio device) amplifier, as a disco voice-over or 'ducking' unit, or as a precision dynamic range compressor or expander.

The Signetics NE571 is identical to the NE570, but has a slightly relaxed specification. *Figure 3.4* lists the basic characteristics of the two ICs. Each IC is housed in a 16-pin DIL package, as shown in *Figure 3.5*, which also shows the block diagram of one IC channel. Note in the block diagram (and in all following NE570/571 circuits) that pin numbers relating to the left-hand

| Parameter | NE570 | NE571 |
|---|---|---|
| Supply voltage range | 6 V to 24 V | 6 V to 18 V |
| Supply current | 3.2 mA | 3.2 mA |
| Output current capability | ±20 mA | ±20 mA |
| Output slew rate | 0.5 V/$\mu$s | 0.5 V/$\mu$s |
| Gain block distortion: | | |
| Untrimmed | 0.3% | 0.5% |
| Trimmed | 0.05% | 0.1% |
| Internal reference voltage | 1.8 V | 1.8 V |
| Output DC shift | ±20 mV | ±30 mV |
| Expander output noise | 20 $\mu$V | 20 $\mu$V |

**Figure 3.4** *Basic characteristics of the NE570 and NE571 'compander' ICs*

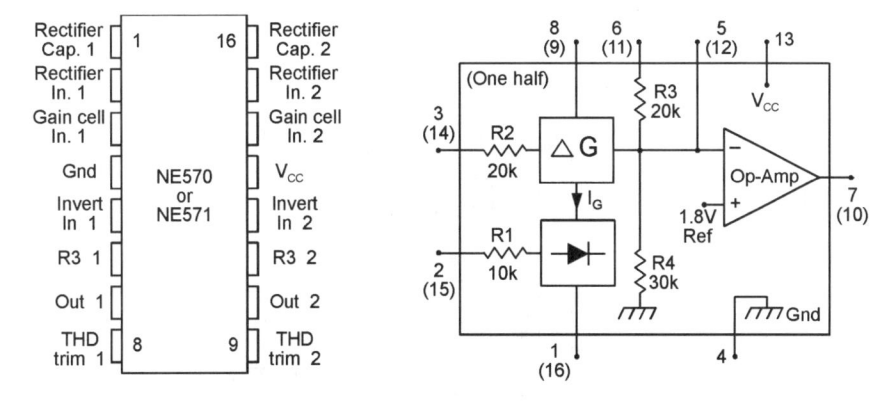

**Figure 3.5** *Outline, pin notation and block diagram of one of the two identical channels of the NE570/571 dual compander IC*

channel of the IC are shown in plain numbers, and those relating to the right-hand half are shown in bracketed numbers.

## NE570/571 circuit description

The operation of the individual elements of the *Figure 3.5* block diagram are fairly easy to understand. Dealing first with the 'rectifier' block, input signals that are ac coupled to pin 2 (or 15) are full-wave rectified by this block and fed – in the form of a proportional current – to pin 1 (or 16), where they can be smoothed by an external capacitor. The resulting dc current is then applied to a built-in current mirror, which directly controls the gain of the IC's 'variable-gain' block.

Dealing next with the variable-gain block, input signals that are ac coupled to pin 3 (or 14) are fed to the input of this block, which is a precision temperature-compensated OTA with its gain controlled directly by the rectifier block's current mirror, and thus indirectly via the pin 1 (or 16) voltage; the gain block's output takes the form of a current, but is converted into a proportional output voltage via the IC's op-amp stage. The gain block's signal distortion is quite low, and can be minimized by feeding a 'trim' voltage to pin 8 (or 9).

The channel's op-amp is internally compensated and has its non-inverting input tied to a 1.8V precision band-gap reference; the inverting input is connected to the gain block output, and is externally available. The inverting input is also connected to the R3–R4 resistor network, which can be used (either directly or with the aid of external resistors) to set the op-amp's ac and/or dc gain, using normal op-amp output-to-input feedback techniques. The op-amp output is available at pin 7 (or 10).

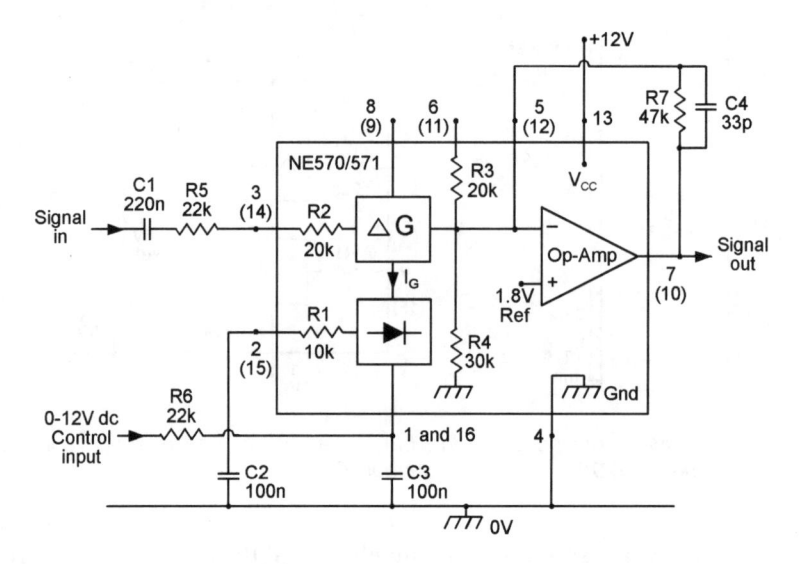

**Figure 3.6** *NE570/571 stereo voltage-controlled amplifier/attenuator (only one channel shown)*

## A stereo VCA

*Figure 3.6* shows how a NE570 or NE571 can be used to make a stereo voltage-controlled amplifier/attenuator. Here, the internal rectifier is disabled via C2, and a 0 to 12V dc control voltage is fed to pins 1 and 16 via R6 and C3, to give direct control of the rectifier's internal current mirror and thus of the variable-gain block. The output of the gain block is fed to pin 7 (or 10) via the op-amp, which has its ac and dc gain set at ×2.56 via R4–R7 and thus generates a quiescent output of 4.62V (= 2.56 × 1.8V). Both channels of the circuit are identical (the control voltage is fed to pins 1 and 16), and give about 6dB of gain with a control input of 12V, or 80dB of attenuation with a control input of zero volts.

## A stereo VOGAD unit

*Figure 3.7* shows a NE570/571 IC used to make a stereo constant-volume amplifier or VOGAD unit, in which the mean audio output amplitude varies by only ±1dB when the input signal amplitude is varied over the range +14dB to −43dB (the '0dB' reference value is 0.9Vrms). This type of circuit is often used to feed amplified microphone signals to the inputs of telephonic (wire or radio) communication units or sound distribution or recording systems, and eliminates the need to fiddle with amplitude-level controls. The circuit

*Dedicated audio processing IC circuits* 63

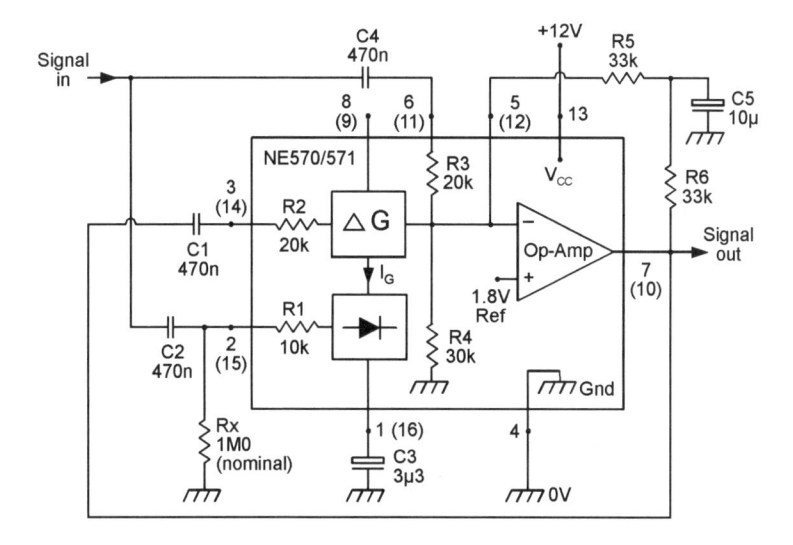

**Figure 3.7** *A NE570 or NE571 IC used to make a stereo constant-volume amplifier or VOGAD unit*

operation is quite simple. The pre-amplified input signal is ac-coupled directly to the input of the internal rectifier, and to the op-amp's inverting input via pin 6 (or 11) of the IC, but the gain block is wired in series with the op-amp's output-to-input negative feedback loop, thus making the overall gain inversely proportional to the input level. Consequently, an '$x$'dB fall in input level causes an identical dB *increase* in gain, thus giving zero change in the circuit's mean output amplitude. Resistor Rx is used to limit the unit's maximum gain, so that the unit does not generate an excessive noise output in the absence of a useful input signal. The Rx value can vary between 100k and 10M, the 'ideal' value (usually about 1M0) being found by trial-and-error.

## A voice-over (ducking) unit

*Figure 3.8* shows the basic circuit of a NE570/571 disco voice-over or 'ducking' unit that automatically fades the music down when the DJ talks into his microphone and gently restores the music again when the chatting is finished. In this design, each channel's op-amp is used as a 2-input audio mixer that has one input taken directly from the microphone input signal and the other taken from the music input signal via the channel's gain cell. Note that each channel's rectifier unit is disabled via C3, and the gain cell is controlled via transistor Q1, which is used as a simple electronic switch that

64　*Dedicated audio processing IC circuits*

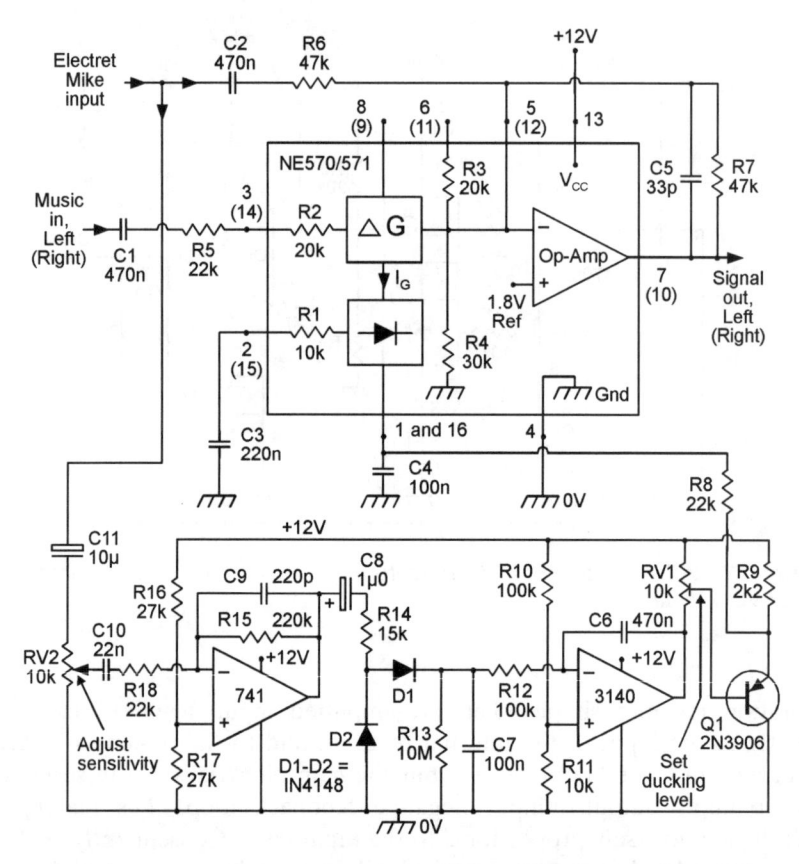

**Figure 3.8**　Basic circuit of an NE570/571 disco voice-over or 'ducking' unit

is activated via the microphone input signal. In the absence of a strong microphone signal, Q1 is cut off and the gain cell is driven fully on via R8–R9, giving maximum amplification to the music signal, which appears at full volume at the pin 7 (10) output terminals. In the presence of a strong microphone signal, however, Q1 is driven on, and the gain cell attenuates the music signal, causing the microphone signal to dominate the pin 7 (10) output.

In practice, the actual ducking or music-attenuation level of the circuit can be fully controlled – from near-zero to –80dB – via RV1, enabling the 'ducked' music and the microphone signal amplitudes to be mixed in any desired ratio, and the microphone-derived gain-control signals are designed to give a smooth fade-over, rather than a sharp switching action. These signals are derived from the microphone input via RV2, are given 20dB of gain via the 741 speech-band (350Hz–3.5kHz) amplifier, are then peak-rectified and filtered via the D1–D2 network, and are used to activate the 3140 voltage comparator, which has a 1.1V reference applied to its non-inverting terminal and is given a 'slow swing' output action via its C6–R12 integrating network. The output of the 3140 op-amp is normally high (and Q1 is thus cut off), but swings low in the presence of a strong microphone signal, this pulling Q1 emitter down (thus attenuating the music signal) by an amount determined by the RV1 control setting.

## Compander theory

The NE570/571 is designed primarily to control the dynamic ranges of various circuits. In acoustics, the term 'dynamic range' can be simply described as the difference between the loudest and the quietest sound levels that can be perceived or recorded. Typically, a healthy human adult has a useful dynamic 'hearing' (acoustic perception) range of about 90dB (= 50 000:1). This range greatly exceeds that of most recording systems. All practical recording systems generate inherent noise, which limits the minimum strength of signals that can be usefully recorded, and this factor (in conjunction with practical limits on maximum signal strength) places a limit on the useful dynamic range of the recording system. Thus, if a recording medium can handle maximum signal amplitudes of 1V r.m.s., but produces a 'noise' output of 1mV r.m.s in the absence of a recorded signal, the system is said to have a signal-to-noise ratio or maximum dynamic range of 1000:1, or 60dB.

Simple tape recorders typically have a useful dynamic range of less than 50dB, and thus cannot directly record and replay high-quality music or other *analogue* signals (these restrictions do not, of course, apply to digitally encoded signals). One way around this problem is to use a compander system to compress the 90dB dynamic range of the analogue input signal down to 45dB when recording it (thus giving a 2:1 compression ratio), and then use a matching 1:2 expander to restore its dynamic range to 90dB when replaying the signals. This same basic technique can be used to improve the quality of many types of analogue audio signal, and the NE570/571 ICs were originally designed specifically for use in low-fi and medium-fi versions of such systems.

## An expander circuit

*Figure 3.9* shows a practical NE570/571 'expander' circuit and its performance table. Here, the input signal is fed to both the rectifier and the variable gain block, and their action is such that circuit gain is directly proportional to the average value of the input. Thus, if the input rises (or falls) by 6dB, the gain also rises (or falls) by 6dB, so the output rises (or falls) by 12dB, giving a 1:2 expansion ratio. Note in this circuit that (because

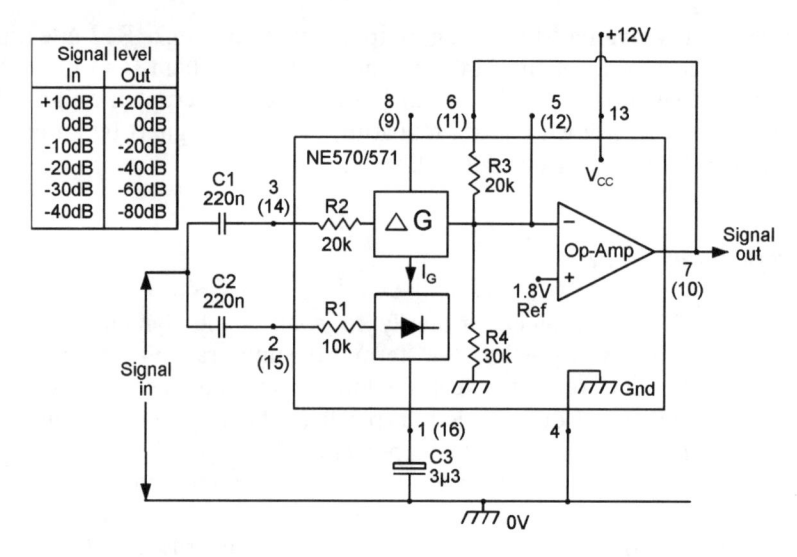

**Figure 3.9** NE570/571 'Expander' circuit and performance table

of the R3 and R4 ratios) the op-amp output takes up a quiescent value of 3V, and can thus supply only modest peak output signals. If desired, the output can be raised to 6V (giving a corresponding increase in peak output levels) by wiring a 12k resistor in parallel with R4 via pins 5 (or 12) and 4, or a 51k resistor in series with R3 via pin 6 (or 11).

### A compressor circuit

*Figure 3.10* shows a practical NE570/571 'compressor' circuit and its performance table. Here, the input signal is fed to the op-amp's inverting input via C4 and R3, but the variable-gain block and rectifier circuitry are connected in exactly the same way as in the above expander design and are ac-coupled into the op-amp's output-to-input negative feedback loop, and the circuit consequently gives a performance that is the exact inverse of the expander, i.e. it gives a 2:1 compression ratio. R5 and R6 form a dc feedback loop (ac decoupled via C5) that biases the op-amp output at a quiescent value of about 6 volts.

### Compander circuit variations

Simple compander system suffer from a number of practical limitations and defects. They usually generate rather high levels of noise, THD, tracking distortion, dc-tracking shifts, and suffer from a phenomena known as

Dedicated audio processing IC circuits 67

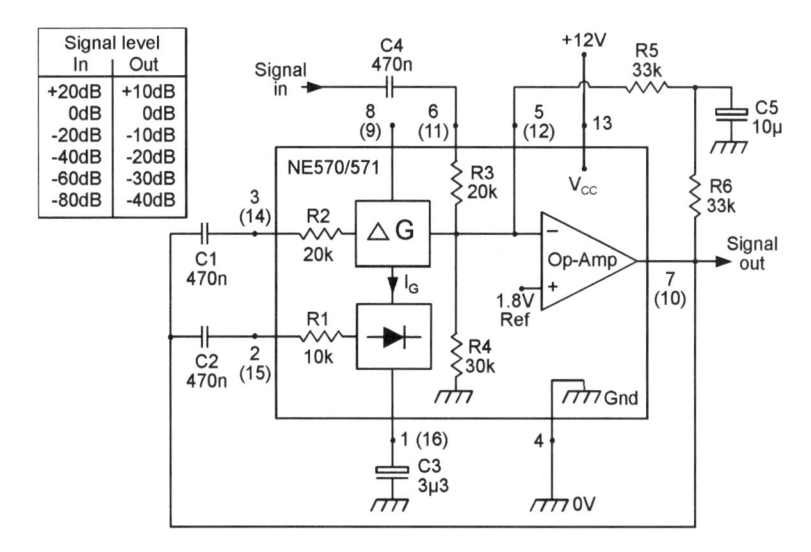

**Figure 3.10** NE570/571 'Compressor' circuit and performance table

'breathing' or 'pumping' (see Chapter 1). These problems may reach intolerably high levels if the system's 'compression' or slope ratio is raised significantly above the basic 2:1 ratio, but can be greatly reducing by using slope ratios of less than 2:1. Anyone wishing to experiment with high-slope compander systems can do so by simply cascading NE570/571 compression or expander circuits. Each IC contains the basics of two of these circuits, and can thus provide an overall 4:1 compression ratio or 1:4 expansion ratio and can, for example, thus compress an 80dB (10 000:1) range of input signals into a 20dB (10:1) output range.

*Figure 3.11* shows how one half of an NE570/571 IC can be used to make a variable-slope compressor–expander in which the slope is fully variable from 2:1 compression to 1:2 expansion via dual-gang 10k pot RV1, which has its two sections wired in anti-phase. When the pot is in its central position the circuit has a 1:1 slope, and acts as a simple amplifier that gives neither compression or expansion.

The circuits shown in *Figures 3.6* to *3.11* are simple designs which can all be improved with the addition of various trim controls, such as those shown in *Figure 3.12*. The THD trimmer networks shown in *(a)* or *(c)* can be used to minimize an NE570/571 circuit's total harmonic distortion figures. To use this trimmer, feed a fairly strong 1kHz sinewave to the input of the main circuit, and then adjust RV1 for minimum output distortion. Note that, if the THD trim facility is not used, pins 8 and 9 of the IC should be decoupled to ground via 220pF capacitors, to eliminate HF instability. The dc offset trimmer networks shown in *Figure 3.12(b)* or *(c)* can be used to minimize

68  Dedicated audio processing IC circuits

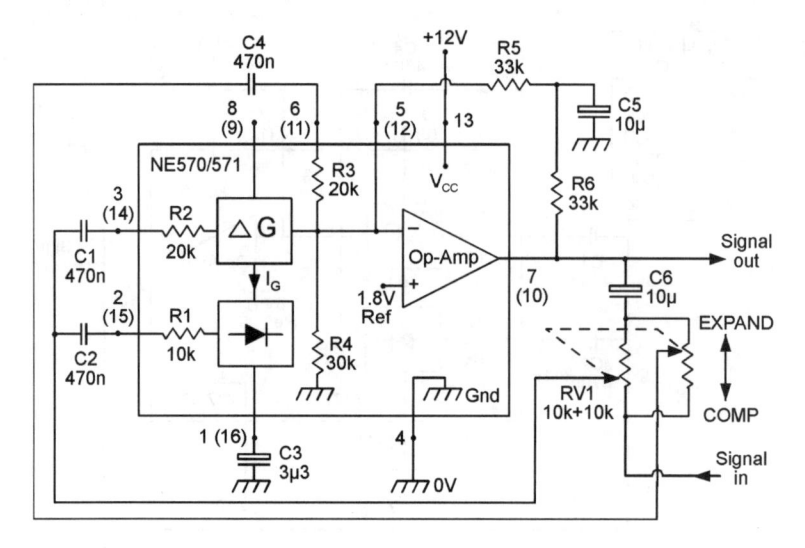

**Figure 3.11**  Variable-slope (2:1 to 1:2) compressor–expander

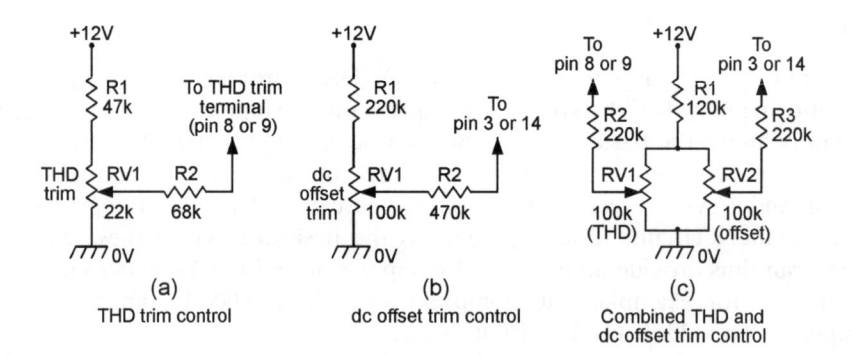

**Figure 3.12**  NE570/571 trimmer networks for minimizing THD and dc offset shifts

any dc output voltage shifts that occur when a circuit's input signal voltages are varied between their maximum and minimum values.

The NE570/571 IC's rectifier elements each consume input bias currents of about 100nA. In the simple *Figure 3.6* to *3.11* circuits this current is derived from the rectifier's input signal, thus limiting the *actual* dynamic range of the rectifier (and also the IC's gain cells) to about 60dB. This snag can be overcome, thus expanding the rectifier's actual dynamic range to its full 80dB+ value, with the help of the *rectifier bias current cancellation network* shown in *Figure 3.13*, which also shows the rectifier's basic performance graph with and without cancellation.

Finally, before leaving the NE570/571, note that this IC's greatest weakness lays in its internal op-amp, which is a very simple and rather noisy

Dedicated audio processing IC circuits   69

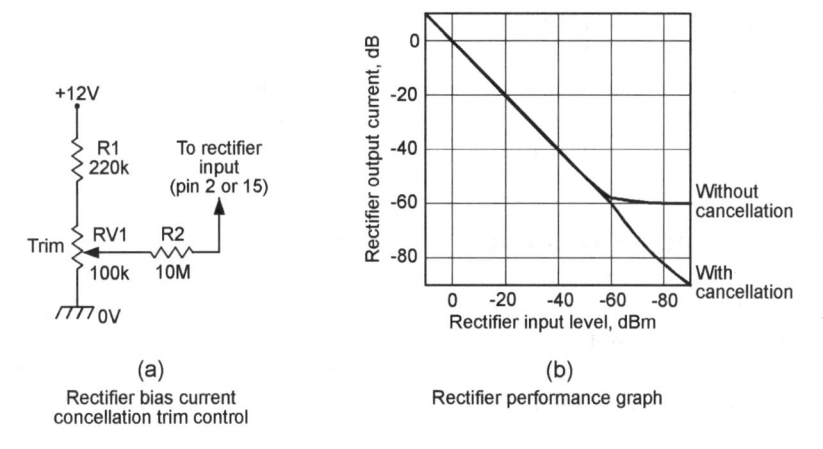

**Figure 3.13** *Rectifier bias current cancellation network (a), and rectifier performance graph with and without cancellation (b)*

mid-fi design. If you wish to use a compander IC in a hi-fi application, therefore, you can either use a rather expensive dedicated hi-fi compander IC such as the SSM2120 (available from Analog Devices) or can simply ignore the NE570/571 IC's internal op-amp and use an external high-performance op-amp instead, using the basic connections shown in *Figure 3.14*. Here, the external op-amp's non-inverting input pin is biased at about 1.8V by

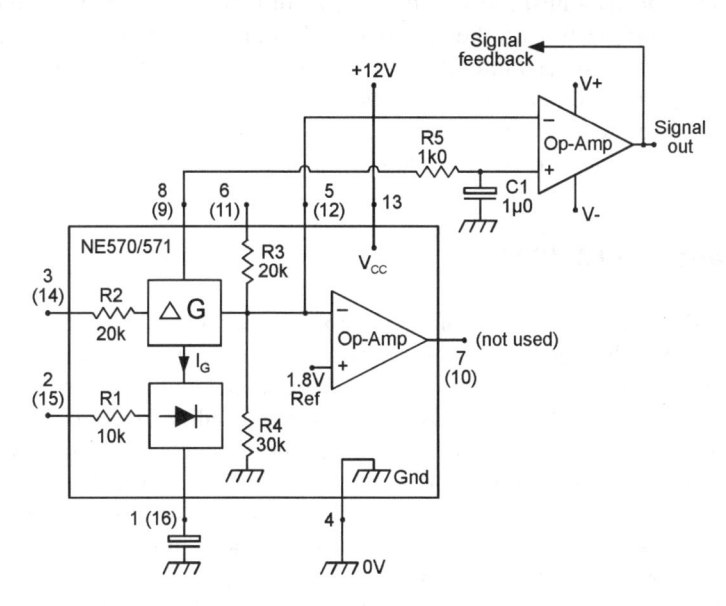

**Figure 3.14** *Basic way of using the NE570/571 with an external (rather than internal) op-amp, to give an improved overall performance*

connecting it to the compander's pin 8 or 9 'THD trim' terminal via the R5–C1 noise filter, and the inverting input pin is tied directly to that of the internal op-amp. The external op-amp can use a dual or single-ended power supply, but in the latter case the op-amp must have an input common mode range that extends down to less than 1.8V.

## Dynamic noise reduction principles

The NE570/571 is just one of many dynamic range manipulation ICs that are designed for use in systems that aim to improve the acoustic reproduction quality of material that is recorded on – or transmitted via – inherently noisy media. Most such systems (including dBx, ANRS, and Dolby) are 'double-ended' and achieve this noise-reduction aim by encoding the material – using dynamic range compression and/or pre-emphasis techniques – at the input end of the system, and using matching decoding (dynamic range expansion and/or de-emphasis) circuits at the output end of the system. There are, however, two 'single-ended' noise reduction systems that can improve the sound quality of virtually *all* recorded (tape or disc) or transmitted (AM or FM) non-coded audio material. The two systems in question are the Philips DNL (dynamic noise limiter) system and the National Semiconductor DNR (dynamic noise reduction) system ('DNR' is a trademark of National Semiconductor Corporation).

The DNL and DNR systems both work by using psychoacoustic techniques that automatically adjust the system's bandwidth and gain to match the instantaneous characteristics of the audio signals that are being processed. The DNR system is of special interest, and is described in detail in the next section of this chapter, together with application details of a special IC, the LM1894, that is designed to implement the system. Note, however, that this is a custom IC, and is available only to approved professional 'bulk purchase' consumers.

### DNR and the LM1894

The DNR system makes use of two simple psychoacoustic facts. The first is that the audibility of white noise (the dominant type of system noise) is proportional to the mean energy level of the noise, which in turn is proportional to the bandwidth of the system. Noise audibility can thus be reduced by reducing the system bandwidth. The second psychoacoustic fact is that, if a simple tone signal and a white noise signal are present at the same time, the tone signal will mask (swamp) the noise signal if the tone's power level is significantly greater than that of the noise signal. Thus, if a low-frequency tone signal is masked by noise in an audio system it can – if the two signals have similar power levels – usually be unmasked by simply reducing the system's bandwidth.

*Dedicated audio processing IC circuits* 71

In National Semiconductor's DNR system, these two sets of facts are utilized by feeding normal audio signals through a filter-amplifier unit that dynamically self-adjusts its bandwidth and gain in sympathy with the instantaneous mean frequency and amplitude of the input signal, thus effectively reducing noise levels by an average of about 10dB, i.e. by a factor of three. All of the active components of a stereo version of this system are contained in the LM1894 IC, and *Figure 3.15* shows the full functional diagram of this device, together with its basic application circuit, and *Figure 3.16* shows the outline and pin notation of the IC. The system functions as follows.

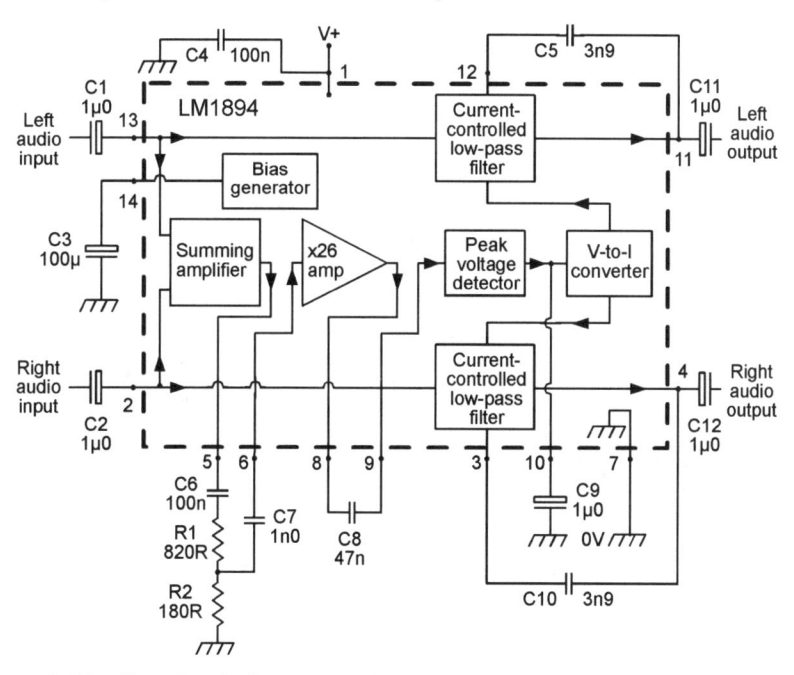

**Figure 3.15** *Functional diagram and basic application circuit of the LM1894 stereo dynamic noise reduction IC*

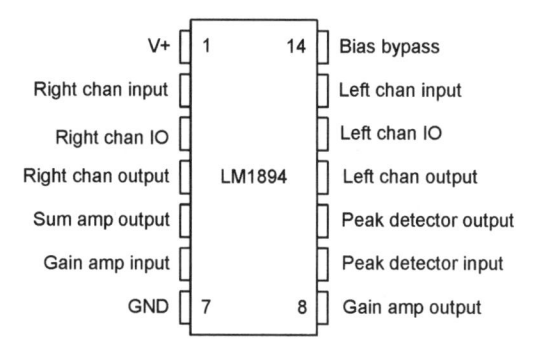

**Figure 3.16** *Outline and pin notation of the LM1894*

72   Dedicated audio processing IC circuits

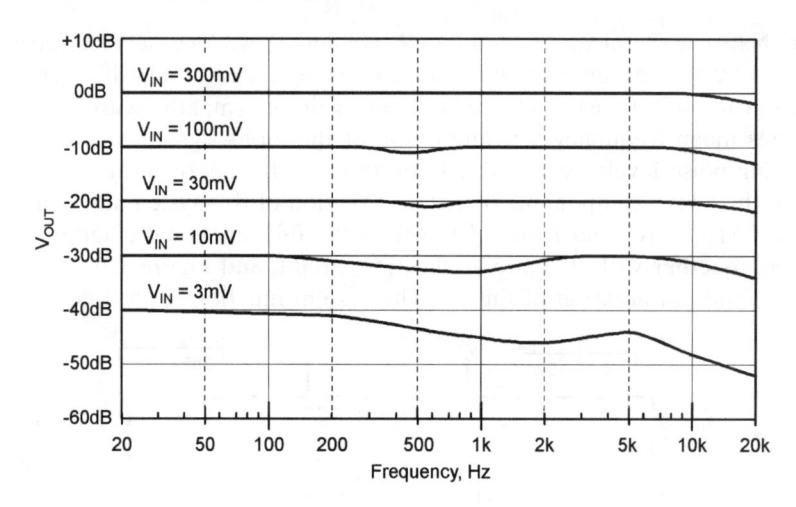

**Figure 3.17**  *Frequency response graph of the LM1894 DNR system*

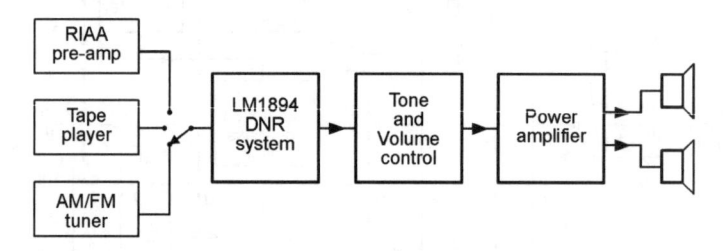

**Figure 3.18**  *The DNR system should be inserted between a hi-fi unit's pre-amplifier and tone/volume control sections*

On entering the IC, the stereo audio channel signals are each passed from input to output via a current-controlled low-pass filter that has its gain controlled via an input-driven bandwidth-control generator circuit. In the latter circuit, the two input signals are added together and then attenuated and filtered via the C6–R1–R2–C7 network; the resulting signal is then amplified, filtered via C8, peak-amplitude detected (rectified), filtered (smoothed) via C9, and finally converted into a proportional current that is used to control the gains of the IC's two current-controlled low-pass filters. Each of the filters in fact consists of an OTA gain cell (of the type used in the NE570/571) plus an op-amp output stage that has its frequency response tailored via C5 or C10. The net result of all this is that each stereo channel exhibits the input-to-output frequency response shown in *Figure 3.17*.

Note in *Figure 3.17* that the frequency response is almost linear when input signals have amplitudes greater than 30mV (and can thus easily swamp system noise), but is subject to fairly heavy noise-attenuating top cut when the input signal amplitudes are less than 10mV. Finally, note that the LM1890 DNL system is intended to be inserted in the middle section of a hi-fi system, between its pre-amplifier and tone/volume control sections, as indicated in *Figure 3.18*, where it will be driven by reasonably strong input signals.

## 'Analogue switch' IC basics

Unscreened cable has a natural tendency to pick up radiated audio and RF signals. Consequently, when conventional control switches are mounted on an audio unit's front panel and are interconnected to various signal-carrying parts of the unit's circuitry, they and their wiring must be very carefully screened to avoid unwanted signal pick-up. Voltage-controlled 'analogue switch' ICs offer a very efficient solution to this particular problem, and are simply placed directly in the signal-carrying paths of the unit's main circuitry (thus eliminating the need for long signal-carrying cables) and are activated, when required, by dc voltages derived from the unit's front panel switches or via a normal remote control system.

All practical voltage-controlled analogue switch ICs conform to one or other of the basic forms represented by the '4-switch' units shown in *Figures 3.19* and *3.20*. In the *Figure 3.19* unit, the input and output of each switch is separately accessible, and each switch can be independently controlled. The signal path of each switch is known as a 'channel', and the *Figure 3.19* unit is thus known as a basic 4-channel analogue switch IC. In the *Figure 3.20* unit, the outputs of the four switches are internally shorted together, and the switches are (in this example) controlled via a 2-bit binary voltage that allows only one switch to be closed at any time. This unit thus acts as a 4-way analogue selector switch or multiplexer. Real-life analogue switch ICs have up to sixteen channels or 'ways'.

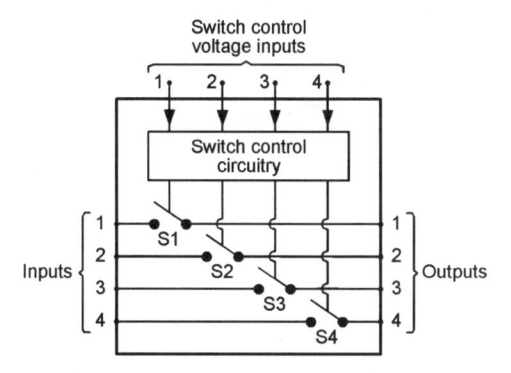

**Figure 3.19** *Diagram representing a basic 4-channel analogue switch IC*

74  Dedicated audio processing IC circuits

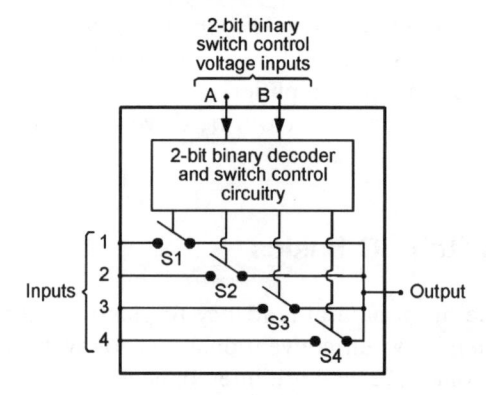

**Figure 3.20** *Diagram representing a basic 4-way analogue selector switch or multiplexer IC*

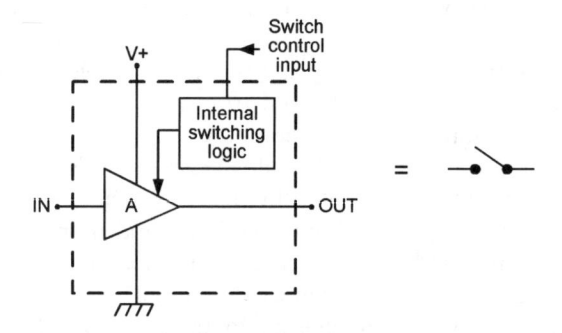

**Figure 3.21** *Basic form and equivalent circuit of a unilateral switch*

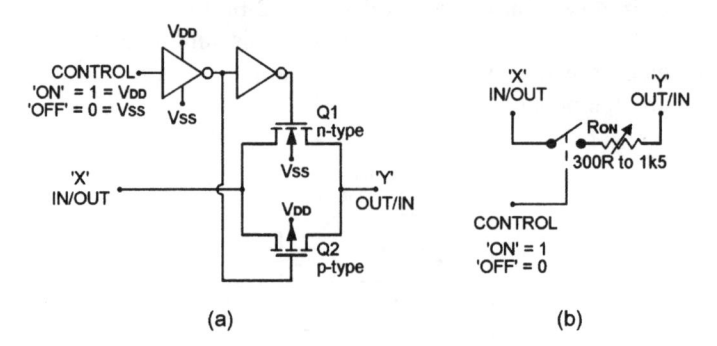

**Figure 3.22** *Basic circuit (a) and equivalent circuit (b) of a simple CMOS bilateral switch*

The actual 'switch' part of the IC may be either a unilateral or a bilateral element. Unilateral switches can pass signals in only one direction, from input to output, and take the basic form shown in *Figure 3.21*, in which '*A*' is a linear amplifier that gives unity gain when acting as a closed switch and

*Dedicated audio processing IC circuits* 75

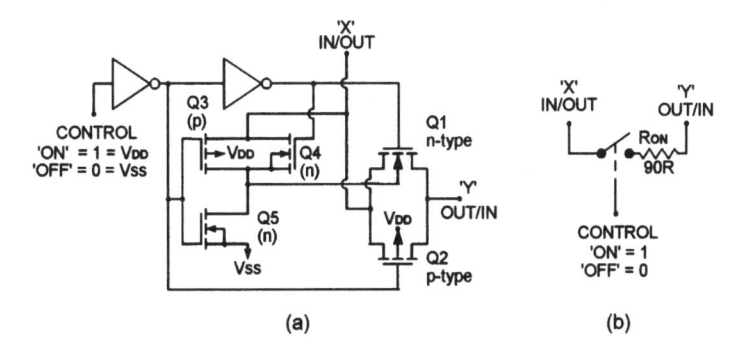

**Figure 3.23** *Basic circuit (a) and equivalent circuit (b) of an improved CMOS bilateral switch*

(ideally) zero gain when acting as an open switch. Usually, the amplifier takes the form of a simple op-amp that has its input and/or feedback connections controlled via one or more JFET switches, or a simple OTA that has its gain controlled via the OTA's tail current.

Bilateral analogue switches can (like an ordinary electrical switch) pass signals in either direction, from input-to-output and from output-to-input. All analogue switches of this type are based on either the simple CMOS bilateral switch circuit shown in *Figure 3.22*, or on an improved version of the basic switch, such as that shown in *Figure 3.23*. In both cases, n-type MOSFET Q1 and p-type MOSFET Q2 are wired in inverse parallel and are driven in antiphase via the CONTROL input signal. When the control signal is low, both MOSFETs are cut off and the circuit acts – between the X and Y points – as an open-circuit switch, but when the control signal is high both MOSFETs are driven fully on and the circuit acts as a closed switch that can pass current in either direction between the X and Y points but has an ON resistance that, in practice, may range from a few ohms to several hundred ohms.

Ignoring the basic architecture of an analogue switch IC (i.e. its type of construction and number of channels), its two most important basic parameters are its signal distortion and channel separation figures. The signal distortion (THD) figures are usually specified at a particular frequency (typically 1kHz) and amplitude level (typically in the range 100mV to 1Vrms. Channel separation defines the amount of signal breathrough (in dB) that occurs between used and unused channels of the IC, and is measured using the basic technique shown in *Figure 3.24*. Here, the ac test signal is applied to the 'S1' input, and all other inputs are decoupled to ground; the magnitudes of the 'unused channel' output signals are then measured and the largest of these is compared to that of the used channel output to determine the IC's channel separation figure. Thus, if the Channel 1 and 2 output signals have r.m.s. magitudes of 1V and 0.1mV, respectively, the channel separation figure works out at –80dB.

76  Dedicated audio processing IC circuits

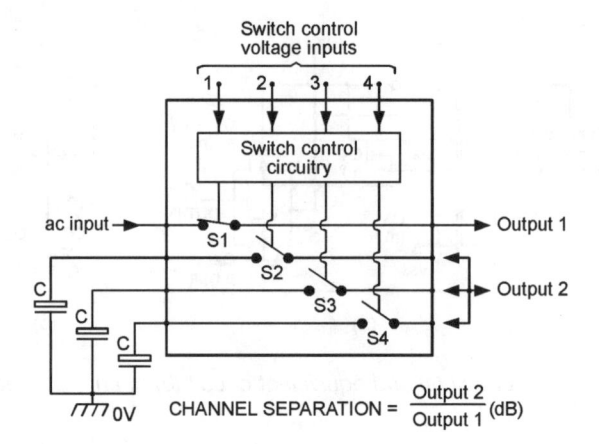

**Figure 3.24** *Basic way of measuring the channel separation of an analogue switch IC*

$$\text{CHANNEL SEPARATION} = \frac{\text{Output 2}}{\text{Output 1}} \text{ (dB)}$$

## Practical analogue switch ICs

For many years, the most popular analogue switch IC for use in hi-fi applications has been National Semiconductor's LM1037 dual 4-channel IC, and the most popular types for use in mid-fi and low-fi applications are the low-cost 4016B and 4066B 4-channel members of the '4000-series' CMOS family. In recent times, however, changes in hi-fi fashions and advances in semiconductor technology have caused a shift in this situation. In the hi-fi field, for example, it is now fashionable to make all major user-variable hi-fi functions fully remote controllable via one sophisticated IC that houses full analogue switching and tone/volume control circuitry. This trend has greatly reduced the market for dedicated 'hi-fi' analogue switch ICs such as the LM1037 type.

Concurrent with the above development, advances in CMOS technology have resulted in the development of very-high-performance low-cost variants of the 4016B and 4066B 4-channel switches, the 74HC4016 and 74HC4066, which each give lower distortion than the LM1037. As a result of these changes, the LM1037 has now ceased production, but will continue to be available from individual distributors for several years. Thus, if you wish to use a dedicated analogue switching IC in an audio project, you can still use an LM1037 or a low-cost CMOS IC such as the 4016B, 4066B, 74HC4016 or 74HC4066 for the purpose. *Figure 3.25* lists basic details of these five devices.

## The LM1037 IC

The LM1037 is a dual 4-channel unilateral analogue switch that allows any one of four stereo input signals to be selected via the appropriate one of four control

*Dedicated audio processing IC circuits* 77

| Device number | Supply voltage range | Switch 'ON' impedance | THD, at 1kHz (12V supply) | Channel separation | IC description |
|---|---|---|---|---|---|
| LM1037 | 5V - 28V | 10R | 0.04% at 1Vrms | -95dB at 1kHz | Dual 4-channel analogue switch. |
| 4016B | 3V - 15V | 400R (15V supply) | 0.4% at 1Vrms | -90dB at 1kHz | Quad bilateral analogue switch. |
| 4066B | 3V - 15V | 80R (15V supply) | 0.1% at 1Vrms | -90dB at 1kHz | Quad bilateral analogue switch. |
| 74HC4016 | 2V - 12V | 20R (12V supply) | 0.01% at 1Vrms | -100dB at 1kHz | Quad bilateral analogue switch. |
| 74HC4066 | 2V - 12V | 15R (12V supply) | 0.01% at 1Vrms | -100dB at 1kHz | Quad bilateral analogue switch. |

**Figure 3.25** *Basic details of five popular analogue switching ICs*

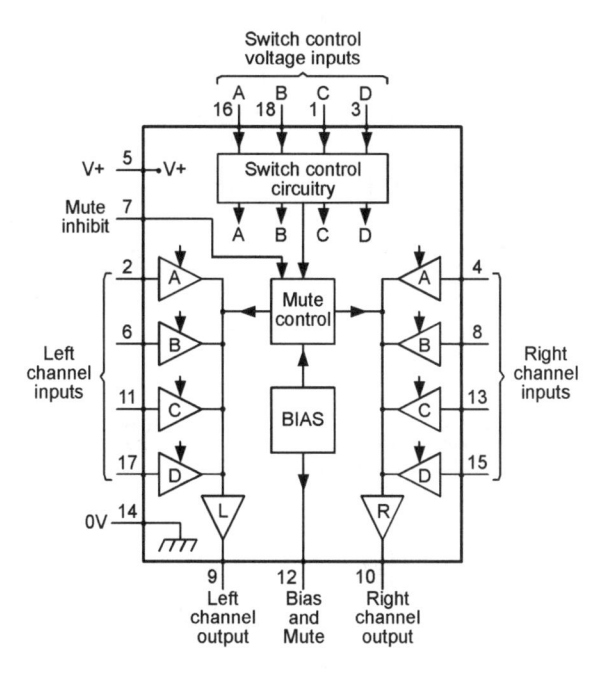

**Figure 3.26** *Block diagram of the LM1037 dual 4-channel analogue switch*

terminals (pins 1, 3, 16 or 18), and which incorporates a 'mute' facility. *Figure 3.26* shows the internal block diagram of the IC, *Figure 3.27* shows the IC's outline and pin notation, and *Figure 3.28* shows the IC's basic usage circuit. Stereo output signals are available on pins 9 and 10, and input channel selection

78  Dedicated audio processing IC circuits

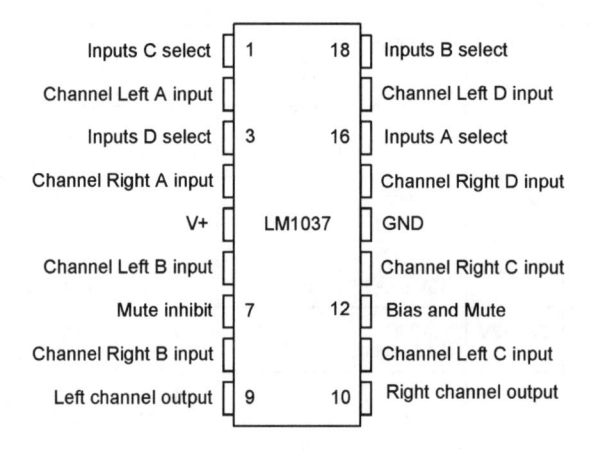

**Figure 3.27** *Outline and pin notation of the LM1037 IC*

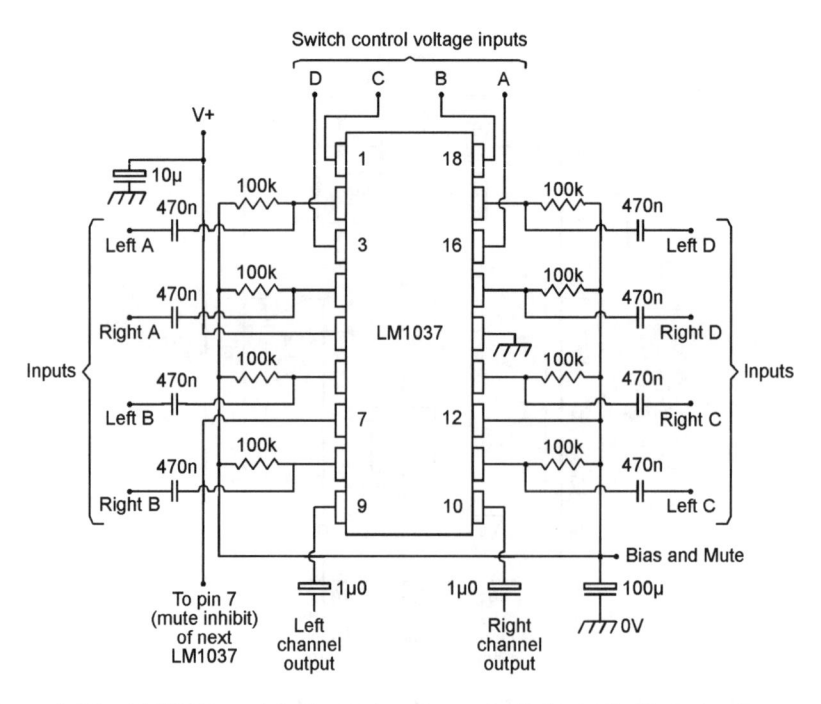

**Figure 3.28** *LM1037 dual 4-channel analogue switch application circuit*

is achieved by taking the appropriate switch control terminal high (only one terminal must be high at any given moment). Each signal input is applied to the IC via a 470n capacitor, and each signal input pin is biased into its linear region by connecting it to the pin-12 'bias' terminal via a 100k resistor.

*Dedicated audio processing IC circuits* 79

LM1037 ICs can be wired in parallel to increase the available channel switching capacity, e.g. two ICs can select up to eight stereo channels. In this application, the pin-7 mute inhibit terminals of both ICs should be direct coupled to each other, and the stereo output terminals of the ICs should be shorted together (pin 9 to pin 9, and pin 10 to pin 10) and made externally available via a single pair of 1μ0 capacitors.

## *The 4016/4066 IC family*

The 4016/4066 family of CMOS ICs are designed primarily for use in high-speed digital applications (the 74HC4066, for example, has a switch signal bandwidth of 120MHz and a switch turn ON or OFF time of 10nS), but are – as shown in the table of *Figure 3.25* – perfectly suitable for use in audio switching applications. The 74HC4066 is particularly useful in hi-fi switching applications, having an 'ON' switch impedance of only 15R and a 1kHz THD figure of 0.01% at 1Vrms.

The 4016B, 4066B, 74HC4016 and 74HC4066 ICs all have identical outlines and functional diagrams, as shown in *Figure 3.29*, and each act as four independent single-pole single-throw (SPST) switches that are open when their control voltage is low (at logic-0) and closed when the control voltages are high (at logic-1). The four ICs are – except for their supply voltage limits – used in exactly the same way, and although the rest of this section deals specifically with the 74HC4066 IC, all of the points mentioned are equally applicable to the 4016B, 4066B and 74HC4066.

The 74HC4066 is very easy to use, provided that basic CMOS usage rules are obeyed. Specifically, input and switching signals must not be allowed to rise above $V_{DD}$ or fall below $V_{SS}$, and all unwanted switches must be disabled by tying their input, output, and control pins to $V_{SS}$. The IC can be used with single-ended or split (dual) supplies by using the basic connections shown in

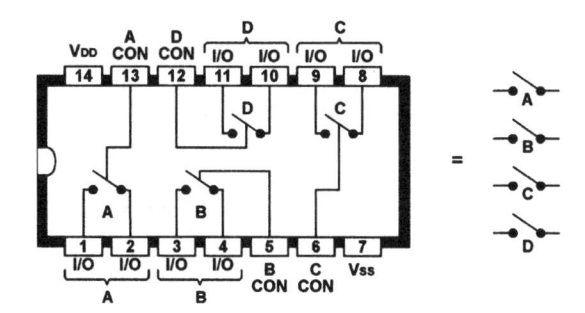

**Figure 3.29** Outline and functional diagram common to the 4016B, 4066B, 74HC4016 and 74HC4066 quad bilateral switch ICs, which each act as four independent SPST switches

80  Dedicated audio processing IC circuits

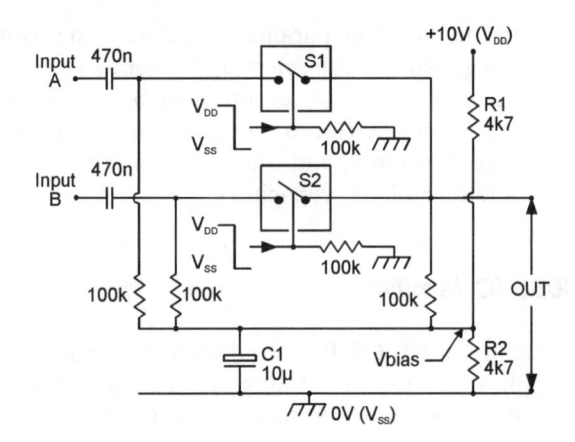

**Figure 3.30**  Basic way of using the 74HC4066 IC with a single-ended +10V power supply

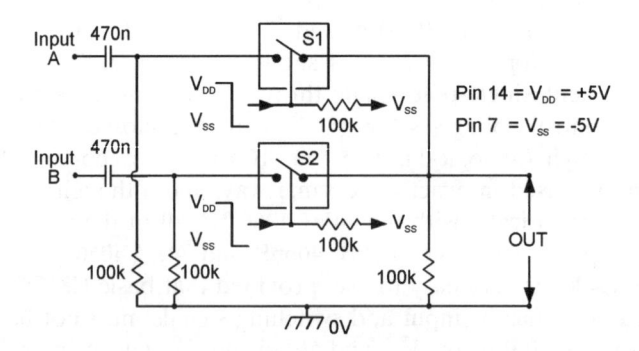

**Figure 3.31**  Basic way of using the 74HC4066 with a split power supply

*Figures 3.30* or *3.31*, which show two of the IC's switches used with their outputs tied together and with the IC powered from a 10V supply. In the *Figure 3.30* single-ended circuit the switch inputs and outputs are biased at half-supply volts via the decoupled R1–R2–C1 bias network and via 100k isolating resistors, and the switch control voltage swings between 0V ($V_{SS}$) and +10V ($V_{DD}$). In the *Figure 3.31* split-supply circuit the switch inputs are tied to zero volts via 100k resistors, and the switch control voltage swings between –5V ($V_{SS}$) and +5V ($V_{DD}$).

The 74HC4066 IC's internal switches can be used in a wide variety of configurations, and *Figure 3.32* shows how it can be used to implement the four basic switching functions of SPST, SPDP, DPST and DPDT. *Figure 3.32(a)* shows the SPST connection, which has already been described. The SPDT function (*Figure 3.32(b)*) is obtained by wiring a simple inverter stage

*Dedicated audio processing IC circuits* 81

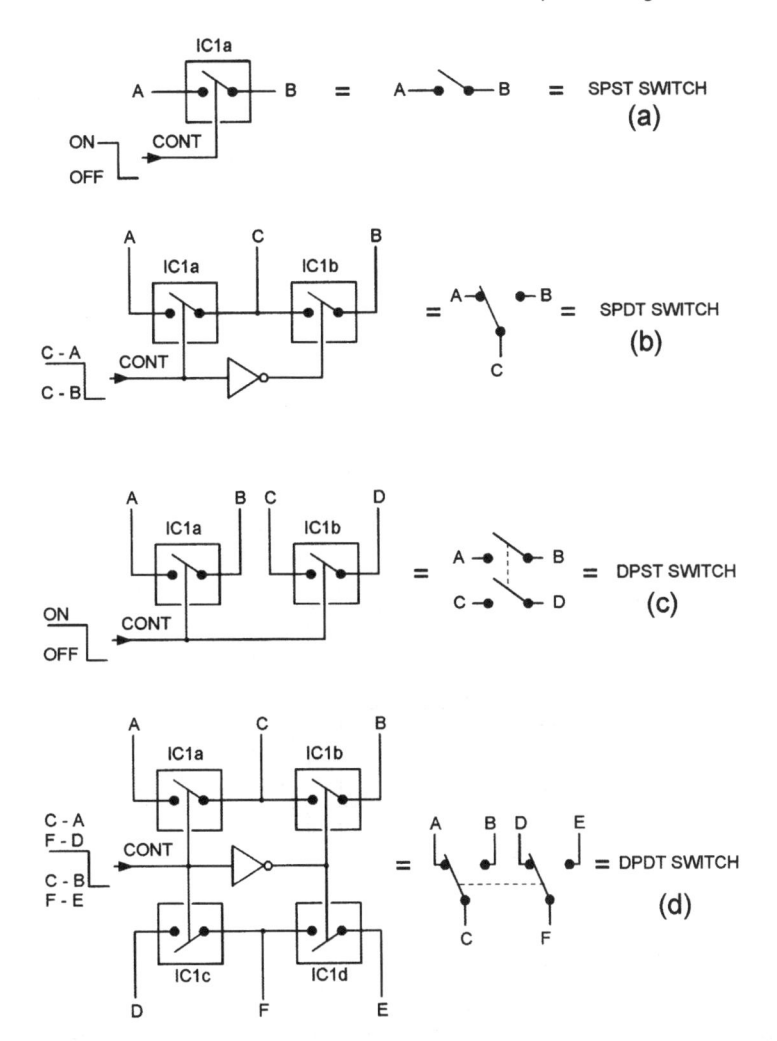

**Figure 3.32** *Implementation of four basic switching functions via the 74HC4066 IC (IC1)*

between the IC1a and IC1b control terminals. The DPST switch (*Figure 3.32(c)*) is simply two SPST switches sharing a common control terminal, and the DPDT switch (*Figure 3.32(d)*) is two SPDT switches sharing an inverter stage in the control line.

In audio applications, the most useful application of the 74HC4066 is as a voltage-controlled multi-input signal selector or switch. *Figure 3.33* shows the basic way of using the 74HC4066 as a single-pole 4-way selector switch in which the desired input signal is selected by driving the appropriate control input high. The 74HC4066 can provide a maximum of four 'ways' per IC,

82  Dedicated audio processing IC circuits

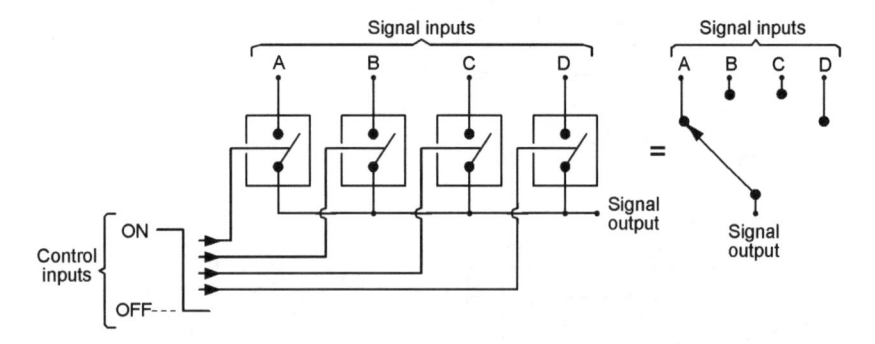

**Figure 3.33**  Basic way of using the 74HC4066 as a single-pole 4-way selector switch

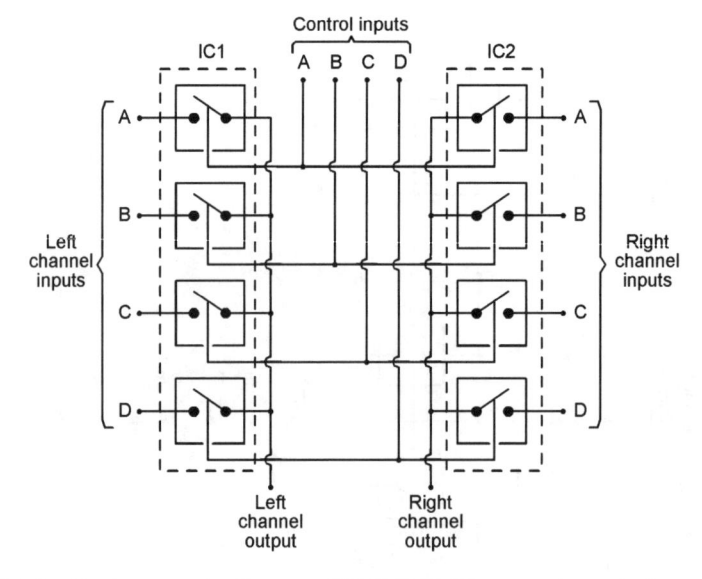

**Figure 3.34**  Basic way of using two 74HC4066 ICs to make a 4-channel stereo input selector switch

and an 8-way switch can thus be made by using two ICs and connecting all of their switch outputs together, or a 4-way stereo selector can be made by using two ICs, configured as in *Figure 3.33*, but with the control inputs of the two ICs wired in parallel, as shown in *Figure 3.34*.

## The 74HC4052 IC

Another low-cost and easy-to-use CMOS IC that can be used to make a 4-way voltage-controlled low-distortion stereo selector switch is the popular 74HC4052 dual 4-channel analogue multiplexer IC. *Figure 3.35* shows this

*Dedicated audio processing IC circuits* 83

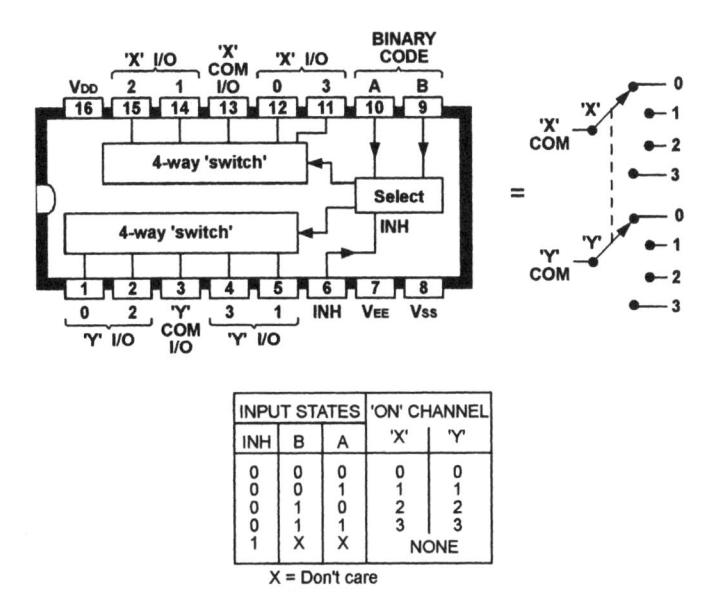

**Figure 3.35** *Functional diagram and truth table of the 74HC4052 dual 4-channel analogue multiplexer IC, which acts like a ganged 2-pole 4-way analogue switch*

IC's functional diagram and truth table, etc. The IC is used in the same basic way as the 74HC4066, except that its channel selection is executed by applying two simultaneous voltages in the form of a 2-bit binary code, as shown in the truth table. One very useful feature of this IC is that the selection process can be inhibited by applying a logic-1 (high) voltage to pin 6, which can thus be used as an audio 'mute' control.

## Variable tone/volume IC basics

### Voltage-controlled ICs

Prior to the advent of modern signal-processing IC technology, all audio tone and volume control circuits were designed around simple but inherently noisy electro-mechanical 'pots', which were usually mounted on the audio unit's front panel and were thus prone to pick up unwanted signals via their signal-carrying interconnecting leads. During the late 1970s, however, the ever-increasing demands for improved audio quality and remote controllability of tone and volume in hi-fi and TV sets resulted in the development of new types of IC that housed all-electronic tone and volume control circuits that were variable via interference-immune input signals applied to the appropriate 'control' pins of the IC.

84  Dedicated audio processing IC circuits

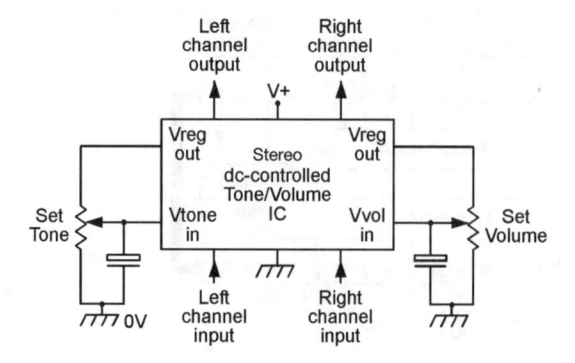

**Figure 3.36** Dc-controlled tone/volume IC basic usage circuit, using pot-generated control voltages

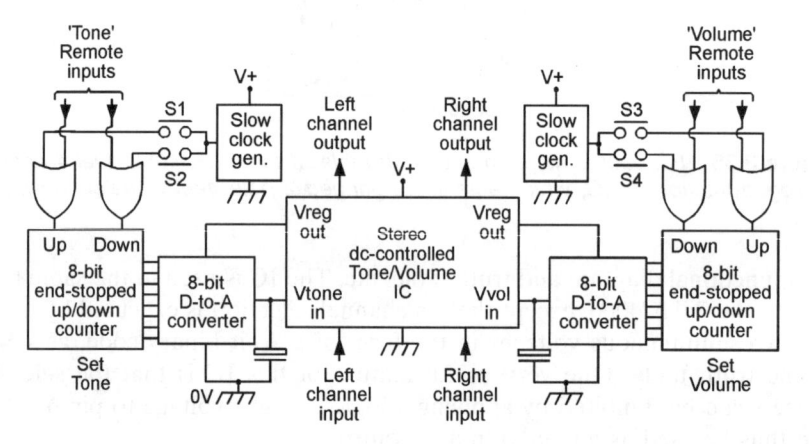

**Figure 3.37** Dc-controlled tone/volume IC basic usage circuit, using digitally-generated control voltages

Throughout the 1980s and the early 1990s, most of these 'variable tone/volume' ICs were designed around OTA types of analogue circuit elements and used variable dc input voltages as their primary control medium. One of the best known of these ICs is National Semiconductor's LM1036 dual (stereo) dc-controlled tone/volume/balance IC, which is very easy to use and typically generates a THD figure of only 0.05%. This IC is described in greater detail later in this chapter.

*Figure 3.36* shows a typical basic circuit for a dc-controlled stereo variable tone/volume IC, using pot-generated dc tone and volume control voltages, and *Figure 3.37* shows a 'remote-controlled' version of the circuit, in which the dc control voltages are derived from digital voltage generator circuitry that can be activated via panel-mounted push buttons or via a conventional remote control system. In the digital circuit, each dc control voltage is derived from the output of an 8-bit digital-to-analogue converter that is driven via a locally or remotely controlled 8-bit end-stopped up/down counter, thus enabling the dc control voltage to be varied from zero to maximum in 256 discrete steps.

## Digitally controlled ICs

During the early 1990s, the continuing demand for ever-better audio quality and ever-greater circuit sophistication resulted in the development of a new generation of variable tone/volume ICs that relied heavily on a mixture of analogue and digital – rather than purely analogue – circuit techniques, and which were specifically designed to be digitally controlled via simple microprocessor or microcontroller systems. Currently, the best known of these ICs is National Semiconductor's LMC1983, a 28-pin device that can select any one of three stereo inputs, can control the selected channel's bass and treble response and its volume, and typically generates a THD figure of only 0.008%. This IC – which, requires the use of a specially designed microprocessor interface – is described in greater detail later in this chapter.

*Figures 3.38* and *3.39* illustrate some of the basic techniques that are used in the circuitry of digitally controlled tone/volume ICs. *Figure 3.38(c)* shows how a simple electromechanical pot can be simulated by a multi-stage potential divider and a set of digitally controlled electronic switches similar to those used in the 74HC4066 IC described earlier. The diagram shows a unit using only five divider and switch sections, but in practice the unit can be given any desired number of sections. If, for example, the digitally controlled pot is to be used as a bass or treble tone control element, a total of only

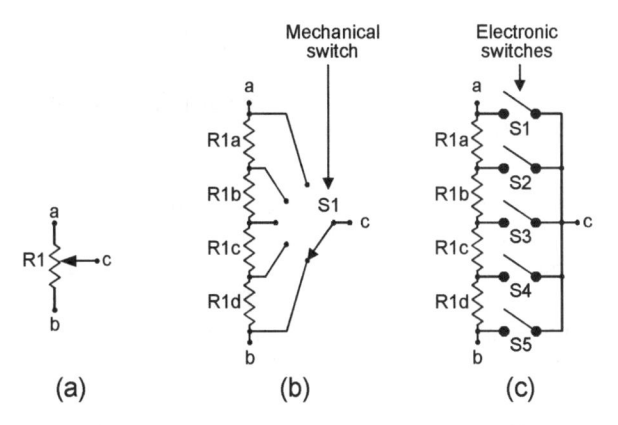

**Figure 3.38** *A simple electromechanical pot (a) can be simulated by a multi-stage potential divider and an electromechanical switch, as in (b), or by a multi-stage potential divider and a set of electronic switches, as in (c)*

86  Dedicated audio processing IC circuits

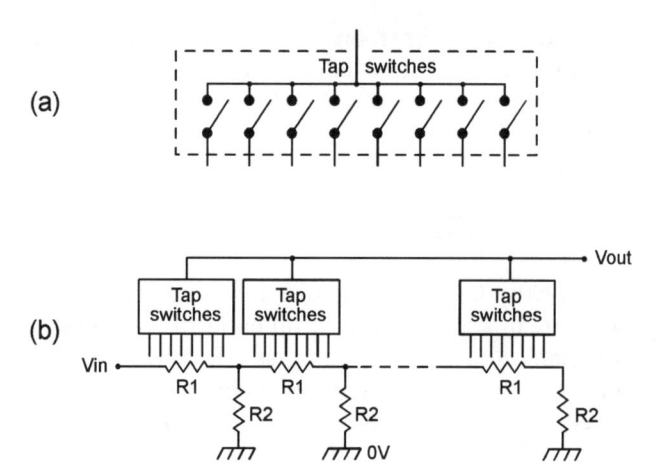

**Figure 3.39** *Groups of multi-section electronic 'tap' switches (a) used to make a variable ladder attenuator (b)*

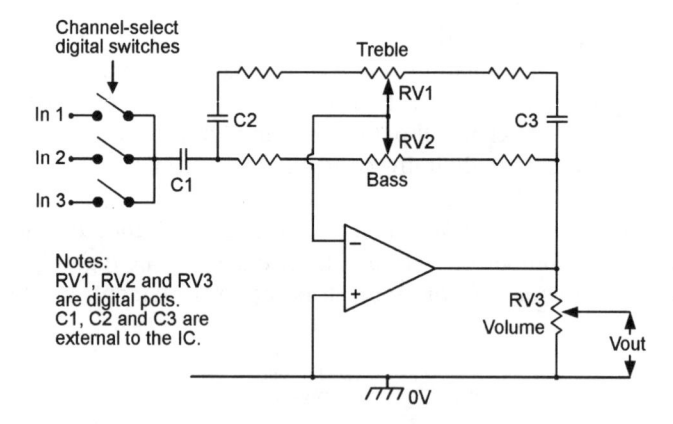

**Figure 3.40** *Typical basic circuitry of one complete channel of a stereo digitally controlled 3-input variable tone/volume IC*

twelve sections are normally considered adequate, and allow the tone response to be varied from +12dB to −12dB in 2dB steps. If the pot is to be used as a volume control, however, at least forty sections are needed, to allow the volume to be varied over an 80dB range in 2dB steps.

In practice, a 'digital' volume control may take the form of a ladder attenuator, rather than a pot, as shows in *Figure 3.39*. Here, groups of 8-way electronic 'tap' switches are used to tap into the R1 parts of the R1/R2 ladder sections, and only one switch is ever closed at any given moment, to select the desired attenuation value. This type of attenuator presents a constant input impedance at all attenuation settings.

*Figure 3.40* shows, in slightly simplified form, the typical basic circuitry of one complete channel of a stereo digitally controlled 3-input variable tone/volume IC. Here, the desired input can be selected digitally and passed on to the main tone-control circuitry via C1, and the treble, bass and volume levels can be varied digitally via the RV1 to RV3 digital pots. Note that C1, C2 and C3 are the only components that are external to the IC, and that all other components are high-precision very-low-distortion elements.

## Practical variable tone/volume ICs

### The LM1036 dc-controlled IC

National Semiconductor's LM1036 is an easy-to-use 20-pin dual (stereo) dc-controlled tone/volume/balance IC that can use any supply in the 9V to 16V range and that typically generates a THD of only 0.05%. *Figure 3.41* lists the IC's basic specification, and *Figure 3.42* shows the IC's outline (not to scale), pin notation and basic functional diagram. Note in each channel that the input signal is applied to pin 2 or 19 and then flows through the volume control network before being processed by the bass/treble tone control network; the signal is then passed to the pin 8 or 13 output terminal via the IC's volume balance network.

*Figure 3.43* shows the LM1036 IC's standard application circuit, using dc control voltages that are derived from simple pots that are powered from the IC's pin-17 5.4V zener output voltage. *Figure 3.44* shows the circuit's tone

| Parameter | Value |
| --- | --- |
| Supply voltage range | 9V - 16V |
| Supply current (typ) | 35mA |
| Zener output | 5.4V at 5mA max |
| Vin max | 1.6Vrms at 12V |
| Vout max | 1.0Vrms at 12V |
| Vgain max | 0dB at 1kHz |
| THD (typ) | 0.05% |
| Channel separation | 75dB |
| Volume control range | 75dB |
| Bass control range | ±15dB at 40Hz |
| Treble control range | ±15dB at 16kHz |
| Balance control range | +1dB to -26dB |

**Figure 3.41** *Basic specification of the LM1036 dual (stereo) dc-controlled tone/volume/balance IC*

88  Dedicated audio processing IC circuits

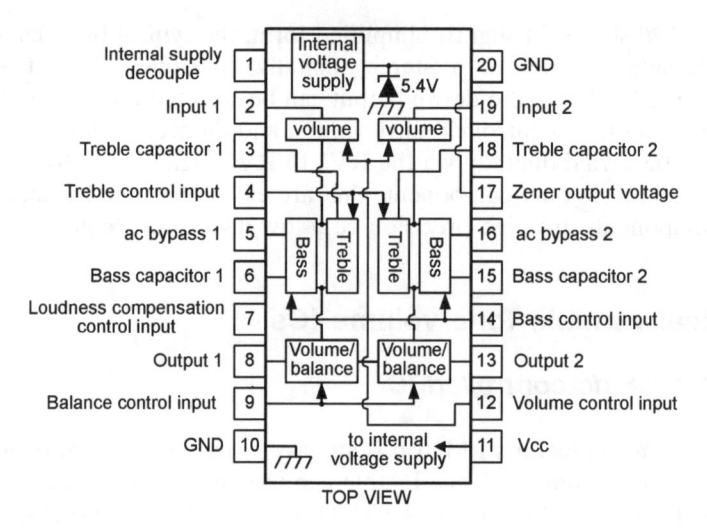

**Figure 3.42** *Outline (not to scale), pin notation and functional diagram of the LM1036*

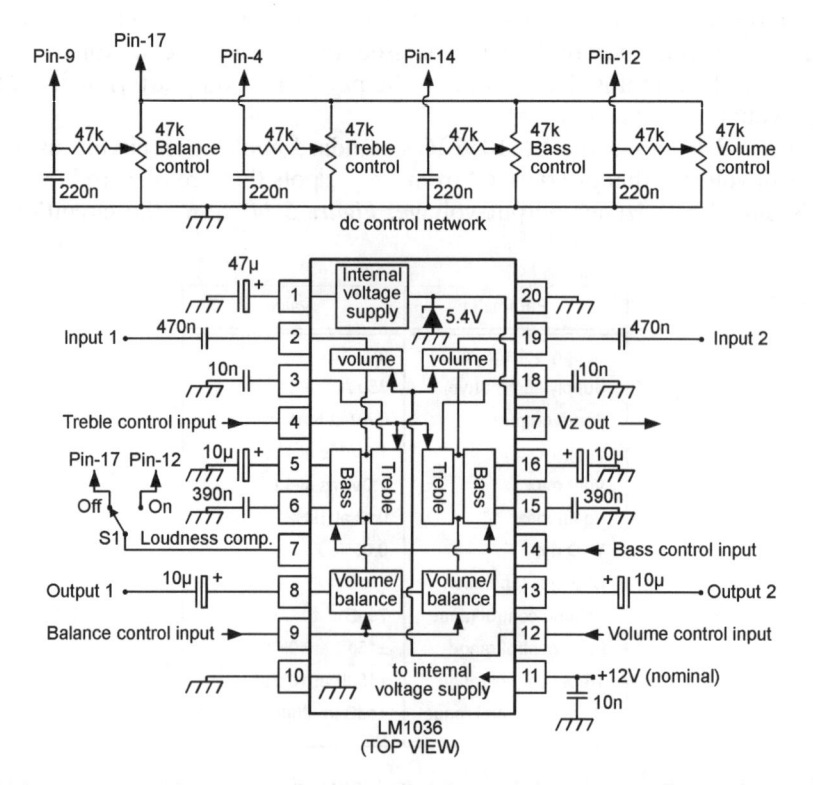

**Figure 3.43** *Standard LM1036 application circuit, using pot-generated dc control voltages*

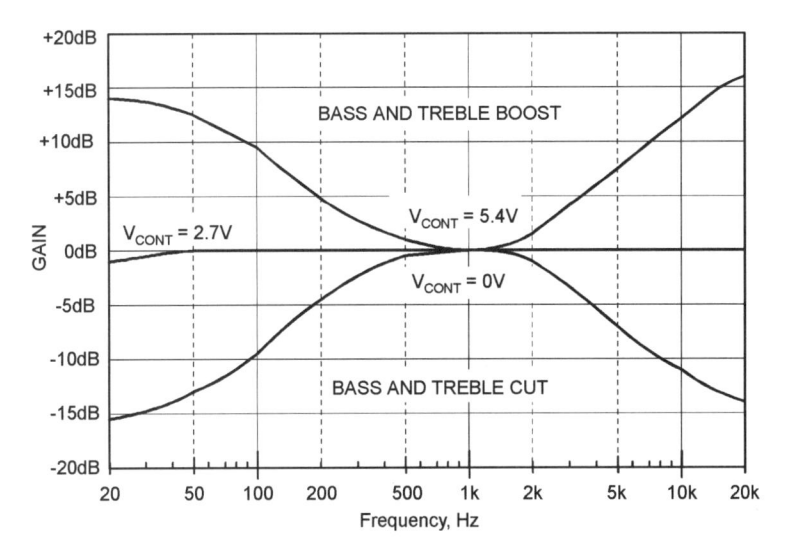

**Figure 3.44** *Tone gain characteristics of the standard LM1036 application circuit*

gain characteristics, at three different values of dc control voltage, $V_{CONT}$. Thus, the circuit gives maximum cut at zero volts, maximum boost at 5.4 volts, and gives a flat frequency response at 2.7 volts.

Normal human hearing has a sensitivity curve that peaks at 3.5kHz and falls off significantly below 300Hz and above 5kHz. The rate of fall off varies, however, with the relative mean amplitude of the signal. At 100Hz, for example, sensitivity is typically down by 10dB (relative to 1kHz sensitivity) at moderate to loud sound levels, but is down by almost 20dB at very low sound levels. Consequently, conventional *volume* control circuits (which have a perfectly flat frequency response at all signal levels) produce low-level outputs that sound distinctly bass and treble cut to the sensitive human ear. To overcome this particular problem, most good-quality hi-fi units are fitted with an optional *loudness* control that modifies the response curve of the normal volume control so that the unit produces an acoustic output that is tailored to match the characteristics of the human ear. The LM1036 is fitted with just such a facility.

The LM1036's volume control circuitry can be made to give normal volume control action by simply tying pin 7 to pin 17, or can be made to give *loudness compensation* action by tying pin 7 to pin 12. In *Figure 3.43* these two options are shown made available via switch S1. *Figure 3.45* shows the actual loudness compensated volume control frequency response curve of the standard LM1036 application circuit, with the normal bass and treble controls set to the 'flat' position. Note that, at low output levels, effective bass and treble boost is attained by simply increasing the midband attenuation.

90  Dedicated audio processing IC circuits

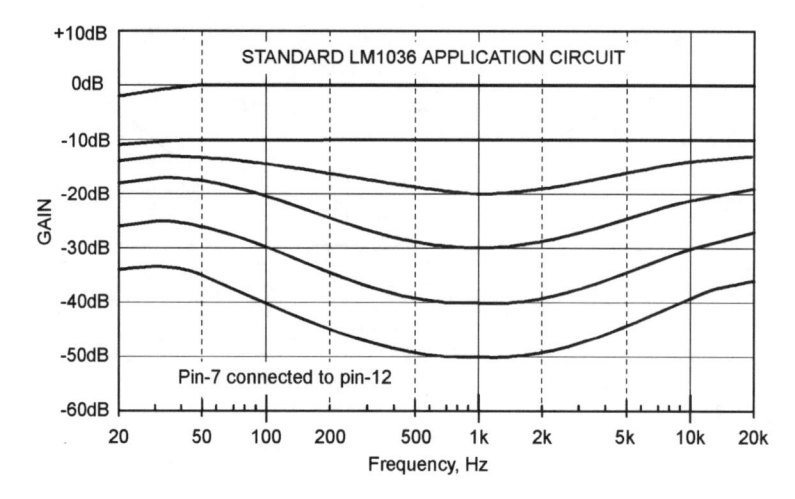

**Figure 3.45** *Loudness-compensated volume characteristics of the standard LM1036 application circuit*

## The LMC1983 digitally controlled IC

National Semiconductor's LMC1983 is a 28-pin digitally controlled 3-channel stereo selector and tone/volume control IC that can use any supply in the 6V to 12V range and that typically generates a THD of only 0.008%. *Figure 3.46* lists the IC's basic specification, *Figure 3.47* shows the IC's outline and pin notations, and *Figure 3.48* shows the IC's basic functional diagram together

| Parameter | Value |
| --- | --- |
| Supply voltage range | 6V - 12V |
| Supply current (typ) | 15mA |
| Vin max | 2.0Vrms at 12V |
| THD (typ) | 0.008% at Vin = 0.3Vrms |
| Channel separation | 80dB |
| Mute attenuation | 105dB |
| Volume control range | 80dB, in 2dB steps |
| Bass control range | ±12dB, in 2dB steps, at 100Hz |
| Treble control range | ±12dB, in 2dB steps, at 10kHz |
| Channel balance (typ) | 0.2dB |
| Signal-to-noise ratio | 95dB (0dB = 1Vrms) |
| Maximum clock frequency | 5MHz |

**Figure 3.46** *Basic specification of the LMC1983 digitally controlled 3-channel stereo selector and tone/volume control IC*

Dedicated audio processing IC circuits 91

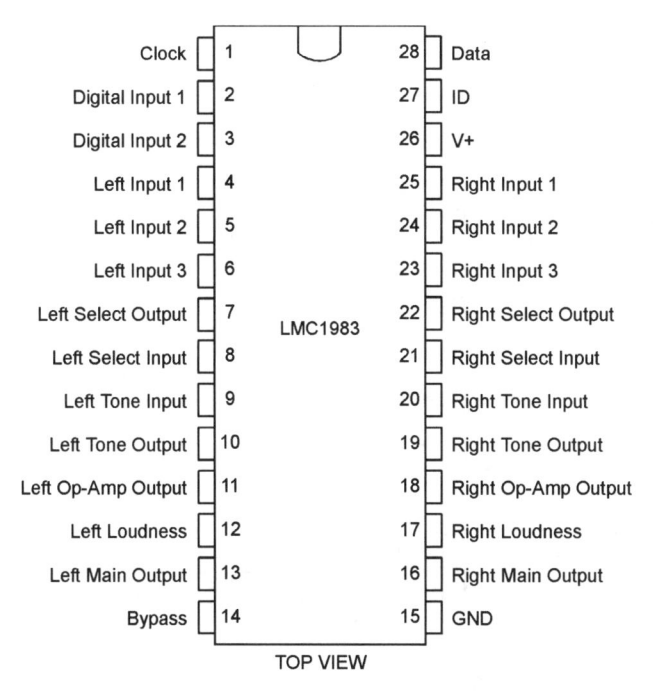

**Figure 3.47** *Outline and pin notation of the LMC1983 IC*

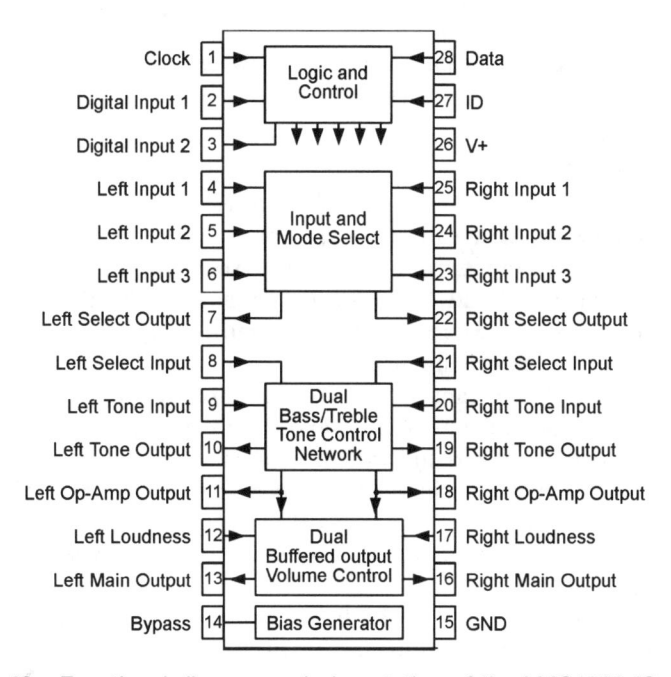

**Figure 3.48** *Functional diagram and pin notation of the LMC1983 IC*

92  Dedicated audio processing IC circuits

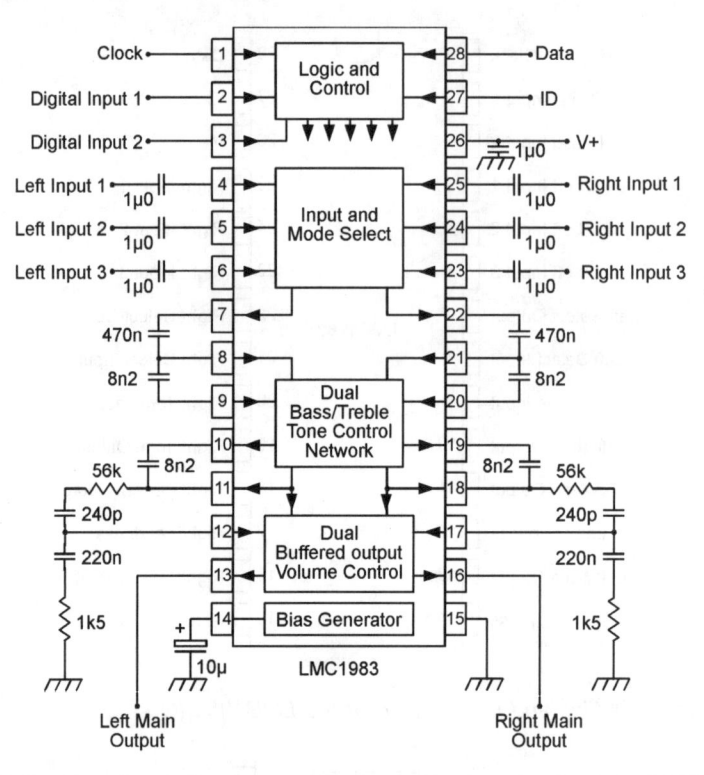

**Figure 3.49** *Functional diagram and basic application circuit of the LMC1983 IC*

with (for clarity) its pin notation. Referring to the functional diagram, the IC operates as follows.

The LMC1983's internal *Input and Mode Select* section acts like a 2-pole 4-way selector switch that can select any one of three stereo inputs or a 'mute' (zero signal) input, and makes the selected input available (in buffered form) at pins 7 and 22 via a selector that enables the signals to be presented in Left Mono, Stereo, or Right Mono form. From pins 7 and 22, the signals may be externally processed (via a dynamic noise reduction system) or can be simply capacitively coupled directly back into pins 8 and 21 of the IC. Signals fed into pins 8 and 21 are passed through the IC's tandem controlled tone and individually controlled left and right volume control networks, which are similar to the basic circuit shown in *Figure 3.40* except that the volume control network can be subjected to loudness compensation and the outputs are buffered, and are then passed out of the IC via pins 13 and 16.

*Figure 3.49* shows the LMC1983's basic application circuit, together with (for clarity) the IC's functional diagram. The 8n2 capacitors wired between pins 8 and 9 (or 20 and 21) and between pins 10 and 11 (or 18 and 19) control

Dedicated audio processing IC circuits 93

the IC's *tone response*, and the four components (56k, 240p, 220n, and 1k5) wired between pins 11 and 12 (or 17 and 18) and ground control the *loudness* response of the volume control. The settings of the selector and mute switches and the tone and volume units are all controlled via digital input signals applied to pins 1, 27 and 28 via a three-wire INTERMETAL (IM) bus interface. When the supply voltage is initially applied to the IC, however, the volume is automaticaly set to minimum and the tone controls are set to flat. Digital input pins 2 and 3 do not form a direct part of the control system, but allow peripheral devices to send single-bit data to the unit's external microprocessor control system via the DATA pin (pin 28).

Note at this point that, since it uses a microprocessor-based digital control system, the LMC1983 is not really suitable for use in one-off or short production-run audio systems, but *is* highly suitable for use in long production-run audio systems, where (since it uses no external pots or switches and uses very few other external components) it offers great economies in system manufacturing costs. Design engineers considering serious use of the LMC1983 should note the following data regarding the IC's digital control system.

Control instructions are sent to the LMC1983 via digital signals applied to pins 1, 27 and 28 via a three-wire IM bus interface. Pin 28 is the serial DATA input. Signals arriving here take the form of a 16-bit word in which the first eight bits represent an Address, which specifically selects the LM1983 IC and one of its eight basic functions, and the remaining eight bits represent a Data word that sets the selected function to the desired value. Each one of the 16 bits is clocked into the IC on the rising edge of a clock signal applied to pin 1, and various signals on the pin-27 ID (IDENTITY) terminal allows the IC to identify the 'Address' and 'Data' part of the 16-bit pin-28 word and to identify an EOT (end of transmission) condition. *Figure 3.50* shows the basic timing details of the LM1983 INTERMETAL bus system, and *Figure 3.51* lists the 16-bit programming codes that are used by the LM1983.

Note in *Figure 3.51* that, in the tone and volume control sections, the table lists only various 'spot' data codes and their corresponding function values, and that intermediate values can be worked out by interpolation. The tone and volume settings are variable in 2dB steps, and the volume can, for

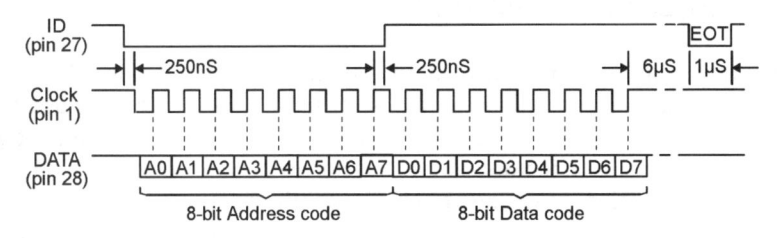

**Figure 3.50** *Basic timing details of the LM1983 INTERMETAL (IM) bus system*

94  Dedicated audio processing IC circuits

| Address (A7-A0) | Function | Data (D7-D0) | Function Selected |
|---|---|---|---|
| 01000000 | Input Select + Mute | XXXXXX00<br>XXXXXX01<br>XXXXXX10<br>XXXXXX11 | INPUT 1<br>INPUT 2<br>INPUT 3<br>MUTE |
| 01000001 | Loudness | XXXXXXX0<br>XXXXXXX1 | Loudness OFF<br>Loudness ON |
| 01000010 | Bass | XXXX0000<br>XXXX0011<br>XXXX0110<br>XXXX1001<br>XXXX11XX | -12dB<br>-6dB<br>FLAT<br>+6dB<br>+12dB |
| 01000011 | Treble | XXXX0000<br>XXXX0011<br>XXXX0110<br>XXXX1001<br>XXXX11XX | -12dB<br>-6dB<br>FLAT<br>+6dB<br>+12dB |
| 01000100 | Left Volume | XX000000<br>XX010100<br>XX101XXX<br>XX11XXXX | 0dB<br>-40dB<br>-80dB<br>-80dB |
| 01000101 | Right Volume | XX000000<br>XX010100<br>XX101XXX<br>XX11XXXX | 0dB<br>-40dB<br>-80dB<br>-80dB |
| 01000110 | Mode Select | XXXXX100<br>XXXXX101<br>XXXXX11X | Left Mono<br>Stereo<br>Right Mono |
| 01000111 | Read Digital Input 1 or 2 on IM Bus | XXXXXXD1D0 | D0 = Digital Input 1<br>D1 = Digital Input 2 |

Note: X = don't care

**Figure 3.51** *List of the main 16-bit IM bus programming codes that are used by the LM1983, together with their functions*

example, thus be set at –20dB by a Data code of XX001010 (i.e. ten binary steps greater than the XX000000 '0dB' value), or at –60dB by a Data code of XX011110 (thirty steps up on the 0dB code), and so on.

## Switched-capacitor filter IC basics

In the great majority of practical audio systems, filter circuits take the form of fairly simple 'active' (usually op-amp based) 1st- or 2nd-order low-pass, band-pass or notch types such as speech-band selectors, tone controls, graphic equalizers and hum or tone rejectors. There are, however, a few rare occasions when special high-precision frequency-tracking or 4th-order or greater filters are required, and in such cases the filters may be best designed around a special type of IC known as a switched-capacitor filter. The most obvious application

Dedicated audio processing IC circuits 95

(a) Basic switched capacitor unit (a) simulates a resistor (b) that is variable via the clock period

(c) In a real switched capacitor unit, the capacitor is switched via MOSFETs

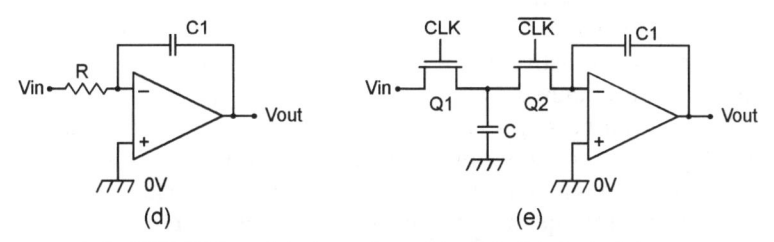

In the MF10 IC, the switched capacitor unit is used within the basic integrator circuit show in (d), using the actual connections shown in (e)

**Figure 3.52** *Diagram illustrating the basic operating principles of the MF10 switched-capacitor filter IC*

of such a filter is as an anti-aliasing auto-tracking low-pass filter in analogue or digital delay line systems, such as those described in Chapter 8, and the most popular switched-capacitor filter IC for use in this type of application is a device known as the MF10, which is available from several manufacturers.

A switched-capacitor filter is a device in which the filter's turnover frequency is directly proportional to that of an external high-frequency clock signal, and can thus be set or varied via the clock signal. This type of action is achieved by replacing the resistive elements of conventional R–C active filters with resistor simulators made from switched-capacitor units. The basic operating principle can be understood with the help of *Figure 3.52*.

Thus, referring to *Figure 3.52*, in the basic switched-capacitor unit shown in *(a)* the capacitor is alternately switched between *V*1 and *V*2 via a symmetrical clock waveform, tranferring a charge (*Q*) of C(*V*1–*V*2) in each complete

96  Dedicated audio processing IC circuits

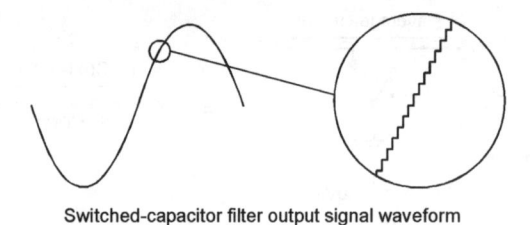

**Figure 3.53** *The signal output waveform of a switched-capacitor filter is quantized, and takes the form of a large number of discrete digital steps*

clock cycle (*T*), thus passing an averaged current of $C(V1-V2)/T$ as shown in *(b)* and acting as a simulated resistance that has a value that is directly proportional to the '*T*' value (i.e. inversely proportional to the clock frequency). The circuit's 'switch' actually takes the form of two MOSFET switches that are clocked in antiphase, as shown in *(c)*. In the MF10 IC, the switched capacitor units are used in the basic integrator circuit shown in *(d)*, using the connection shown in *(e)*. The integrator functions as a 1st-order low-pass filter that has a fall-off slope of 6dB per octave (20dB per decade).

Switched-capacitor active filter circuits produce outputs that are *almost* identical to those obtained from similar types of conventional active filters, except that their outputs appear in quantized – rather than purely analogue – form. In other words, the filter output waveforms appear as a number of small digital steps, rather than as a smoothly varying waveform. The number of steps (*S*) equals the clock frequency ($f_c$) divided by the signal frequency ($f_s$). In ICs such as the MF10, these steps are – as illustrated in *Figure 3.53* – usually too small to be seen on an oscilloscope, and their basic frequency is too high to create problems in most audio applications. If, for example, the MF10 is used as a 12kHz low-pass filter, it will use a clock frequency of 600kHz, thus producing 600 steps on a 1kHz filter signal or 100 steps on a 6kHz filter signal.

### *The MF10C switched-capacitor filter IC*

The MF10C is the most popular commercial version of the MF10 IC, and is supplied in a standard 20-pin dual-in-line plastic package. Like all MF10 ICs, it is a dual switched-capacitor filter IC that is configured as two identical general-purpose 2nd-order active filter blocks that can each be used in a variety of modes (high-pass, low-pass, band-pass, notch, or all-pass) at frequencies up to 20kHz. The two blocks can be used independently or can be cascaded to give 3rd-order or 4th-order filter action from a single IC, or can give an even greater number of orders from two or more cascaded MF10 ICs. *Figure 3.54* shows the outline (not to scale), pin notation, and simplified block diagram of the MF10C IC, and *Figure 3.55* shows the full functional block diagram of one of the two

Dedicated audio processing IC circuits 97

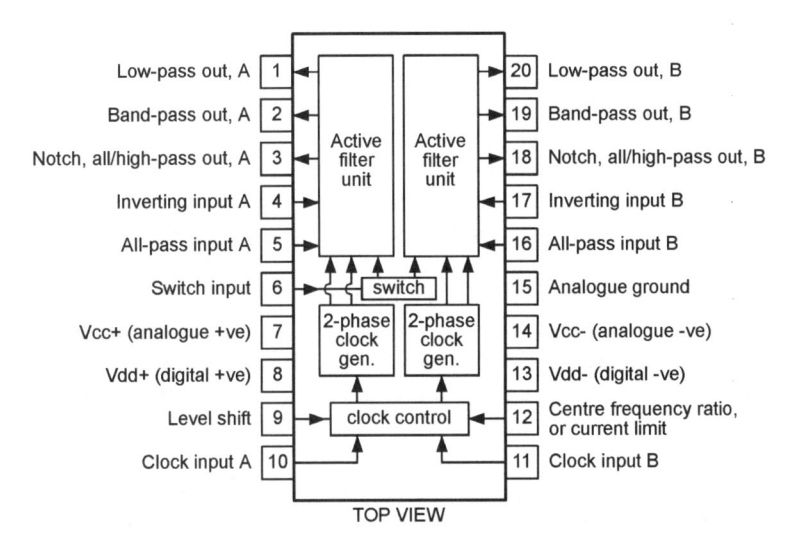

**Figure 3.54** Outline (not to scale), pin descriptions, and simplified block diagram of the MF10C dual switched-capacitor filter IC

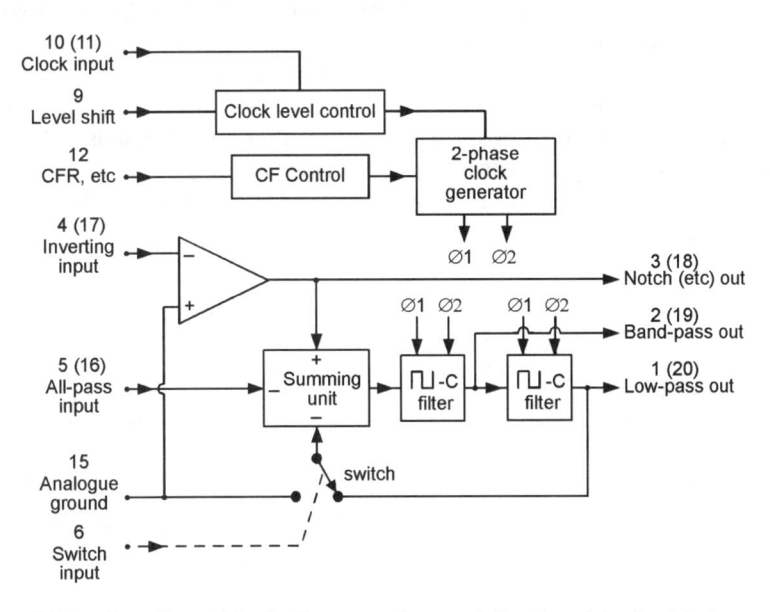

**Figure 3.55** Functional block diagram of one of the two identical halves of the MF10 IC

identical halves of the IC, which operates as follows.

Referring to *Figure 3.55*, each active filter unit consists of an op-amp, a special 3-input 'summing' unit that subtracts two of the inputs from the third, two cascaded non-inverting switched-capacitor filter units (which are similar

## 98 Dedicated audio processing IC circuits

to those shown in *Figure 3.52(e)* but have a modified switching arrangement), and a feedback control switch. These components can be configured, via the switch and/or by using various feedback connections, to make the unit function as any one of the five basic types of active filter, and – by using suitable feedback resistors – to give any one of several types of response, including Butterworth, Chebyshev, and Bessel. The IC can use single-ended or split (±5V) 10V supplies, and the clock network thus includes 'level shift' circuitry for use with 5V clock inputs. The pin-12 'CFR' terminal gives the filter a clock frequency ($f_c$) to filter turn-over frequency ($f_{to}$) ratio of 50:1 when tied to $V_{dd+}$, or 100:1 when tied to ground with dual supplies or to mid-supply level with single-ended supplies. The IC can accept a maximum $f_c$ input of 1MHz, and can thus give a maximum $f_{to}$ value of 10kHz with a 100:1 $f_c/f_{to}$ value, or 20kHz with a 50:1 value. The IC is thus very versatile.

Note in *Figure 3.55* that functions that are associated with pin numbers that are shown singly (such as 9 and 12) are common to both halves of the IC, and that where pin numbers are shown as a pair (such as 10 and (11)) the plain number applies to the IC's L/H filter, and the number in brackets applies to the R/H filter.

*Figure 3.56* shows the basic way of connecting an MF10C filter network in its most popular mode, as a combined 2nd-order notch, band-pass and low-pass filter. In this mode, the notch centre frequency, $f_o$, equals the clock input frequency divided by the 100:1 or 50:1 CFR value set by pin 12, and the notch $Q$ equals R3/R2. In the low-pass mode, the filter's gain equals R2/R1 and the fall-off slope equals 12dB/octave (40dB/decade).

In hi-fi audio systems, the most important application of the MF10C is as

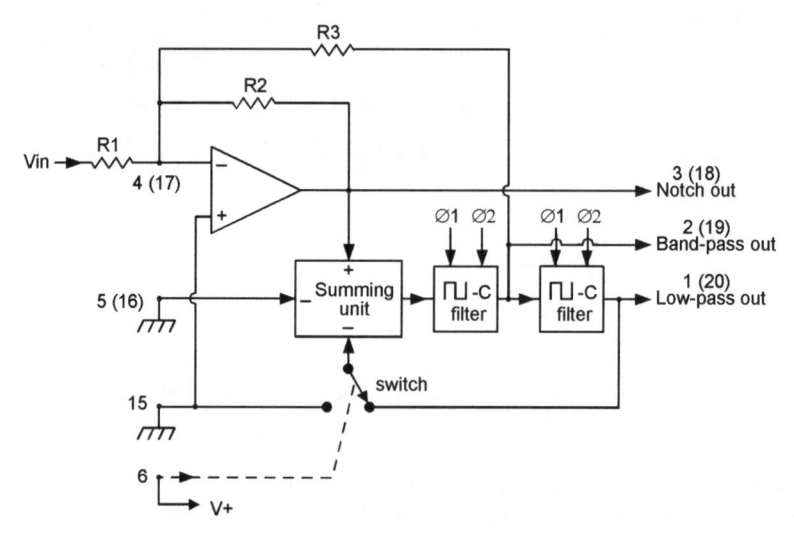

**Figure 3.56** Basic way of connecting an MF10C filter network as a combined 2nd-order notch, band-pass and low-pass filter

Dedicated audio processing IC circuits  99

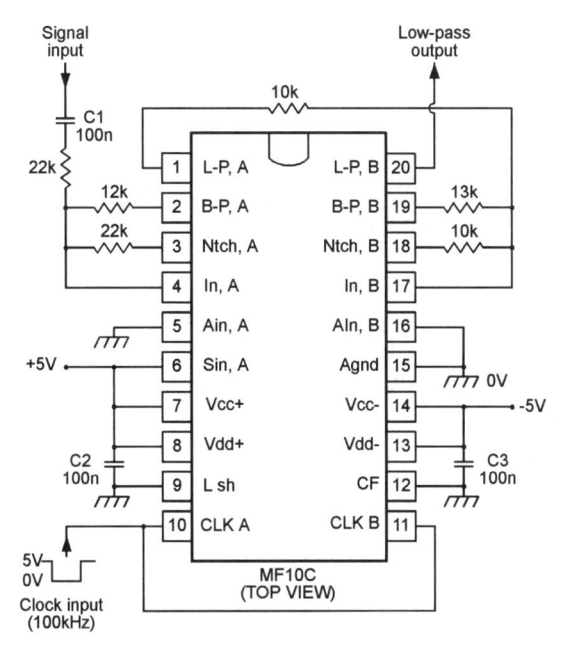

**Figure 3.57**  Connections for using the MF10C as a Butterworth 1kHz 4th-order low-pass filter, using split (±5V) power supplies.

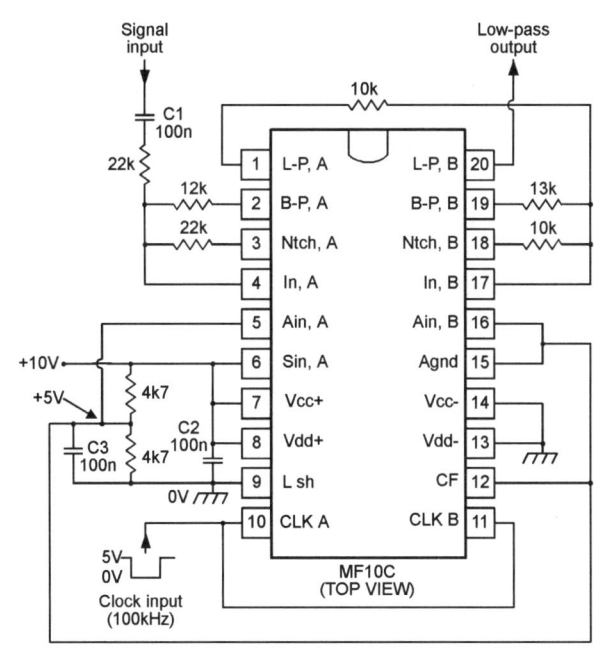

**Figure 3.58**  Connections for using the MF10C as a Butterworth 1kHz 4th-order low-pass filter, using a single-ended 10V power supply

a 4th-order Butterworth low-pass filter, which gives a fall-off slope of 24dB/octave (80dB/decade), and to conclude this chapter *Figures 3.57* and *3.58* show practical examples of the MF10C connected in this mode, as a 1kHz (nominal) filter using a 100kHz clock frequency, using alternative supply connections. In both circuits, the clock input signals are derived from normal +5V TTL or CMOS sources. In *Figure 3.57*, the circuit is shown for use with a split (±5V) power supply, with the analogue ground (pin 15) terminal, etc., tied to zero volts. In *Figure 3.58*, the circuit is shown for use with a single-ended 10V supply, and in this case a simple potential divider is used to generate a +5V 'half supply' voltage, which serves as the IC's analogue ground reference. Note in both filter circuits that the turn-over frequency (1kHz nominal with a 100kHz clock signal) can be varied from less than 100Hz up to at least 10kHz by simply changing the clock signal frequency.

# 4

# Audio pre-amplifier circuits

Each channel of a modern stereo hi-fi audio system comprises a number of interconnected circuit blocks, as indicated in slightly simplified form in *Figure 4.1*. In this example, input signals from either a radio tuner, a tape (cassette) deck, or a phono (phonograph) pre-amplifier are selected via SW1 and then fed to the input of a power amplifier stage via a tone-control system and a volume control. In practice, the tone-control system may include refinements such as 'scratch' and 'rumble' filters, etc.

The tone/volume-control system needs to be driven by input signals with fairly large mean amplitudes (typically of tens or hundreds of millivolts). Signals of suitable amplitude are usually available directly from the output of a tape or tuner unit, but not directly from the output of a device such as a magnetic phono pickup. In the latter case, therefore, the phono signal must be passed to the tone-control input via a suitable pre-amplifier (pre-amp) stage, as indicated in the diagram. An audio pre-amp is a unit that generates negligible signal distortion or noise and has excellent power-supply ripple

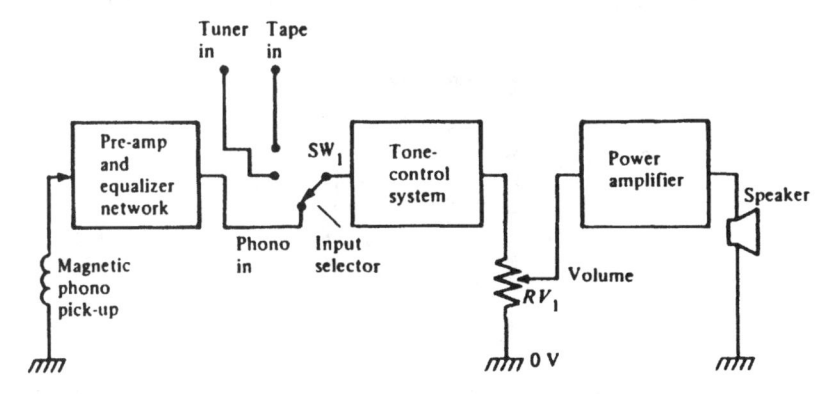

**Figure 4.1** *Block diagram of one channel of a typical hi-fi system*

rejection, and can thus amplify weak input signals up to useful output levels without introducing significant deterioration in the quality of the signal.

IC versions of audio pre-amps have been available for very many years, and for most of the past couple of decades the two most popular of these devices were the LM381 and LM382 dual pre-amp types manufactured by National Semiconductor (NS). In recent times, however, these two particular ICs have been phased out of production, and the most popular dual pre-amp IC is now the LM387, which is really a utility version of the old LM381/LM382 types. Each of the LM387's pre-amps can be regarded as an easy-to-use op-amp that operates from a single-ended power supply, and this 'op-amp' design approach typifies the current trend in IC audio pre-amp design. Thus, the second most popular 'audio pre-amp' IC is another NS product, the LM833, which the manufacturers actually describe as a 'dual audio op-amp'. Both of these ICs are described in detail in this chapter.

## The LM387 dual pre-amp IC

National Semiconductor's LM387 is a high-performance, internally compensated dual pre-amp IC with short-circuit protected outputs, and uses an 8-pin DIL package. The IC is available in a standard LM387 version and in a test-selected premium-grade LM387A version that has improved supply voltage and noise figure ratings. Both versions of the IC are internally and externally identical, and *Figure 4.2* shows the outline, pin notation and simplified equivalent circuit that is common to both versions of the IC which, as shown, can each be regarded as two independent op-amps that share a common power supply. *Figure 4.3* lists the basic performance features of the two versions of the IC.

*Figure 4.4* shows the actual internal circuit that is common to the two halves of the LM387/LM387A IC. This circuit in fact comprises four major sections, these being a 1st-stage amplifier (Q1–Q2), a 2nd-stage amplifier

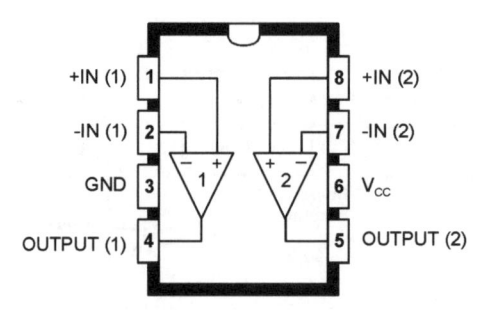

**Figure 4.2** *Outline, pin notation and simplified equivalent circuit of the LM387/LM387A IC*

*Audio pre-amplifier circuits* 103

| PARAMETER | LM387 | LM387A |
| --- | --- | --- |
| Supply voltage range | 9 to 30V | 9 to 40V |
| Supply current (typical) | 10mA | 10mA |
| Maximum voltage gain (at 100Hz) | 104dB | 104dB |
| Unity gain bandwidth | 15MHz | 15MHz |
| Power supply signal rejection (at 1kHz) | 110dB | 110dB |
| Channel separation (at 1kHz) | 60dB | 60dB |
| THD, at 60dB gain, at 1kHz | 0.1% | 0.1% |
| Total equivalent input noise (10Hz - 10kHz) | 1.0µVrms | 0.8µVrms |

**Figure 4.3** *Basic performance features of the LM387/LM387A dual pre-amp IC*

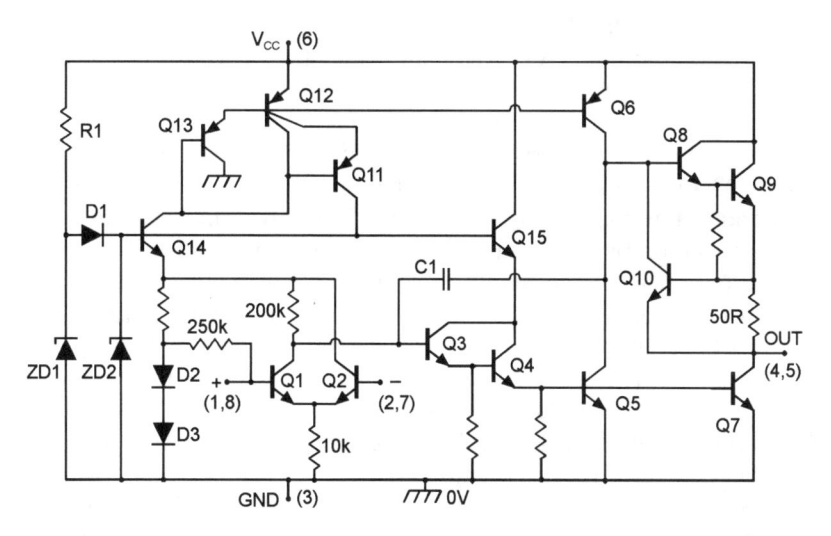

**Figure 4.4** *Internal circuit that is common to the two halves of the LM387/LM387A IC*

(Q3 to Q6), an output stage (Q7 to Q10), and a biasing network (Q11 to Q15). *Figure 4.5* shows a simplified 'equivalent' circuit of the complete pre-amplifier, showing its four major sections.

The Q1–Q2 1st-stage input amplifier of the IC is powered via the internal biasing network, and has a biasing potential of 1.2V permanently applied to Q1 base via a 250k series resistor. This 1st-stage can only be operated as a differential amplifier, and this is achieved by feeding 1.2V to Q2 base via an

104  *Audio pre-amplifier circuits*

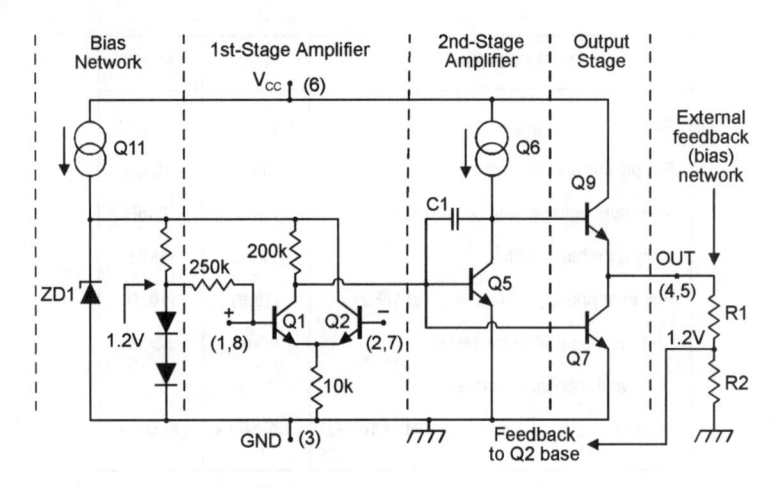

**Figure 4.5**  *Simplified equivalent circuit of the LM387/LM387A amplifier*

external R1–R2 feedback biasing network connected as shown in *Figure 4.5*. This 1st-stage amplifier gives a voltage gain of ×80 when correctly biased into the differential mode.

The 2nd-stage amplifier comprises common-emitter stage Q5 (with constant-current load Q6), which is driven from the output of Q1 via Darlington emitter follower Q3–Q4. This 2nd-stage amplifier gives an overall voltage gain of ×2000, and is internally compensated via C1 to give unity gain at 15 MHz. This compensation provides stability at closed-loop gains of ×10 or greater.

The output stage of the amplifier comprises Darlington emitter follower Q8–Q9, which is provided with an active current sink via Q7. Transistor Q10 and the 50R current-sensing resistor (connected between Q9 emitter and the OUT terminal) provides short-circuit output protection by automatically limiting the peak output current to 12mA.

The biasing network of the amplifier is designed to give a very high supply-signal rejection ratio (120dB), and consists essentially of very high impedance constant-current generator Q11–Q12–Q13, which is used to generate a ripple-free reference voltage across Zener diode ZD2. This reference voltage is then used to power the first two stages of the amplifier via Q14 and Q15, and to provide internal biasing to Q1 base. Components R1–ZD1 and D1 generate initial start-up conditions for the circuit's bias generator, and perform no other useful function.

## Basic LM387 circuits

The LM387 (and LM387A) IC is designed to operate only in the basic differential input mode. To use a LM387 pre-amp in this mode the IC must first

*Audio pre-amplifier circuits* 105

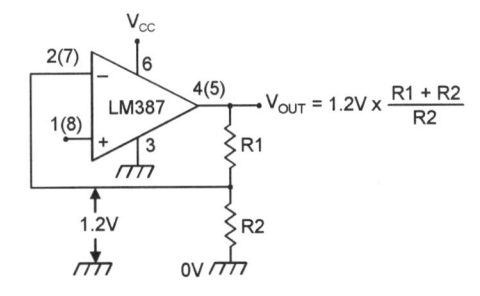

**Figure 4.6** *Differential biasing of the LM387 (or LM387A)*

be biased so that its output takes up a positive quiescent value that is independent of variations in supply voltage, and this can be achieved by connecting potential divider R1–R2 between the output and the inverting input of the IC, as shown in *Figure 4.6*, thus forming a dc negative feedback loop. The non-inverting input terminal of the IC (Q1 base in *Figure 4.5*) is internally biased at about 1.2V above zero; consequently, when R1 and R2 are connected as shown in *Figure 4.6*, dc negative feedback causes the inverting input terminal to take up a value equal to that of the non-inverting terminal (1.2V). The amplifier output therefore attains a dc value of 1V2 × (R1+R2)/R2, and can be set at any desired value by suitable choice of R1/R2 ratio. In practice, R2 should have a value less than 250k.

Note in *Figure 4.6* and all other 'LM387' diagrams shown in the rest of this chapter that the diagram is valid for both versions of the IC, and that the amplifier input and output pin numbers shown without brackets apply to the IC's left-hand (number 1) amplifier, and those within brackets apply to the right-hand (number 2) amplifier.

The *Figure 4.6* circuit can be made to act as a non-inverting ac amplifier by simply ac coupling the input signal to the non-inverting input terminal of the amplifier. In this configuration the circuit has an input impedance of about 250k: input signals must be limited to 300mVrms maximum, to avoid excessive distortion. The dc voltage gain of the circuit is determined by R1 and R2; if the desired ac gain differs from the dc value, the desired ac gain can be obtained by ac shunting one or other of the bias network resistors. *Figure 4.7*, for example, shows the circuit of a low-noise ×100 non-inverting amplifier that is optimized for operation from a 24V supply rail. In this case the dc gain is determined by R1 and R2 and is less than ×10, but the ac gain is determined mainly by R1 and R3, and approximates ×100.

The basic *Figure 4.6* circuit can be made to act as an inverting ac amplifier by ac grounding the non-inverting terminal and feeding the input signal to the inverting terminal via a gain-determining resistor, as shown in *Figure 4.8*. Here, bias resistors R2 and R3 give a dc gain of about ×10, and thus set the quiescent output at +12V. The ac gain, however, is determined by the

# 106 Audio pre-amplifier circuits

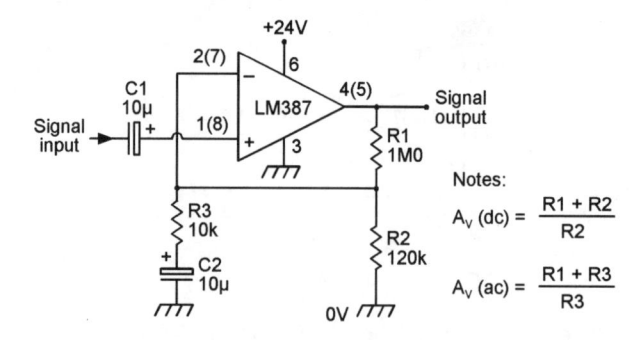

**Figure 4.7** A low-noise ×100 non-inverting amplifier

Notes:

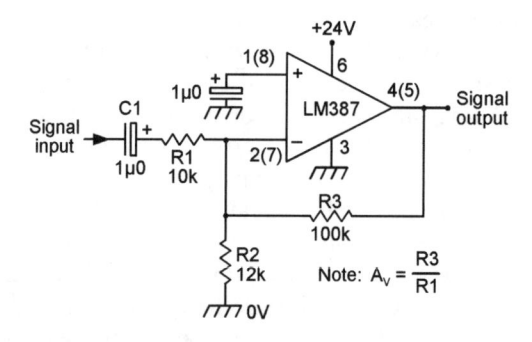

$$A_v \text{ (dc)} = \frac{R1 + R2}{R2}$$

$$A_v \text{ (ac)} = \frac{R1 + R3}{R3}$$

**Figure 4.8** Low-noise ×10 inverting amplifier

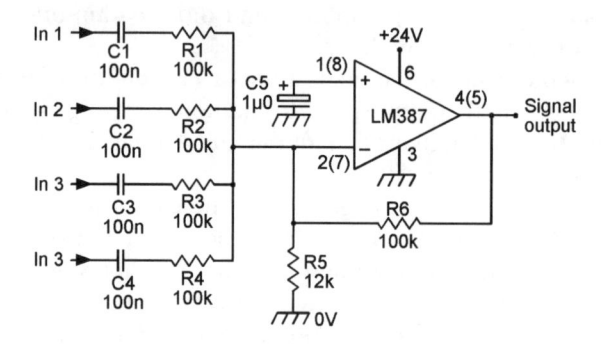

Note: $A_v = \frac{R3}{R1}$

**Figure 4.9** 4-input unity-gain audio mixer

R3/R1 ratio, and has a value of ×10 in this example: the input impedance roughly equals the R1 value. Finally, *Figure 4.9* shows how the above circuit can be made to act as a unity-gain 4-input audio mixer by simply providing each of the four input channels with its own series-input resistor; in practice,

## LM387 filter circuits

In most practical pre-amp IC applications, the device is used to act as both a low-noise amplifier and a filter or equalizer circuit, and with this point in mind *Figures 4.10* to *4.15* show various types of practical LM387 (or LM387A) filter circuit.

*Figure 4.10* shows how to modify the basic *Figure 4.7* non-inverting circuit for use as a phono pre-amplifier with RIAA equalization, and *Figure 4.11* shows how to modify it for use as a NAB tape playback amplifier. In both

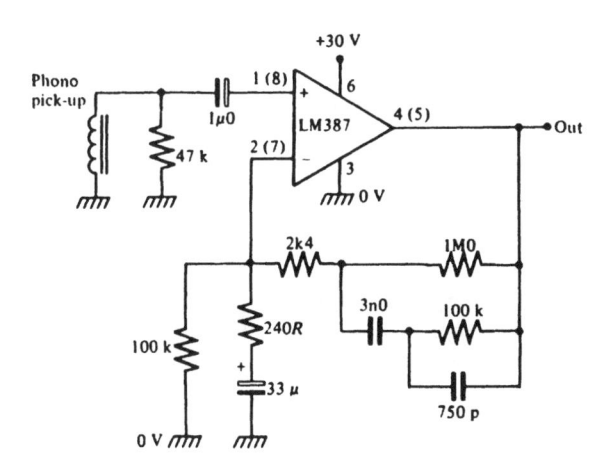

**Figure 4.10** *LM387 phono pre-amp (with RIAA equalization)*

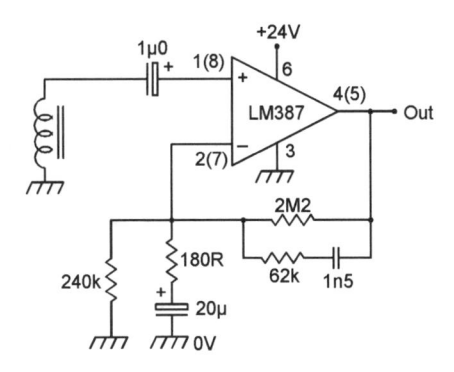

**Figure 4.11** *LM387 tape playback amplifier (with NAB equalization)*

108  *Audio pre-amplifier circuits*

cases, the low-level pickup signal is applied to the non-inverting input terminal of the pre-amp, which has its frequency response tailored by the feedback network wired between the pre-amp's output and its inverting input terminal.

*Figures 4.12* to *4.15* show various ways of using the LM387 in the inverting amplifier mode in active filter applications. The *Figure 4.12* circuit is that of an active tone control that gives unity gain with its controls in the 'flat' position, or 20dB of boost or cut with the controls fully rotated.

The 'rumble' filter of *Figure 4.13* is actually a 2nd-order high-pass active filter that rejects signals below 50Hz and does so with a slope of 12dB/octave. The 'scratch' filter of *Figure 4.14* is a 2nd-order low-pass filter that rejects signals above 10kHz. Finally, the 'speech' filter of *Figure 4.15* consists of a

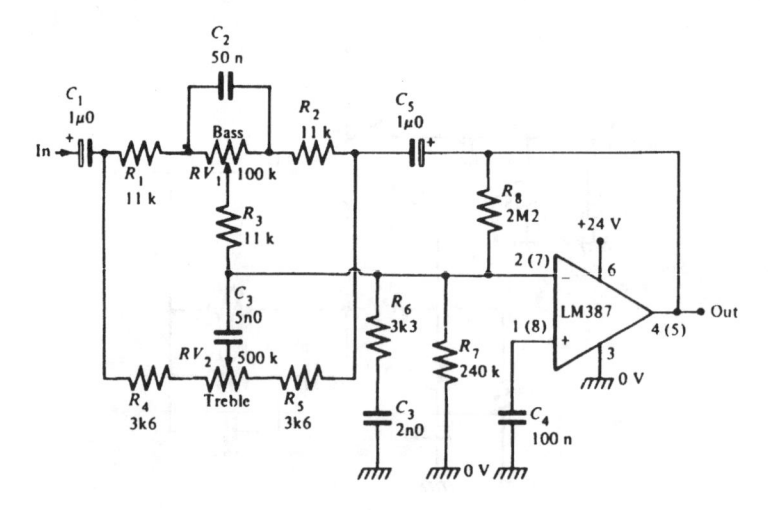

**Figure 4.12**  *LM387 active tone control circuit*

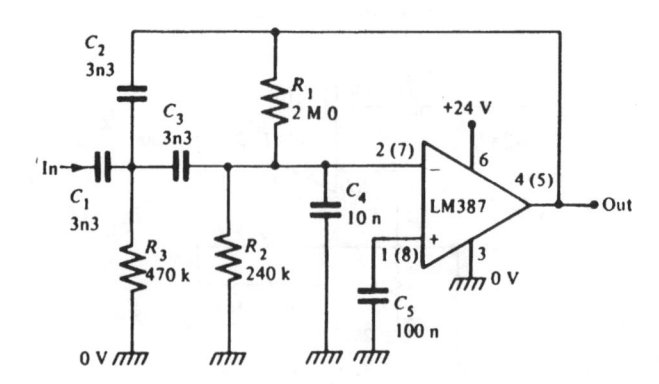

**Figure 4.13**  *'Rumble' filter*

*Audio pre-amplifier circuits* 109

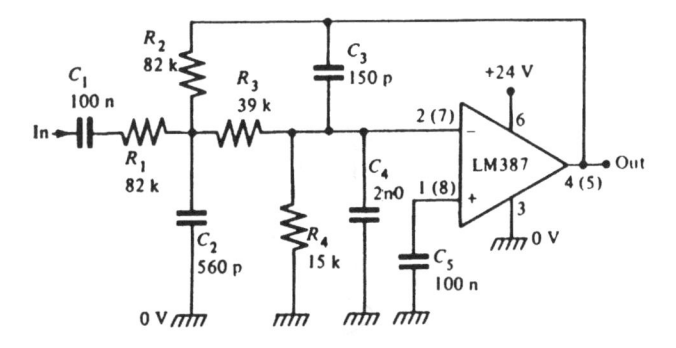

**Figure 4.14** *'Scratch' filter*

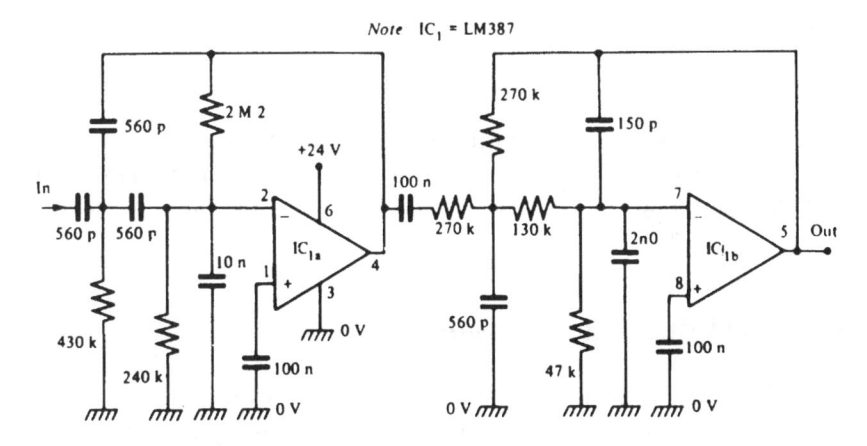

**Figure 4.15** *'Speech' (300Hz to 3kHz) filter*

2nd-order high-pass and a 2nd-order low-pass filter wired in series, to give 12dB/octave rejection to signals below 300Hz or above 3kHz.

## LM387 usage hints

Most pre-amp ICs – including the LM387 types – are high-gain, wide-band devices, and some care must consequently be taken when using them in practical circuits to ensure that they function correctly. The two most frequently encountered practical problems are those of RF-instability and RF 'pickup'.

The RF-instability problem is usually caused by inadequate high-frequency power supply decoupling: note in *all* pre-amp circuits that the power supply to the IC must be RF-decoupled by wiring a 100n ceramic or 1n0 tantalum capacitor directly across the power supply pins of the IC.

110  Audio pre-amplifier circuits

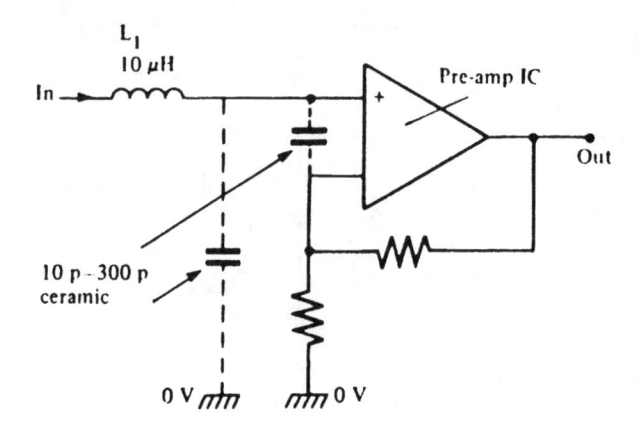

**Figure 4.16** *'RF pickup' elimination circuitry*

The RF pickup problem manifests itself in the pickup and demodulation of AM broadcast signals. This problem can usually be eliminated by wiring a 10μH RF choke in series with the IC input terminal, and perhaps by also decoupling the input terminal (or terminals) with a low-value capacitor, as shown in *Figure 4.16*.

## The LM833 dual audio op-amp IC

The LM387 dual pre-amp IC described in the previous sections of this chapter is a low-cost, low-noise device that uses a single-ended power supply but is based on a design that dates back to the 1970s. The LM833 dual op-amp IC, on the other hand, is a low-cost, ultra-low-noise bipolar device that uses a dual (split) power supply, is specifically designed for use in high-quality audio pre-amp and filter applications, and is based on modern IC technology. The LM833 has a typically noise figure that is less than one half of that of the LM387 (and about sixty times lower than that of a 741 or 747 op-amp), and has a distortion figure that is fifty times lower than that of the LM387. *Figure 4.17* shows the outline, pin notation and simplified equivalent circuit of the LM833 IC, which uses an 8-pin DIL package, and *Figure 4.18* lists the IC's basic performance features.

The LM833 IC is used in exactly the same way as any conventional bipolar op-amp, and can thus be used in any of the 741-type circuits shown in Chapter 2. The LM833 is, however, a 'dual' device in which the two op-amps are powered from the same set of split supply lines, and if – in some special application – only one of the available op-amps is used, the unwanted op-amp can be effectively disabled by simple tying its two input terminals directly to the ground (0V) supply line. Important 'application' points to note

*Audio pre-amplifier circuits* 111

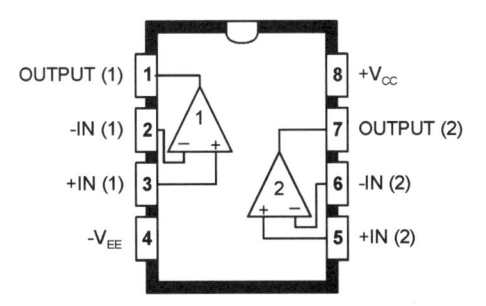

**Figure 4.17** *Outline, pin notation and simplified equivalent circuit of the LM833 IC*

| PARAMETER | LM833 |
|---|---|
| Supply voltage range | ±5 to ±18V |
| Supply current (typical) | 5mA |
| Maximum voltage gain (at 100Hz) | 110dB |
| Unity gain bandwidth (typical) | 9MHz |
| Power supply signal rejection (at 1kHz) | 100dB |
| Channel separation (at 1kHz) | 120dB |
| THD, at 60dB gain, at 1kHz | 0.002% |
| Total equivalent input noise (10Hz - 10kHz) | 0.45µVrms |

**Figure 4.18** *Basic performance features of the LM833 dual audio op-amp IC.*

about the LM833 are that it can use any split supply voltage values in the range ±5V to ±18V, and that it generates a typical THD of only 0.002% and has a typical input noise figure of a mere 0.45µVrms.

*Figures 4.19 to 4.25* show some practical applications of the LM833 IC. Note that all of these circuits are, for simplicity, shown using dual 12V supply rails, and that the op-amp input and output pin numbers shown without brackets apply to the IC's left-hand (number 1) amplifier, and those within brackets apply to the right-hand (number 2) amplifier. Most of the circuits show only one op-amp in use, as one of two identical halves of a 2-channel (stereo) system.

*Figure 4.19* shows the circuit of a general-purpose low-noise switched-gain (×10 and ×100) non-inverting audio pre-amplifier that can be used to boost very weak input signals up to a level suitable for feeding into the tone/volume control section of an audio power amplifier. Here, with S1 closed, the op-amp gives an ac signal gain of (R3+R4)/R4, i.e. very close to ×10 (+20dB),

112  Audio pre-amplifier circuits

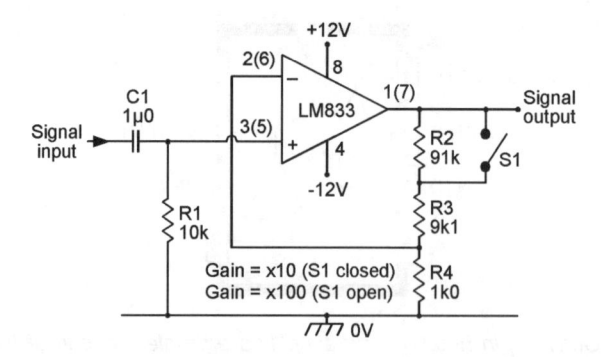

**Figure 4.19** *Circuit of a general-purpose low-noise switched-gain (×10 and ×100) non-inverting audio pre-amplifier*

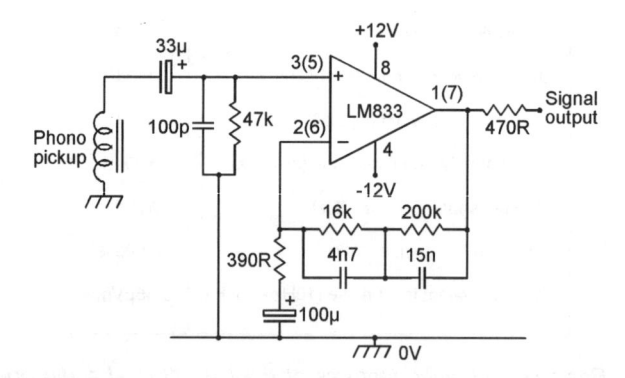

**Figure 4.20** *Low-noise RIAA phono pre-amplifier*

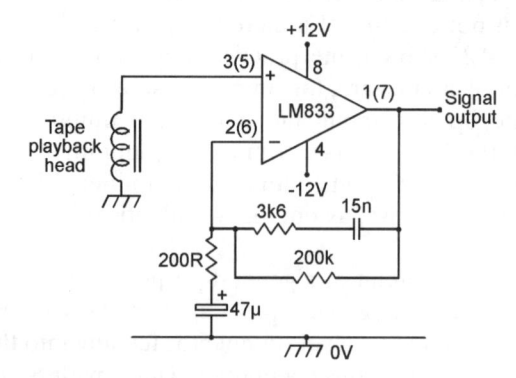

**Figure 4.21** *Low-noise NAB tape playback pre-amplifier*

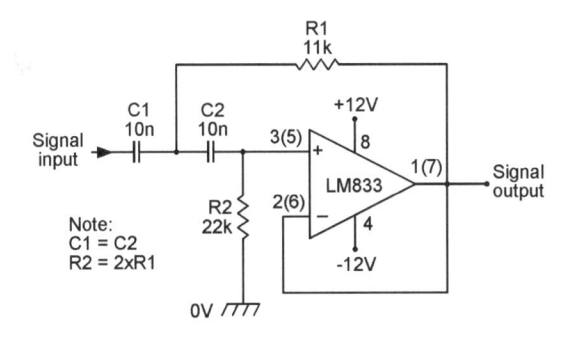

**Figure 4.22**  2nd-order 1kHz Butterworth high-pass filter circuit

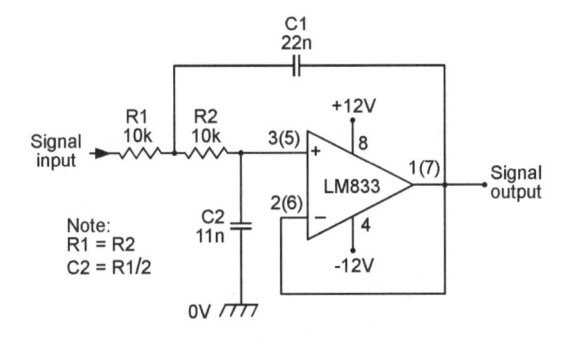

**Figure 4.23**  2nd-order 1kHz Butterworth low-pass filter circuit

but with S1 open gives a gain of (R2+R3+R4)/R4, i.e. very close to ×100 (+40dB) with the component values shown.

*Figures 4.20* and *4.21* show low-noise LM833 versions of an RIAA phono pre-amp and an NAB tape-playback pre-amplifier, respectively. The basic circuits are (with minor modifications) similar to those of *Figures 4.10* and *4.11*, but are adapted for use with conventional split-supply op-amps. Ideally, the electrolytic capacitors used in these two circuits should be non-polarized types, but in practice the bias voltages generated across these components are too small to make the use of non-polarized types absolutely necessary.

*Figures 4.22* and *4.23* show the circuits of low-noise 2nd-order Butterworth-response 1kHz high-pass and low-pass filters respectively. In these circuits, the filter turn-over frequency is inversely proportional to the values of the filter capacitor or resistor values, and can thus be increased (or decreased) by reducing (or increasing) one or other set of these component values. Thus, to double the turn-over frequency, simply halve the values of C1 and C2 or R1 and R2.

114  Audio pre-amplifier circuits

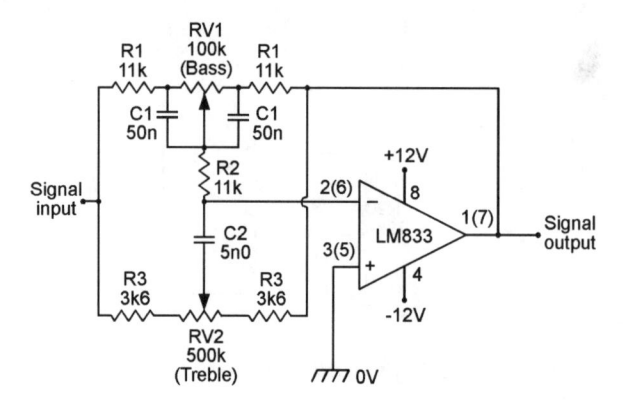

**Figure 4.24**  *Low-noise active tone-control circuit*

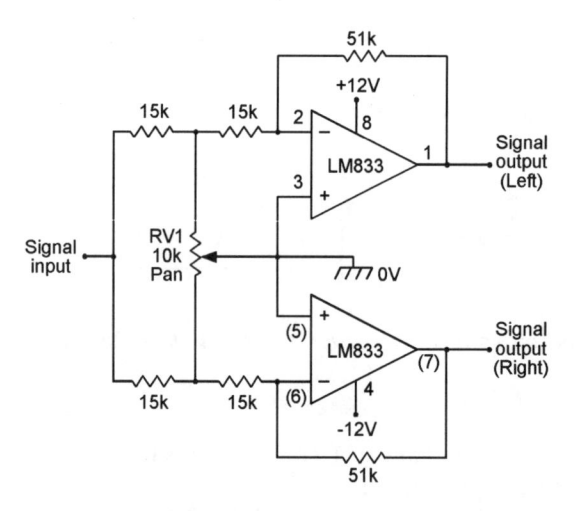

**Figure 4.25**  *2-channel pan pot circuit*

*Figure 4.24* shows the practical circuit of a low-noise active tone control that gives a flat frequency response when RV1 and RV2 are set at their middle values, but enables the bass and treble responses to be independently cut or boosted by up to 20dB. The RV1 bass control network has an upper turn-over frequency of 320Hz, and the RV2 treble control network has a lower turn-over frequency of 1.1kHz.

Finally, to complete this chapter, *Figure 4.25* shows the practical circuit of a so-called 2-channel 'pan pot', which enables a single input signal to be fed, via a pair of power amplifiers, to a pair of loudspeakers in any desired ratio via 'pan control' pot RV1. Thus, with RV1 set to one end of its travel the output signal is fed only to the left speaker, and at the other end of RV1's travel the signal is fed only to the right speaker, and with RV1 in its central position the signal is applied equally to both speakers. Thus, if one speaker is placed in the left corner of a room and the other is placed in the right corner, the apparent source point of the audio system's output sounds can be made to sweep (pan) back and forth across the room by suitably twiddling with the RV1 control.

## 5

# *Audio power amplifier circuits*

An 'ideal' audio power amplifier can be simply defined as a circuit that can deliver audio power into an external load (usually a loudspeaker) without generating significant signal distortion and without overheating or consuming excessive quiescent current. Circuits that come very close to this ideal can easily be built, using modern integrated circuits.

Simple audio power amplifiers with outputs up to only a few hundred milliwatts can be built at low cost by using little more than a standard op-amp and a couple of general-purpose transistors. For higher power levels, a wide range of special-purpose 'single' or 'dual' audio power amplifier ICs are readily available, and can provide maximum outputs ranging from a few hundred milliwatts to roughly 68W. The specific IC chosen for a given application depends mainly on the constraints of the available power supply voltage and on the required output power level or levels. The present chapter explains various audio power amplifier operating principles and then goes on to present a wide selection of practical IC-based circuits with maximum power-output ratings up to almost 5.5W; a further range of practical circuits, with power ratings in the range 6W to 68W, are described in Chapter 6.

### Audio power amplifier basics

The circuit techniques used within modern audio power amplifier ICs have evolved from those used in simple transistor circuits. Transistors can be used as reasonably linear power amplifiers by operating them in either of two basic modes, known as the 'class-A' and the 'class-B' modes. *Figures 5.1* and *5.2* outline the basic details of these two operating modes.

A basic class-A amplifier may consist of a single transistor, wired in the common-emitter mode, with the speaker acting as its collector load, as shown in *Figure 5.1(a)*. The main feature of this type of amplifier is that its input

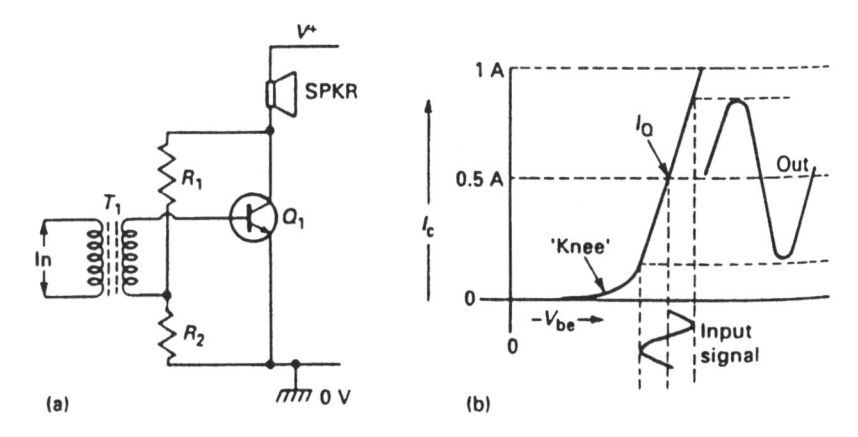

**Figure 5.1** *Basic circuit (a) and transfer characteristics (b) of a class-A amplifier*

(base) is heavily forward biased so that the collector current takes up a quiescent value roughly half way between the desired maximum and minimum swings of output current, as shown in *Figure 5.1(b)*, so that maximal undistorted output signal swing can be obtained. From this description, it can be seen that if the ac and dc impedances of the speaker load are the same, the transistor collector voltage takes up a quiescent value of roughly half of the supply voltage.

The class-A amplifier is simple and produces good low-distortion audio signals, but consumes a very high quiescent current and is relatively inefficient. Amplifier 'efficiency' can be regarded as the ratio of ac power feeding into the load, compared with the dc power consumed by the circuit. At maximum output power, the efficiency of a class-A *audio* amplifier is typically 30 percent, falling to 3 percent at one tenth of maximum output and to near-zero at very low ouput power levels.

A basic class-B amplifier consists of a pair of unbiased transistors driven in antiphase but driving a common ouput load, as shown in *Figure 5.2(a)*. In this particular design the two transistors are wired in the common-emitter mode and drive the speaker via push-pull transformer T2, and the antiphase input drive is obtained via phase-splitting transformer T1. The basic features of this type of amplifier are that neither transistor is driven on under quiescent conditions, and that one transistor is driven on by the input signal's positive half cycles and the other by its negative half cycles, with only one transistor being turned on at any given moment.

Major advantages of the class-B amplifier are that it consumes near-zero quiescent current and has good efficiency (up to 78.5 percent) under all operating conditions, but its big disadvantage is that it produces very high levels of signal distortion, as shown in the transfer characteristics graph of *Figure 5.2(b)*. Note in *Figure 5.2* that, since both transistors are operated with

118  *Audio power amplifier circuits*

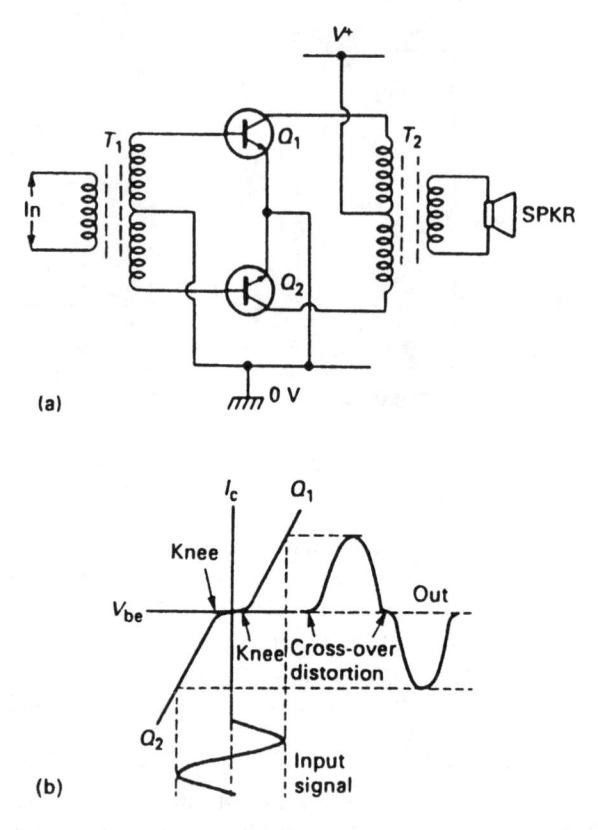

**Figure 5.2** *Basic circuit (a) and transfer characteristics (b) of a class-B amplifier*

zero bias, neither transistor conducts until its input drive signal exceeds the transistor's base–emitter 'knee' voltage (about 600mV), and this factor results in severe cross-over distortion in the amplifier's output signal, as shown in the diagram. Cross-over distortion is very objectionable to the audio listener, and the basic class-B amplifier is thus not suited to audio amplifier use. The circuit's basic 'cross-over distortion' defect can, however, easily be eliminated by modifying the circuit slightly, to convert it into what is known as a 'class-AB' amplifier.

## Class-AB amplifier basics

The cross-over distortion of the basic class-B amplifier can be virtually eliminated by applying slight forward bias to the base of each transistor via potential divider R1–RV1 as shown in *Figure 5.3*, so that each transistor passes a modest quiescent current. Such a circuit is known as a class-AB amplifier,

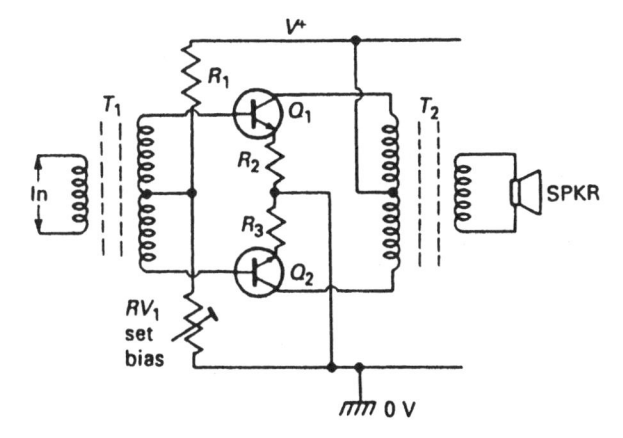

**Figure 5.3** *Basic circuit of a class-AB amplifier*

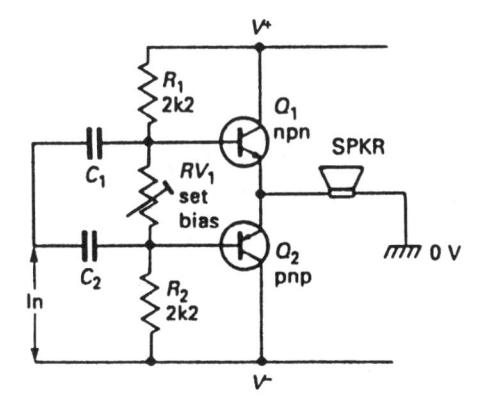

**Figure 5.4** *Basic transformerless class-AB amplifier with complementary emitter follower output and dual power supply*

and its basic features are that it operates in the push-pull load-driving mode, with its transistors biased into the linear operating mode under quiescent conditions. The precise type of circuit shown in this diagram was widely used in early transistor power amplifiers, but, due to its use of transformers for input phase-splitting and output load driving, became obsolete many years ago, and was replaced by the basic 'transformerless' type of class-AB amplifier shown in *Figure 5.4*.

The *Figure 5.4* class-AB amplifier uses a complementary pair of transistors (one pnp and one npn) wired in the emitter follower mode, and uses a split (dual) power supply. The two emitter followers are biased (via R1–RV1–R2) so that their outputs are at zero volts and zero current flows in the speaker load under quiescent conditions, but have slight forward bias applied (via

## 120  Audio power amplifier circuits

RV1), so that they pass modest quiescent currents and thus do not suffer from cross-over distortion problems. Identical input signals are applied (via C1 and C2) to the bases of both emitter followers; the circuit operates as follows.

When an input signal is applied to the *Figure 5.4* circuit the positive parts drive Q2 off and Q1 on. Q1 is an npn transistor and acts as a current source with a very low output (emitter) impedance; it feeds a faithful unity-voltage-gain copy of the input signal directly to the speaker under this condition, almost irrespective of Q1's actual parameter values. Similarly, the negative parts of the input signal drive Q1 off and Q2 on. Q2 is a pnp device and acts as a current sink with a very low emitter impedance; it sinks a faithful unity-voltage-gain copy of the input signal from the speaker under this condition, almost irrespective of Q2's actual parameter values.

Thus, the basic *Figure 5.4* transformerless class-AB circuit consumes a low quiescent current, generates good low-distortion audio signals in the output

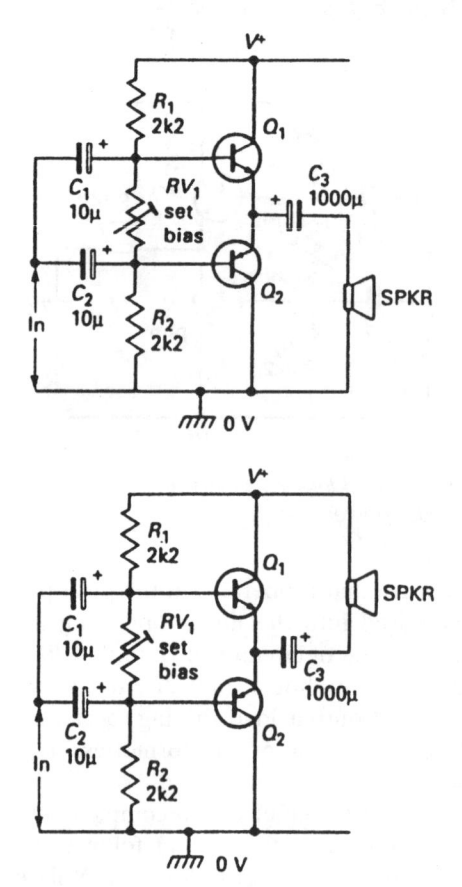

**Figure 5.5**  *Alternative versions of the transformerless class-AB amplifier with a single-ended power supply*

## Audio power amplifier circuits

speaker, and does not require the use of transistors with closely matched characteristics. The basic circuit can be modified for use with a single-ended power supply by simply connecting one end of the speaker to either the zero or the positive supply rail, and connecting the other end to the amplifier output via a high-value blocking capacitor, as shown in *Figure 5.5*.

The basic *Figure 5.4* and *5.5* circuits form the basis of virtually all modern audio power amplifier ICs. Many modifications and variations can, however, be made to the basic circuits.

### Class-AB circuit modifications

The basic *Figure 5.4* circuit gives zero overall voltage gain, and one obvious circuit modification is to provide it with a voltage-amplifying driver stage, as in *Figure 5.6*. Here, Q1 is wired as a common-emitter amplifier and drives the Q2–Q3 complementary pair of emitter followers via collector load resistor R1. Note that Q1's base bias is derived from the circuit's output via R2–R3, thus providing dc feedback to stabilize the circuit's operating points and ac feedback to minimize signal distortion.

*Figure 5.6* also shows auto-bias applied to Q2 and Q3 via silicon diodes D1 and D2, which inherently have thermal characteristics almost identical to those of the Q2 and Q3 base–emitter junctions, thus giving (when the components are mounted on a single IC chip) excellent thermally compensated auto bias. Low-value resistors R4 and R5 are wired in series with Q2 and Q3 emitters to provide a degree of dc negative feedback to the auto-biasing network.

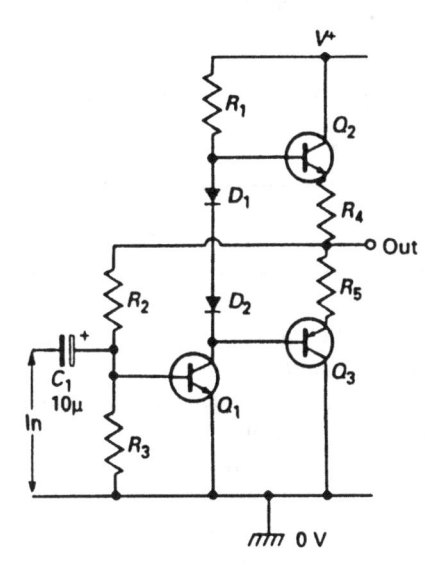

**Figure 5.6** *Complementary amplifier with driver and auto-bias*

122  Audio power amplifier circuits

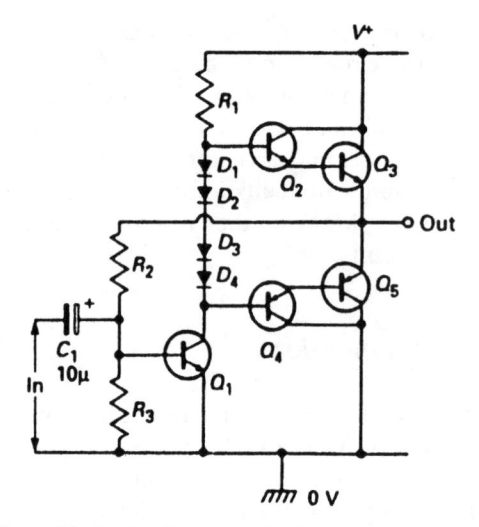

**Figure 5.7** Amplifier with Darlington output stages

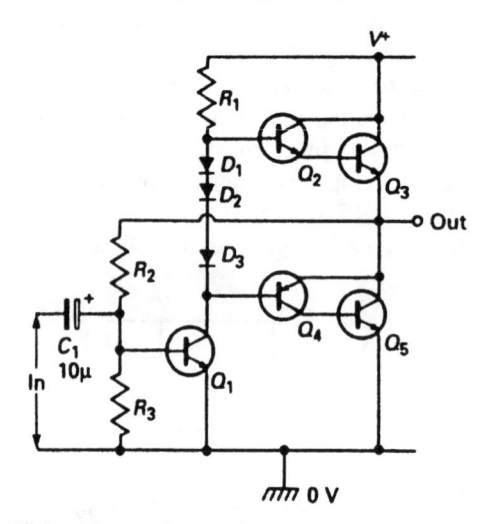

**Figure 5.8** Amplifier with quasi-complementary output stages

The input impedance of the basic *Figure 5.4* circuit equals the product of the speaker load impedance and the current gain values of Q1 or Q2. An obvious circuit improvement is to replace the individual Q1 and Q2 transistors with Darlington or Super-Alpha pairs of transistors, thereby greatly increasing the circuit's input impedance and enabling it to be used with a driver with a high-value collector load. *Figures 5.7* to *5.9* show three alternative ways of modifying the *Figure 5.6* circuit in this way.

In *Figure 5.7*, Q2–Q3 are wired as a Darlington npn pair, and Q3–Q4 as a Darlington pnp pair; note that four base–emitter junctions exist between

Q2 base and Q4 base, so this output circuit must be biased via a chain of four silicon diodes.

In *Figure 5.8*, Q2–Q3 are wired as a Darlington npn pair, but Q3–Q4 are wired as a complementary pair of common-emitter amplifiers that operate with 100% negative feedback and provide unity voltage gain and a very high input impedance. This design is known as a 'quasi-complementary' output stage, and is possibly the most popular of all class-AB amplifier configurations; it calls for the use of three biasing diodes.

Finally, in *Figure 5.9*, both Q2–Q3 and Q4–Q5 are wired as complementary pairs of unity-gain common-emitter amplifiers with 100% negative feedback, but are virtual 'mirror images' of each other. This circuit thus has a complementary output stage; it calls for the use of only two biasing diodes.

Note that the *Figure 5.6* to *5.9* circuits all call for the use of a chain of silicon biasing diodes. In practice, each of these chains can be replaced by a single transistor and two resistors wired in the 'amplified diode' configuration shown in *Figure 5.10*. Note that if R1 is shorted out the circuit acts like

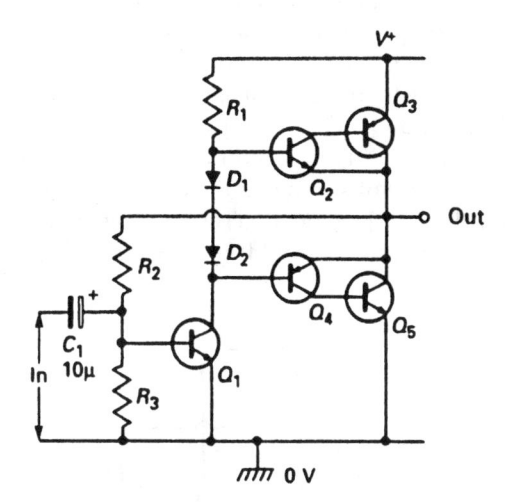

**Figure 5.9** *Amplifier with complementary output stages*

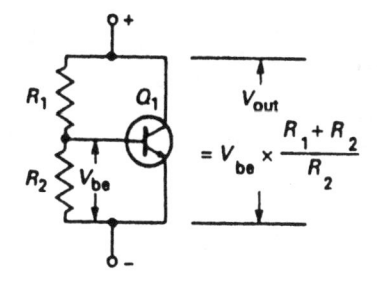

$$V_{out} = V_{be} \times \frac{R_1 + R_2}{R_2}$$

**Figure 5.10** *Fixed-gain 'amplified diode' circuit*

a single base–emitter junction 'diode', and that the circuit can be made to simulate any desired number of series-connected diodes by suitably adjusting the R1/R2 ratios.

### Driver circuit variants

In the complementary amplifier circuit of *Figure 5.6*, the main purpose of the Q1 driver stage is to provide significant overall ac voltage gain, and this is determined by the effective *signal* impedance of collector load R1. The dc resistive value of R1 is, however, usually dictated by dc current-setting biasing requirements, and there is thus sometimes a conflict between these two different R1 value requirements. These 'R1 value' problems can, in practice, easily be overcome in either of two ways. One solution is to 'bootstrap' R1, so that it presents a greater ac impedance than its true dc value. *Figures 5.11* and *5.12* show examples of bootsrapped class-AB power amplifier circuits.

Note in *Figure 5.11* that Q1's collector load comprises R1 and R2 in series, and that the circuit's output signal (which also appears across SPKR) is fed back to the R1–R2 junction via C2. This output signal is a near-unity-voltage-gain copy of that appearing on Q1 collector. Suppose that R1 has an actual value of 1k0 and that the Q2–Q3 stage gives a voltage gain of 0.9. It can be seen that, under actual amplifying conditions, $X$ signal volts appears on the low end of R2 and $0.9X$ volts appears at the top end of R2, i.e. only one tenth of $X$ signal volts are developed across R2, which thus passes only one tenth of the signal current that would be expected from a 1k0 resistor. In

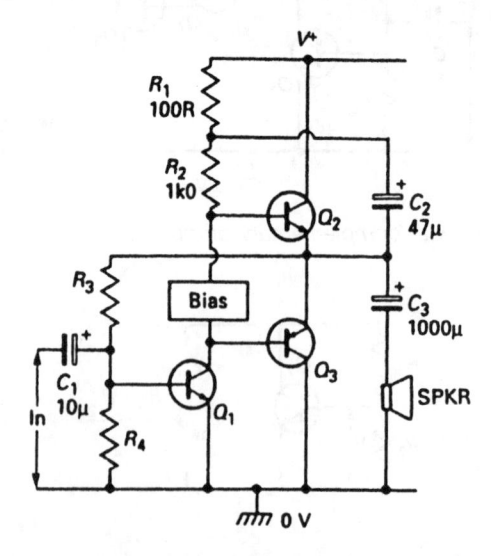

**Figure 5.11** *Amplifier with bootstrapped driver stage*

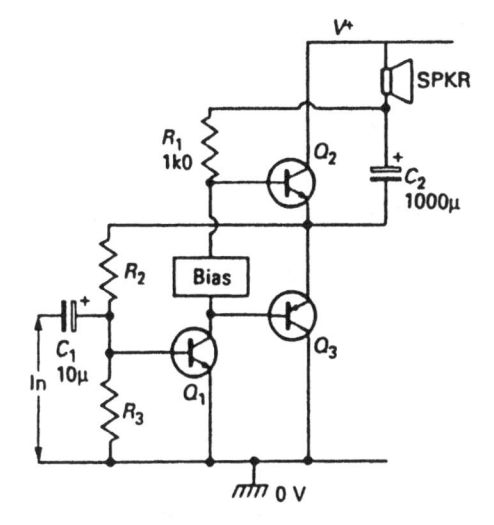

**Figure 5.12** *Alternative amplifier with bootstrapped driver stage*

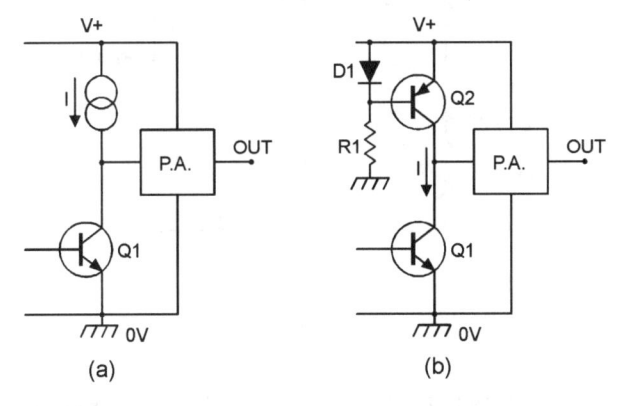

**Figure 5.13** *Symbolic (a) and actual (b) circuits of a driver stage that uses a constant-current collector load*

other words, R2's ac impedance is ten times greater (10k) than its dc value (1k0), and the signal voltage gain is similarly increased.

In practice, the bootstrapping technique enables the effective voltage gain and collector load impedance of Q1 to be increased by a factor of about ×20. *Figure 5.12* shows an alternative version of the circuit, which saves two components; the SPKR forms part of Q1's collector load, and is bootstrapped via C2.

An alternative solution to the basic 'R1 value' problem is (instead of using bootstrapping) to replace R1 with a constant-current generator that sets the desired operating current level but at the same time acts as a very high impedance load. *Figure 5.13* shows (a) the symbolic and (b) actual circuits

126  *Audio power amplifier circuits*

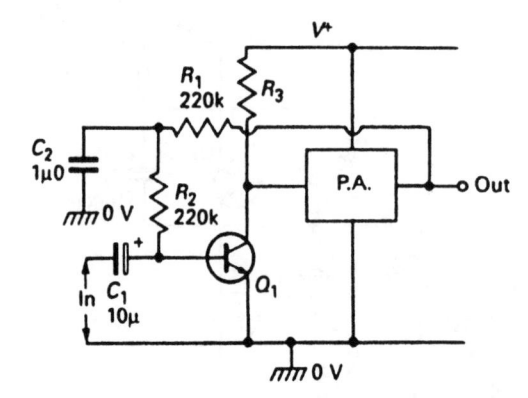

**Figure 5.14**  *Driver stage with decoupled parallel dc feedback*

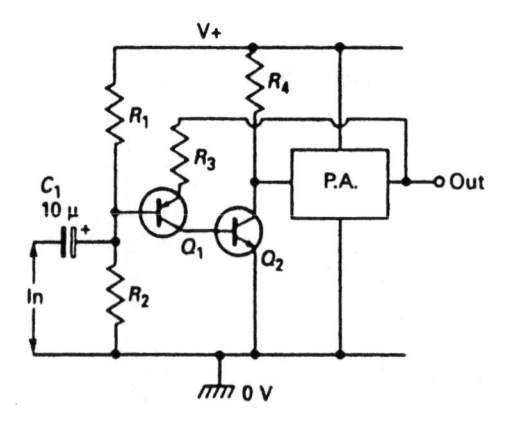

**Figure 5.15**  *Driver stage with series dc feedback*

of a driver stage that uses this technique and drives a power amplifier (P.A.) output stage. Note in *Figure 5.13(b)* that Q2 collector acts as the high-impedance constant-current source point.

Returning again to the basic *Figure 5.6* complementary amplifier circuit, note that the Q1 driver stage uses parallel dc and ac voltage feedback via the R2–R3 divider network, which in practice gives the circuit rather low values of input impedance and gain and enables it to be used over only a very limited range of supply voltages. A simple, and perhaps better, variant of this basic circuit is shown in *Figure 5.14*. It uses current feedback via series resistors R1–R2, thus enabling the circuit to be used over a wide range of supply voltages. The feedback resistors can be ac decoupled (as shown) via C2 to give increased gain and input impedance, at the expense of increased signal distortion. Q1 can be a Darlington type if a very high input impedance is required.

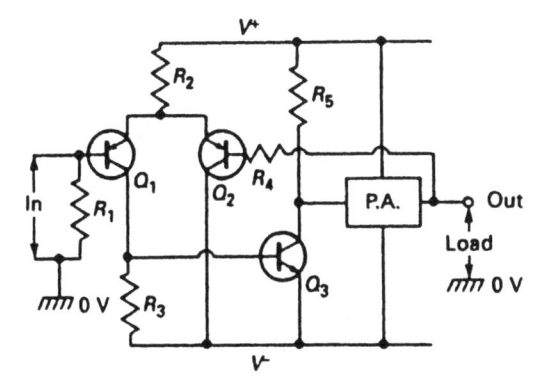

**Figure 5.16** *Driver stage with long-tailed pair input*

*Figure 5.15* shows an alternative configuration of driver stage. This design uses dc and ac feedback, and gives greater gain and input impedance than the basic *Figure 5.6* circuit, but uses two transistors of opposite polarities.

Finally, *Figure 5.16* shows a driver circuit meant for use in amplifiers that use dual power supplies and have direct-coupled ground-referenced inputs and outputs. It uses a long-tailed pair input stage, and the input and output both centre on zero volts if R1 and R4 have equal values. The circuit can be used with a single-ended power supply by simply grounding one supply line and using ac coupling of the input and the output signals. This basic circuit is used in many audio power amplifier ICs.

## An audio power amplifier IC

Practical audio power amplifier ICs use many of the circuit techniques already described in this chapter. To illustrate this point, *Figure 5.17* shows the internal circuit and pin notations/numbers of the very popular LM380 2-watt audio amplifier IC, which can (as shown later in this chapter) be used with single-ended power supplies. The IC operates as follows.

In the LM380, Q1 and Q2 are pnp emitter followers that drive the Q3–Q4 differential amplifier and enable inputs to be dc referenced to the ground line, or to be direct-coupled between ground and input lines. The output of the differential amplifier is direct coupled into the base of Q12, which is wired as a simple common-emitter amplifier with Q11 acting as its high-impedance (constant current) collector load, and the collector signal of Q12 is fed to the IC's output pin via the Q7–Q8–Q9 quasi-complementary emitter follower set of output transistors. The output currents of Q7 and Q9 are limited to 1.3A peak via R6 and R7.

128  *Audio power amplifier circuits*

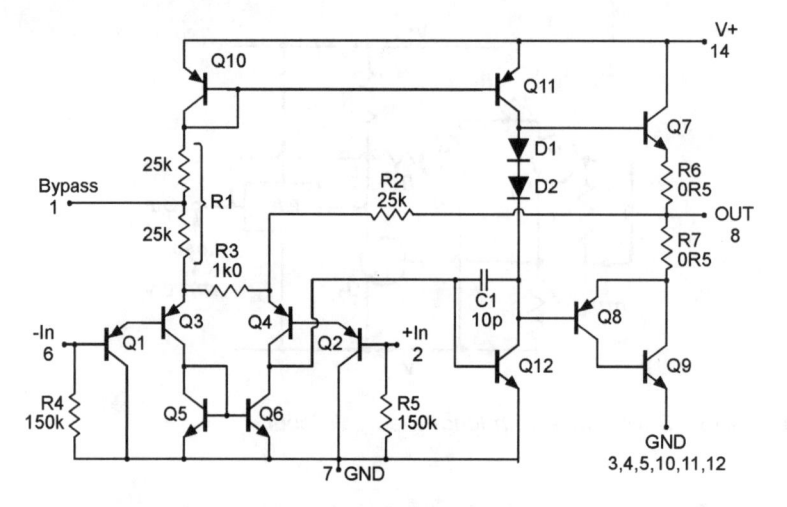

**Figure 5.17**  *Internal circuit of the LM380 2W audio power amplifier IC*

Bias-determining and gain-controlling resistor networks are built into the LM380. Feedback resistor R2 has half the value of R1, and these two resistors cause the amplifier's output to balance at a quiescent value of about half supply-line voltage. The IC's voltage gain is internally set at ×50 (34dB) by the ratios of R2 and R3, but can easily be altered by using external feedback or decoupling networks. The LM380 is thus a versatile and easy-to-use IC.

## Practical audio power amplifiers

### Low-power 'OP-AMP' circuits

The popular 741 general-purpose operational amplifier can supply peak output currents of at least 10mA, and can provide peak output voltage swings of at least 10V into a 1k0 load when powered from a dual 15V supply. This IC can thus supply peaks of about 100mW into a 1k0 load under this condition, and can easily be used as a simple low-power audio amplifier, as shown in *Figures 5.18* and *5.19*.

*Figure 5.18* shows how to use the 741 op-amp as a low-power amplifier in conjunction with a dual power supply. The external load is direct-coupled between the op-amp output and ground, and the two input terminals are ground-referenced. The op-amp is used in the non-inverting mode, and has a voltage gain of ×10 (= R1/R2) and an input impedance of 47k (= R3).

*Figure 5.19* shows how to use the circuit with a single-ended power supply. In this case the external load is ac-coupled between the output and ground,

Audio power amplifier circuits 129

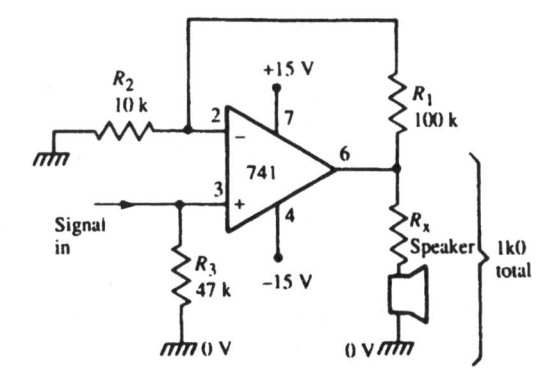

**Figure 5.18** *Low-power amplifier using dual power supplies*

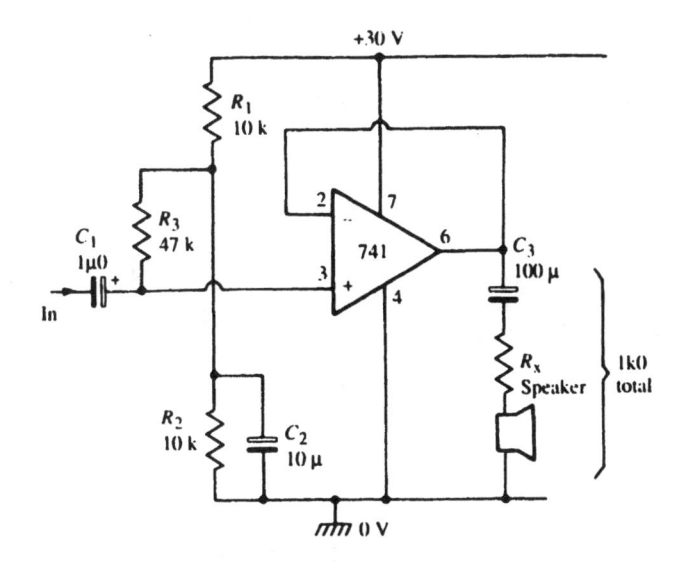

**Figure 5.19** *Low-power amplifier using a single-ended power supply*

and the output is biased to a quiescent value of half-supply volts (to give maximum output voltage swing) via the R1–R2 potential divider. The op-amp is operated in the unity-gain non-inverting mode, and has an input impedance of 47k (= R3).

Note in the above two circuits that the external load must have an impedance of at least 1k0. If the external loudspeaker has an impedance lower than this value, resistor Rx can be connected as shown to raise the impedance to the 1k0 value: Rx inevitably reduces the amount of power reaching the actual loudspeaker.

## Boosted-output OP-AMP circuits

The available output current (and thus power) of an op-amp can be boosted by wiring a complementary emitter follower between its output and its non-inverting input terminal, as shown in *Figure 5.20*. Note that this circuit is configured to give an overall voltage gain of unity, but that the Q1 and Q2 base–emitter junctions are both wired into the circuit's negative feedback loop, so that their effective forward voltage values (about 600mV) are reduced by a factor equal to the open-loop voltage gain of the op-amp. Thus, if this gain is ×10 000 the effective forward voltages of Q1 and Q2 are each reduced to a mere 6μV, and the circuit generates negligible signal distortion.

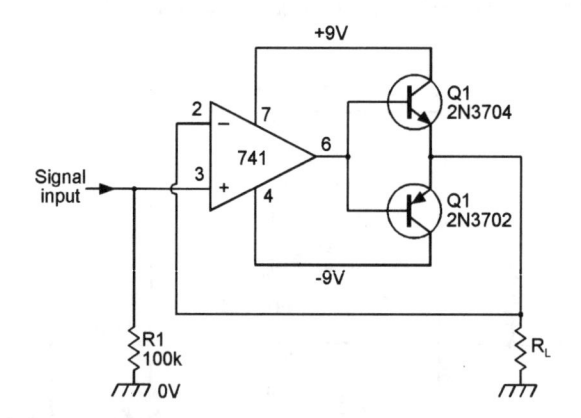

**Figure 5.20** *Basic boosted output current unity-voltage-gain op-amp circuit*

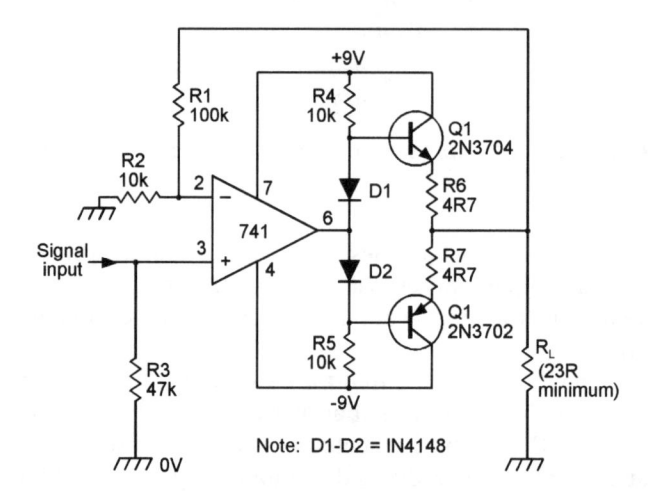

**Figure 5.21** *Op-amp power amplifier using dual supplies (power out = 280mW max)*

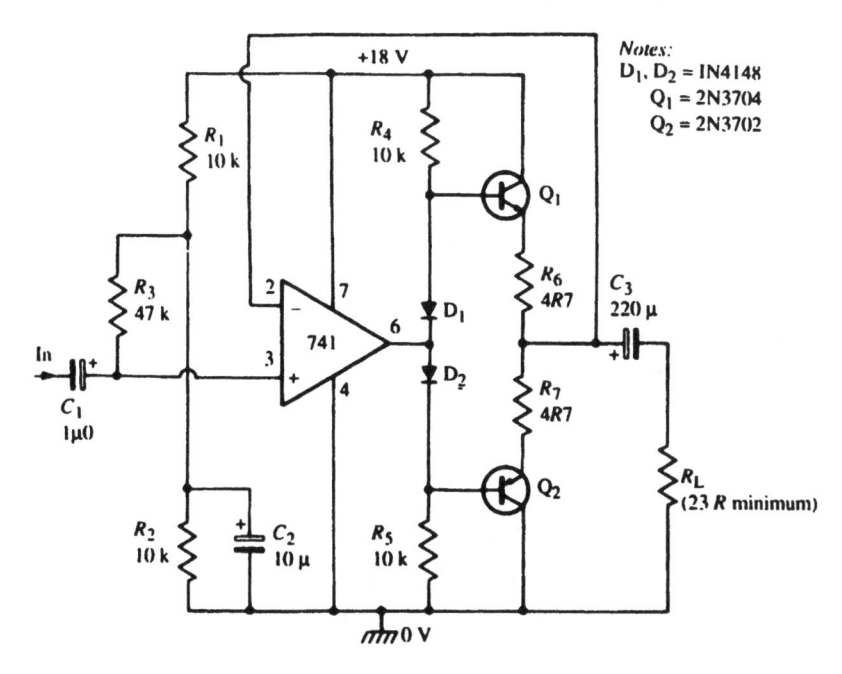

**Figure 5.22** *Op-amp power amplifier using a single-ended supply*

In practice, op-amp open-loop voltage gain falls off at a rate of about 6dB/octave, so although the signal distortion of the *Figure 5.20* amplifier may be insignificant at 10Hz, it can rise to objectionable levels at (say) 10kHz. This problem can be overcome by applying a slight forward bias to Q1 and Q2, as shown in *Figures 5.21* and *5.22*, so that their forward voltage values are reduced to near-zero and distortion is minimized.

The *Figures 5.21* and *5.22* circuits produce output currents up to at least 350mA peak or 50mA rms into a minimum load of 23R, i.e. they produce powers up to 280mW rms into such a load. These limitations are determined by the current/power ratings of Q1 and Q2, and by the power supply voltage values. The *Figure 5.21* circuit is designed for use with dual power supplies, and gives a voltage gain of ×10. The *Figure 5.22* circuit uses a single-ended supply, and gives unity voltage gain.

## IC PA usage notes

If audio output powers in the approximate range 200mW to 68W are needed, the most cost-effective way of getting them is to use a dedicated IC to do the job. A wide range of such ICs are available, in either 'single' or 'dual' form. Many of these ICs take the effective form of an op-amp with a complementary

132  *Audio power amplifier circuits*

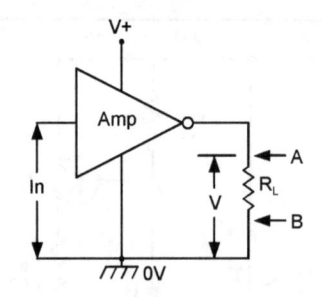

**Figure 5.23**  *An amplifier connected in the single ended output mode gives a peak output of $V^2/R$ watts*

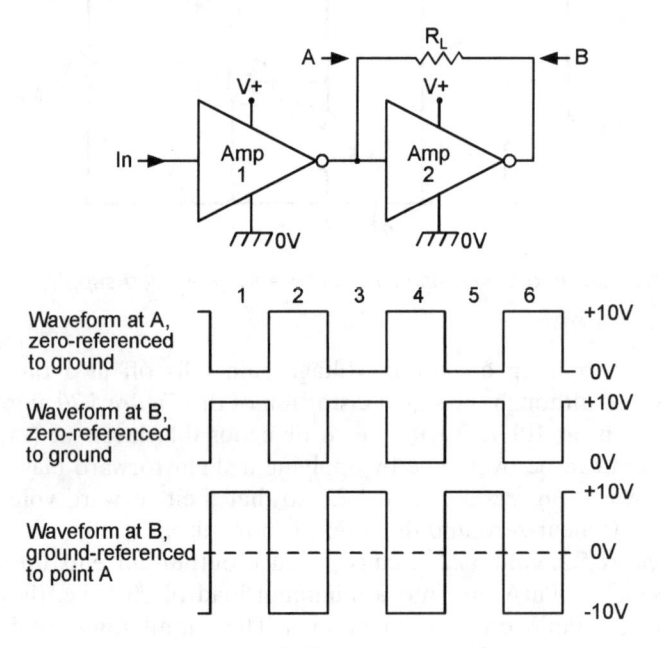

**Figure 5.24**  *A pair of amplifiers connected in the bridge mode give a peak output of $2V^2/R$ watts, i.e. four times the power of a single-ended circuit*

emitter-follower output stage (like *Figures 5.21* and *5.22*); they have differential input terminals and can provide high output current/power, but consume a low quiescent current.

When an IC power amplifier is connected in the single-ended output mode, as shown in *Figure 5.23*, the peak available output power equals $V^2/R$, where $V$ is the peak available output voltage. Note, however, that the available output power can be increased by a factor of four by connecting a pair of amplifier ICs in the 'bridge' configuration shown in *Figure 5.24*, in which the peak available load power equals $2V^2/R$. This power increase can be explained as follows.

In the single-ended amplifier circuit of *Figure 5.23* one end of $R_L$ is grounded, so the peak voltage across $R_L$ equals the voltage value on point A. In *Figure 5.24*, on the other hand, both ends of $R_L$ are floating and are driven in anti-phase, and the voltage across $R_L$ equals the difference between the A and B values.

*Figure 5.24* shows the circuit waveforms that are applied to the load when fed with a 10V peak-to-peak squarewave input signal. Note that although waveforms A and B each have peak values of 10V relative to ground, the two signals are in anti-phase (shifted by 180°). Thus, during period 1 of the drive signal, point B is 10V positive to A and is thus seen as being at +10V. In period 2, however, point B is 10V negative to point A, and is thus seen as being at –10V. Consequently, if point A is regarded as a zero voltage reference point, it can be seen that the point-B voltage varies from +10V to –10V between periods 1 and 2, giving a total voltage change of 20V across $R_L$. Similar changes occur in subsequent waveform periods.

Thus, the load in a 10V bridge-driven circuit sees a total voltage of 20V peak-to-peak, or twice the single-ended input voltage value, as indicated in the diagram. Since doubling the drive voltage results in a doubling of drive current, and power is equal to the *V–I* product, the bridge-driven circuit thus produces four times more power output than a single-ended circuit. A selection of IC-based single-ended and bridge-driven power amplifier circuits are presented in the remaining sections of this and the following chapter (Chapter 6).

## Practical ICs

A large range of audio power amplifier ICs are readily available. Some of these ICs house a single (mono) amplifier, while others house a pair (a dual) of amplifiers. *Figure 5.25* list the basic characteristics of nine popular audio power amplifier ICs with maximum output power ratings in the approximate range 325mW to 5.5W; note that the LM831, TDA2822 and LM1877 are 'dual' types, and that the LM1877, LM380 and LM384 have fully protected (short-circuit proof) output stages. The rest of this chapter is devoted to detailed descriptions of each of the above nine IC types.

### LM386 basics

The LM386 is a very popular audio power amplifier IC (manufactured by National Semiconductor) that is housed in an 8-pin DIL package and uses power supplies in the 4V to 15V range. It consumes about 4mA of quiescent

## 134  Audio power amplifier circuits

| Device number | Amplifier type | Maximum output power | Supply voltage range | Distortion, into 8R0 | $Z_{IN}$ | $A_V$ | Bandwidth | Quiescent current |
|---|---|---|---|---|---|---|---|---|
| LM386 | Mono | 325mW into 8R0 | 4 to 15V | 0.2%, Vs = 6V, Po = 125mW | 50k | 26dB | 300kHz | 4mA |
| LM831 | Dual | 220mW per channel, into 4R0 | 1.8 to 6V | 0.25%, Vs = 3V, Po = 50mW | 25k | 46dB | 20Hz - 20kHz | 6mA |
| TDA7052 | Mono | 1.2W into 8R0 | 3 to 15V | 0.2%, Vs = 6V, Po = 100mW | 100k | 40dB | 20Hz - 20kHz | 4mA |
| LM388 | Mono | 1.5W into 8R0 | 4 to 12V | 0.1%, Vs = 12V, Po = 0.5W | 50k | 26dB | 300kHz | 16mA |
| TDA2822 | Dual | 1.0W per channel, into 8R0 | 1.8 to 15V | 0.3%, Vs = 9V, Po = 0.5W | 100k | 40dB | 120kHz | 6mA |
| TBA820M | Mono | 2.0W into 8R0 | 3 to 16V | 0.4%, Vs = 9V, Po = 0.5W | 5M0 | 34dB | 20Hz - 20kHz | 4mA |
| LM1877 | Protected-output Dual | 2.0W per channel, into 8R0 | 6 to 26V | 0.04%, Vs = 20V, Po = 2W/channel | 4M0 | 34dB | 65kHz | 25mA |
| LM380 | Protected-output Mono | 3.0W into 4R0 | 8 to 22V | 0.2%, Vs = 18V, Po = 2W | 150k | 34dB | 100kHz | 7mA |
| LM384 | Protected-output Mono | 5.5W into 8R0 | 12 to 26V | 0.25%, Vs = 22V, Po = 4W | 150k | 34dB | 450kHz | 8.5mA |

**Figure 5.25**  Basic details of the nine ICs described in this chapter

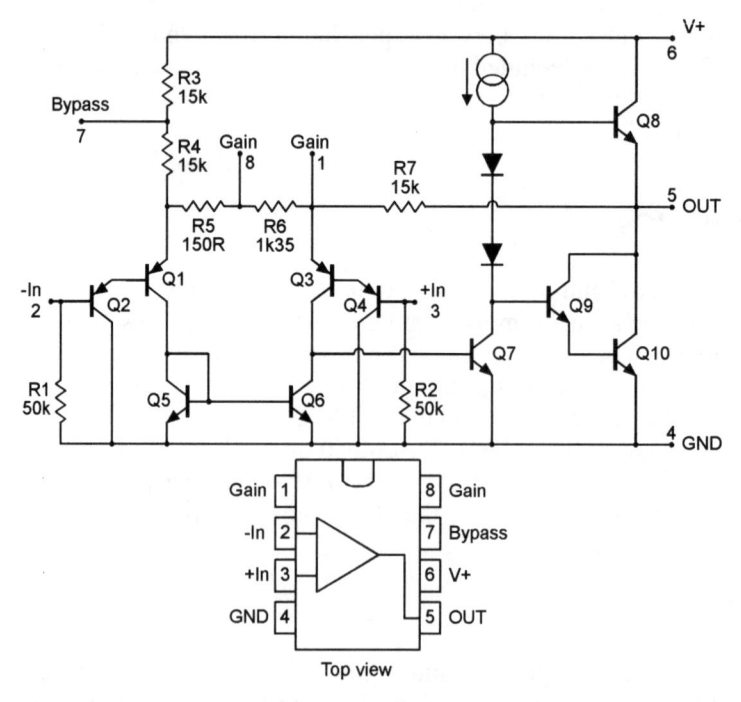

**Figure 5.26**  Internal circuit and pin connections of the LM386 low-voltage audio power amplifier

## Audio power amplifier circuits

current and is ideal for use in battery-powered applications. The IC's voltage gain is variable from ×20 to ×200 via external connections, its output automatically centres on a quiescent half-supply voltage value, and it can feed several hundred milliwatts into an 8R0 load when operated from a 12V supply. Its differential input terminals are both ground-referenced, and have typical input impedances of 50k.

*Figure 5.26* shows the LM386's internal circuit. Here, Q1 to Q6 form a differential amplifier in which both inputs are tied to ground via 50k resistors (R1 and R2) and the output (from Q3) is direct-coupled to the input of common-emitter amplifier Q7. The Q7 collector signal is direct-coupled to the IC's output terminal via class-AB unity-gain power amplifier stage Q8–Q9–Q10 which, to minimize internal volt-drops and maximize the available output power, is *not* provided with overload protection circuitry.

### LM386 applications

The LM386 is very easy to use. Its voltage gain equals double the pin 1 to pin 5 impedance value (15k in *Figure 5.26*) divided by the impedance between the emitters of Q1 and Q3 (=R5+R6). Thus, the IC can be used as a minimum-parts amplifier with an overall voltage gain of ×20 (= 2 × 15k/1.5k) by using the simple connections shown in *Figure 5.27*, where the load is ac-coupled to the IC output via C2, and the input signal is fed to the non-inverting terminal via RV1. Note that C1 is used to RF-decouple the +ve supply pin (pin 6), and R1–C3 is an optional Zobel network that ensures HF (high-frequency) stability when feeding an inductive speaker load.

*Figure 5.28* shows the above circuit modified to give an overall voltage gain of ×200 by using C4 (between pins 1 and 8) to effectively short-circuit the internal 1k35 resistor of the IC. Alternatively, *Figure 5.29* shows how the gain can be set at ×50 by wiring a 1k2 resistor (R2) in series with C4.

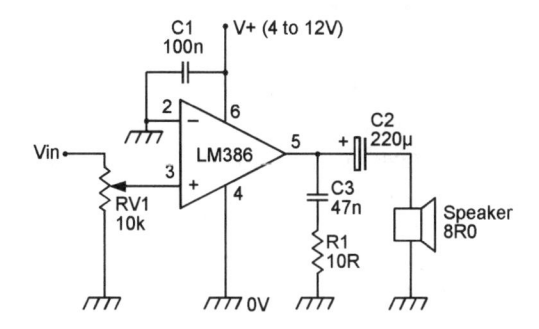

**Figure 5.27** *Minimum-parts LM386 amplifier with $A_V$ = 20*

136  Audio power amplifier circuits

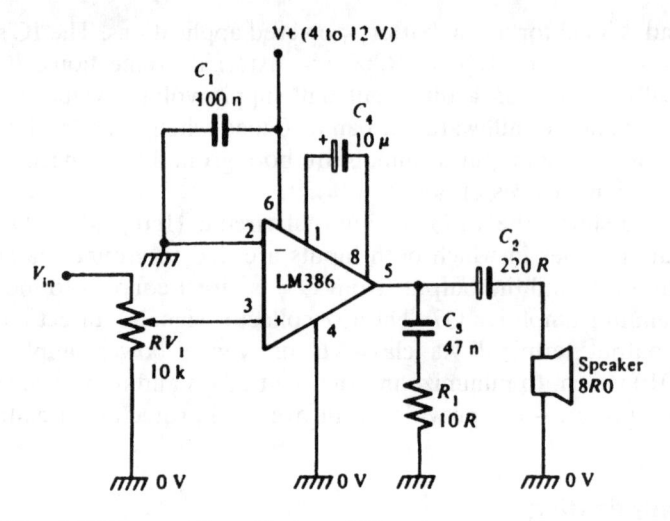

**Figure 5.28**  LM386 amplifier with $A_V = 200$

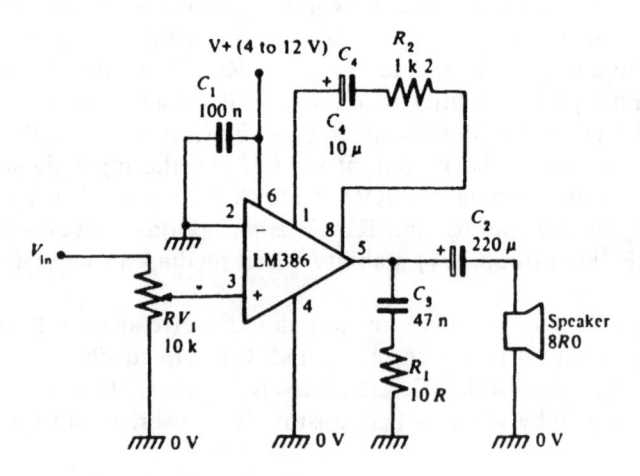

**Figure 5.29**  LM386 amplifier with $A_V = 50$

The voltage gain of the LM386 can also be varied by shunting the internal 15k pin 5 to pin 1 feedback resistor. *Figure 5.30* shows how to shunt this resistor with C4–R2, to give 6dB of bass boost at 85Hz, to compensate for the poor bass response of a cheap loudspeaker.

Finally, *Figure 5.31* shows how the LM386 amplifier can be modified for use as a built-in amplifier in an AM radio. Here, the detected AM signal is fed to the non-inverting input of the IC via volume control RV1, and is RF-decoupled via R1–C3; any residual RF signals are blocked from the load via a ferrite bead. C4 sets the IC's voltage gain at ×200. Note that this circuit is

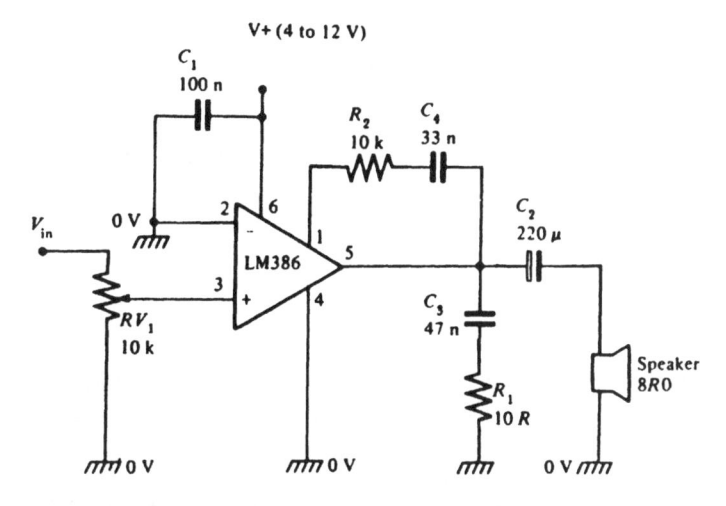

**Figure 5.30** LM386 amplifier with 6dB of bass boost at 85Hz

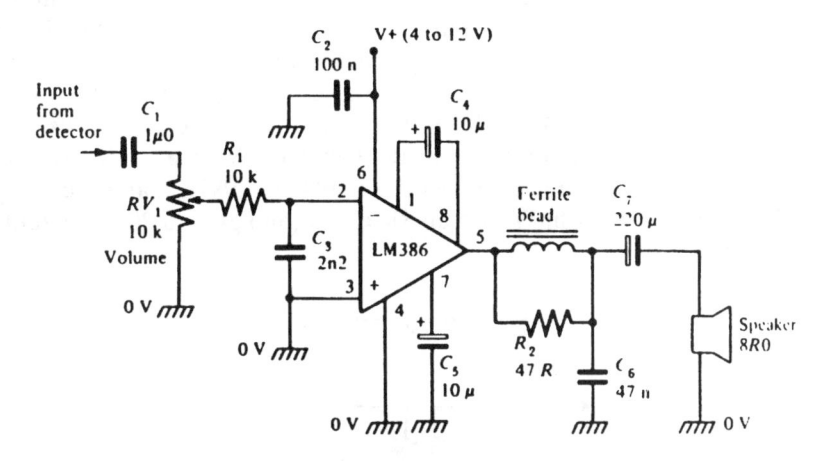

**Figure 5.31** AM-radio power amplifier

provided with additional power-supply ripple rejection by wiring C5 between pin 7 and ground; this ripple-rejection capacitor can also be used with the *Figure 5.27* to *5.30* circuits if required.

Before leaving the LM386, note that this IC is a member of a long-established family of audio power amplifier ICs designed around the basic *Figure 5.26* chip circuit, and that two of these ICs, the LM389 and the LM390, have recently been withdrawn from production through lack of demand. The LM389 contained an array of three independently accessible wide-band npn transistors on the same substrate as the basic amplifier, and

138  *Audio power amplifier circuits*

used an 18-pin DIL package. The LM390 had a modified LM386 output stage and used a 14-pin DIL package that enabled it to feed 1W into a 4R0 load when operating from a 6V supply. One other member of the family, the LM388, which can feed 1.5W into an 8R0 load when using a 12V supply, is still available and is described in greater detail later in this chapter.

## *LM831 circuits*

The LM831 is a dual power amplifier IC designed specifically for very low voltage operation; it can use supplies in the 1.8V to 6V range. Its two independent amplifiers give good low-noise and low-distortion performances, and generate minimal RF radiation, thus enabling the IC to be used in close proximity to an AM receiver. The IC is housed in a 16-pin DIL package, using the pin notation shown in *Figure 5.32*.

The two amplifiers of the LM831 can either be used independently to make a low-voltage stereo amplifier, or can be interconnected in the bridge mode to make a boosted-output mono amplifier. *Figures 5.33* and *5.34* show the circuit connections of these two options. When these circuits are powered from a 3V supply derived from two 1.5V cells, each channel of the stereo amplifier can deliver 220mW into a 4R0 speaker load (and gives a 3dB signal bandwidth of 50Hz to 20kHz), and the bridge amplifier can deliver 440mW into an 8R0 load (and gives a 20Hz to 20kHz bandwidth).

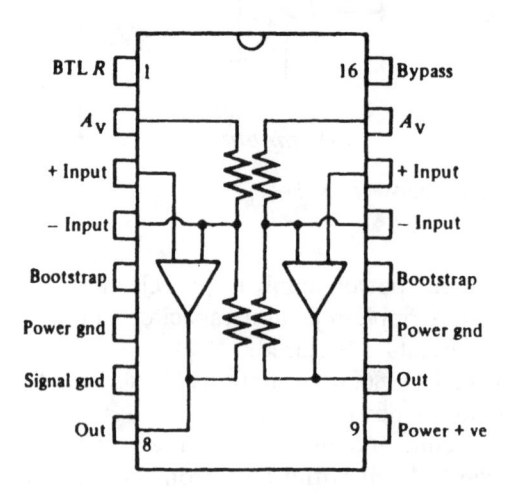

**Figure 5.32** *Outline, basic circuit and pin notation of the LM831 dual low-voltage audio power amplifier*

Audio power amplifier circuits 139

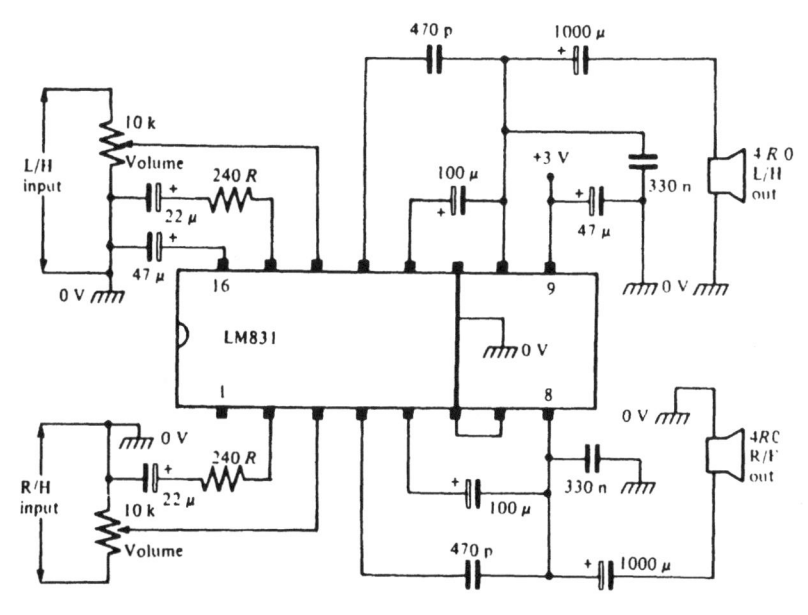

**Figure 5.33** *LM831 stereo amplifier*

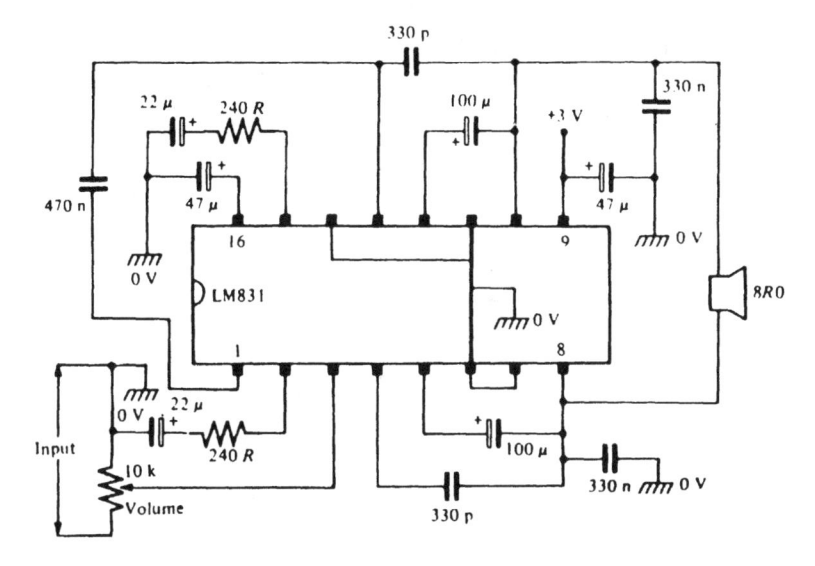

**Figure 5.34** *Bridge-connected LM831 amplifier*

When constructing these two circuits note that, in the interest of adequate circuit stability, the pcb must be laid out with large earth planes, and the pin 9 decoupling capacitor must be as close to the IC as possible and must have

140  *Audio power amplifier circuits*

a minimum value of 47µ; the two 330n decoupling capacitors must also be as close as possible to the IC.

## TDA7052 circuits

The TDA7052 is intended for use in battery-powered equipment and is designed to drive a single loudspeaker in the highly efficient 'bridge' mode, generating up to 1.2W in an 8R0 speaker when using a 6V power supply. The IC, which uses an 8-pin DIL package, is a dedicated fixed-gain (40dB) unit with a protected output and does not require the use of an external heat sink. *Figure 5.35* shows the outline, pin notation and simplified circuit of the TDA7052 IC, and *Figure 5.36* shows the IC's basic application circuit as a 1.2W amplifier.

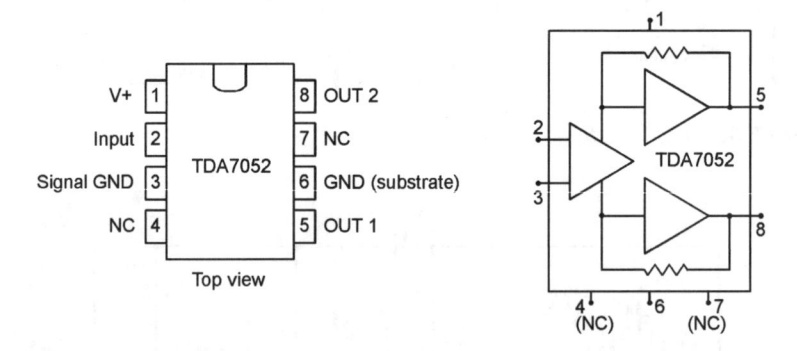

**Figure 5.35**  *Outline, pin notation and simplified circuit of the TDA7052 bridged-output 1.2W audio power amplifier IC*

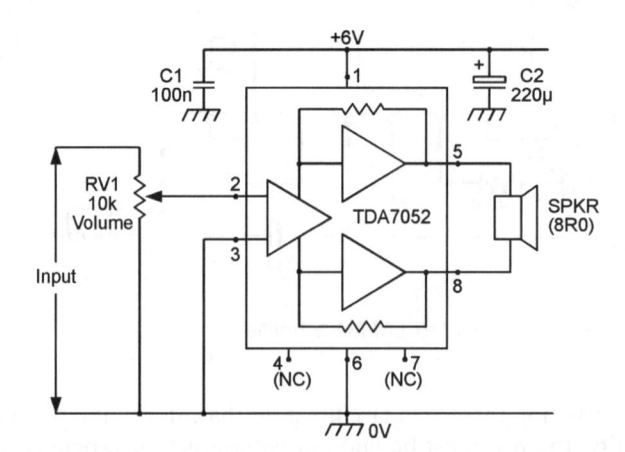

**Figure 5.36**  *Super-simple TDA7052 application circuit as 1.2W amplifier*

Audio power amplifier circuits 141

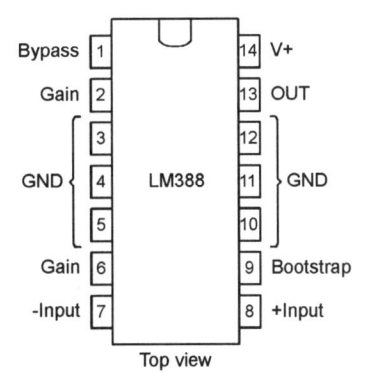

**Figure 5.37** *Outline and pin notation of the LM388 1.5W audio power amplifier IC*

## LM388 circuits

The LM388 is a slightly modified version of the LM386 and uses a 14-pin DIL package with internal heat sink (see *Figure 5.37*) and can feed 1.5W into an 8R0 speaker when powered from a 12V supply. The most significant internal difference between this IC and the LM386 relates to Q7 (see *Figure 5.26*), which uses an internal constant-current collector load in the LM386 but which uses an external load in the LM388. This 'external load' feature greatly increases the versatility of the IC.

*Figure 5.38* shows one way of using the LM388. R1 and R2 are wired in series between the positive supply line and pin 9 of the IC, to provide collector current to the internal Q7, and the R1–R2 junction is bootstrapped from the IC's output via C2, to raise the ac impedance of R2 (and thus the voltage gain of Q7) to a value far greater than its dc value. The LM388's overall voltage gain is determined in the same way as in the LM386, and equals ×20

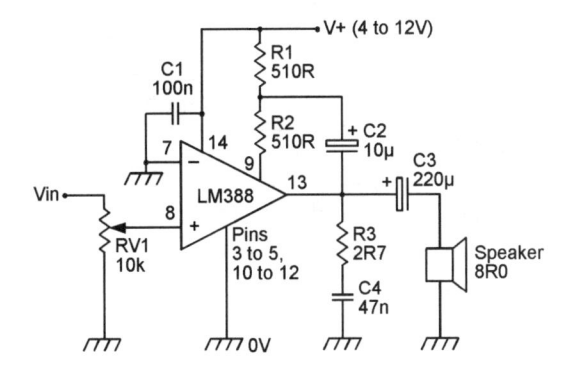

**Figure 5.38** *LM388 with a gain of ×20 and load returned to ground*

142  *Audio power amplifier circuits*

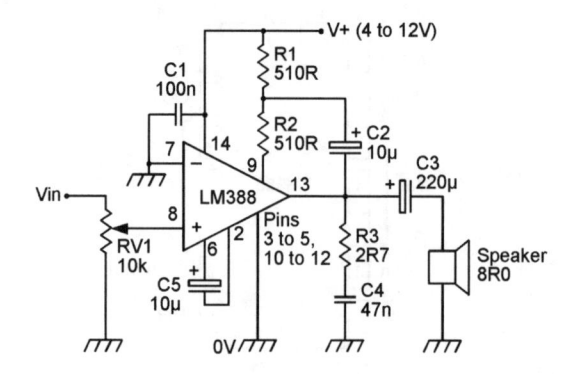

**Figure 5.39** *LM388 with a gain of ×200 and load returned to ground*

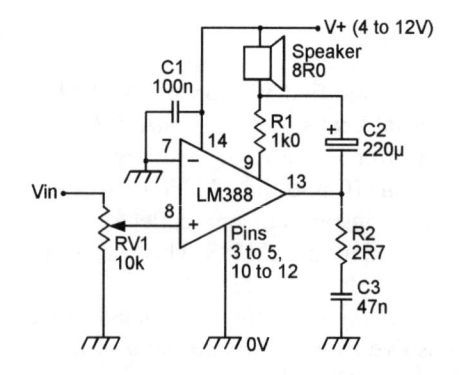

**Figure 5.40** *LM388 with a gain of ×20 and load returned to +ve supply*

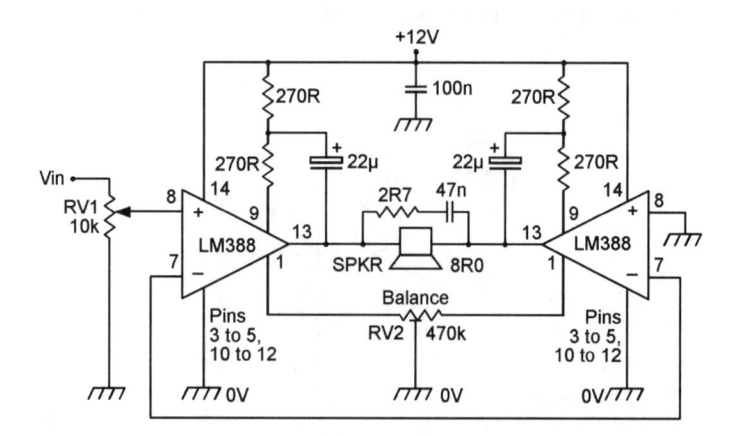

**Figure 5.41** *LM388 bridge amplifier delivering 4W to an 8R0 load*

in *Figure 5.38*. *Figure 5.39* shows how the gain can be increased to ×200, by wiring C5 between pins 2 and 6.

*Figure 5.40* shows another way of using the LM388. Here, dc current is fed to pin 9 via the speaker and R1, and the 'low' end of the speaker is ac driven by the amplifier's output, thus bootstrapping R1 and giving it a high ac impedance value. This circuit thus gives a performance similar to that of *Figure 5.38*, but does so with a saving of two components.

Finally, *Figure 5.41* shows how to connect a pair of LM388 ICs in the bridge configuration, to provide 4W of drive to a direct-coupled 8R0 speaker load when using a 12V power supply. Pre-set pot RV2 is used here to set the quiescent output of the two ICs at identical values, to minimize the circuit's quiescent current consumption.

Before leaving the LM388, note that this IC has a fairly poor supply line ripple rejection performance, and if any problems are met in this respect they can be overcome by wiring a 10μF (or larger) capacitor between pin 1 and ground.

## *TDA2822 circuits*

The TDA2822 is a versatile dual amplifier that can use any DC supply in the 1.8V to 15V range; it can be powered from a 3V supply and used to drive headphones at 20mW per 32-ohm channel, or from a 9V supply and used to drive 8R0 speakers at 1W per channel.

The TDA2822 is housed in an 8-pin DIL package (see *Figure 5.42*), and uses the minimum of external components. *Figure 5.43* shows how it can be used as a stereo speaker or headphone amplifier circuit that is powered from a 6V supply.

## *TBA820M circuits*

The manufacturers describe this device as a low-power amplifier that is capable of generating a few hundred milliwatts in a 4R0 to 16R speaker load,

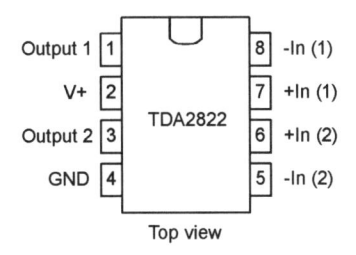

**Figure 5.42** *Outline and pin notation of the TDA2822 dual amplifier IC*

144  Audio power amplifier circuits

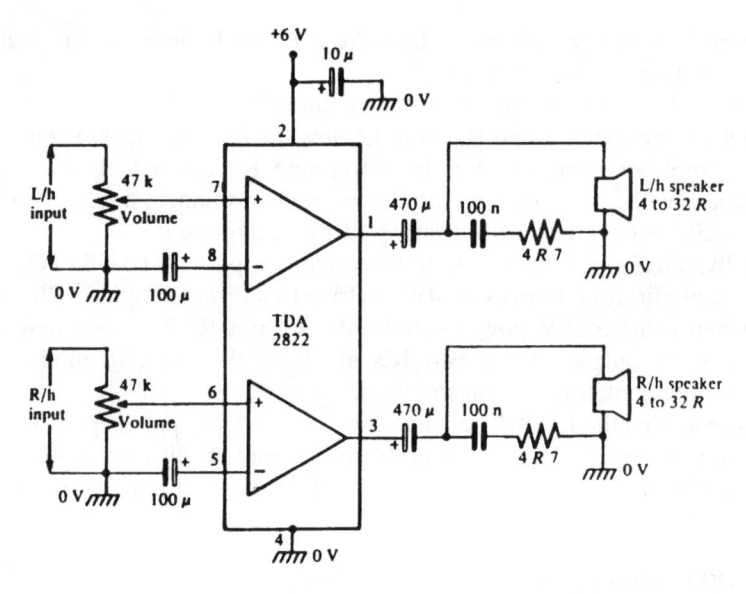

**Figure 5.43**  *TDA2822 stereo amplifier circuit*

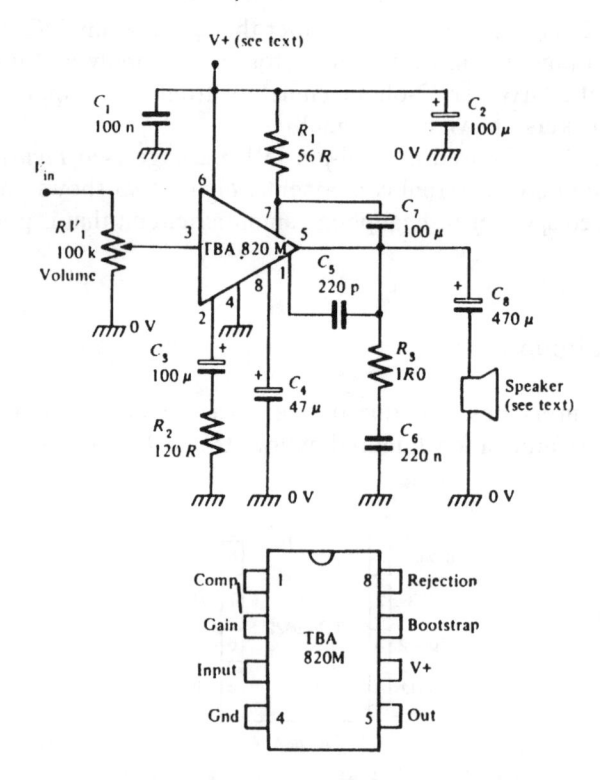

**Figure 5.44**  *TBA820M low-power audio amplifier circuit*

although it can in fact generate as much as 2W in an 8R0 load. The IC uses an 8-pin DIL package, can operate from supplies as low as 3V, and features low quiescent current, good ripple rejection, and low cross-over distortion.

*Figure 5.44* shows the outline and pin notations of the TBA820M IC, plus a practical application circuit for the device. Here, R2 determines the voltage gain of the IC, and R3–C6 form a Zobel network across the loudspeaker. This circuit can use a maximum supply voltage of 16V with a 16R speaker, 12V with an 8R0 speaker, or 9V with a 4R0 speaker.

## LM1877 circuits

The LM1877 is a protected-output dual amplifier that can feed up to 2W into each output channel when using a pair of 8R0 speakers. *Figure 5.45* shows the outline and pin notation of the IC, which uses a 14-pin DIL package; note that, in use, GND pins 3 to 5 and 10 to 12 are meant to be soldered to a large ground plane on the circuit's pcb, which is used as a heat sink.

*Figure 5.46* shows the LM1877's basic application circuit, as a stereo amplifier driving 8R0 speakers and powered from a single-ended supply. The voltage gain of each amplifier is set at ×200 by its 100k/510R negative feedback resistance network, and each amplifier's input signal is applied to its non-inverting terminal, which is wired to the IC's decoupled pin-1 bias terminal via a 1M0 isolating resistor. Other points to note are that the amplifier outputs are ac coupled to the 8R0 speakers via 500µF electrolytics and are fitted with 2R7–100n Zobel networks, and that pins 3 to 5 and 10 to 12 of the IC must be wired to the pcb's ground plane, which is used as a heat sink.

*Figure 5.47* shows the above circuit modified for operation from a split (dual) power supply, and with its voltage gain set at ×50 by the 100k/2k0

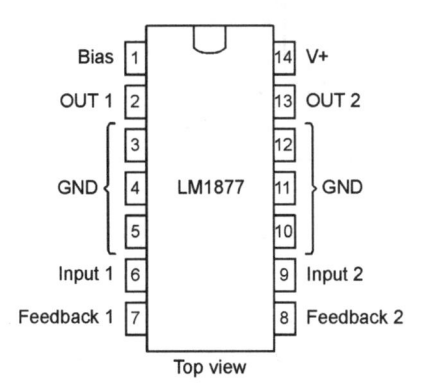

**Figure 5.45** *Outline and pin notation of the LM1877 dual 2W amplifier IC*

146  *Audio power amplifier circuits*

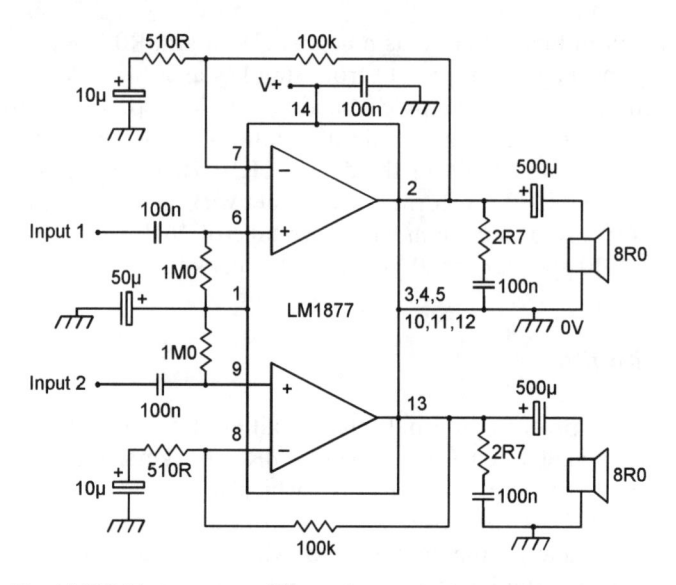

**Figure 5.46** *LM1877 stereo amplifier using a single-ended power supply*

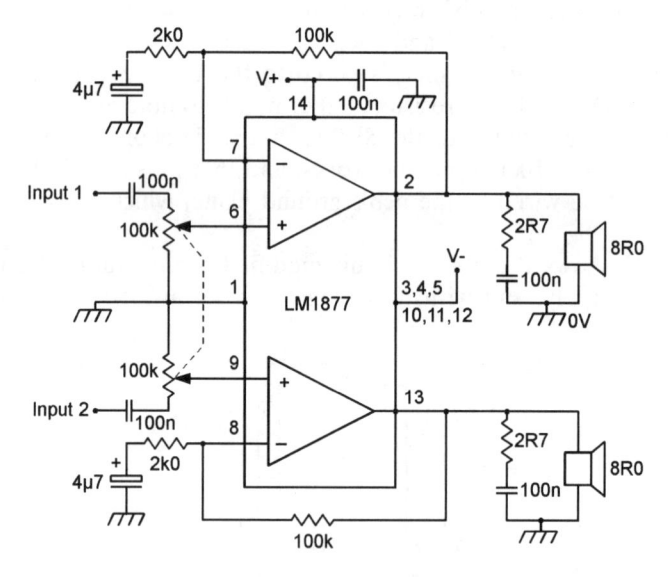

**Figure 5.47** *LM1877 stereo amplifier using a split (dual) power supply*

resistance ratios and with its input signal levels made variable via stereo volume control pot RV1. Note that each amplifier's output is direct coupled to its 8R0 speaker, that the IC's pin-1 bias terminal is connected directly to the power supply 0V (common) line, and that pins 3 to 5 and 10 to 12 of the

## LM380/LM384 circuits

The LM380 (*Figures 5.17* and *5.48*) is probably the best known of all power amplifier ICs. It can work with any supply voltage in the range 8V to 22V, and can deliver 2W into an 8R0 load when operated with an 18V supply (but needs a good external heat sink to cope with this power level). Its differential input terminals are both ground referenced, and the output automatically sets at a quiescent value of half-supply volts. Its voltage gain is fixed at ×50 (= 34dB), the output is short-circuit proof, and the IC is provided with internal thermal limiting.

The LM384 is simply an uprated version of the LM380, capable of operating at supply values up to 26V and of delivering 5.5W into an external load. Both types of IC are housed in a 14-pin DIL package, in which pins 3 to 5 and 10 to 12 are intended to be thermally coupled to an external heat sink.

To conclude this chapter, *Figures 5.49* to *5.52* show some practical applications of these two audio power amplifier ICs. *Figure 5.49* shows how to use either IC as a basic ×50 amplifier with enhanced (via C2) ripple rejection and a simple form of volume control (via RV1). Alternatively, *Figure 5.50* shows how to use either IC as a phono amplifier with RIAA equalization (via R1-C4), and *Figure 5.51* shows how to modify the circuit for use with so-called common mode volume and tone controls. Finally, *Figure 5.52* shows how to use a pair of these ICs in the bridge mode, to give a maximum output of either 4W or 10W.

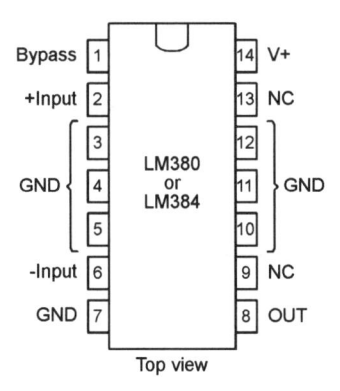

**Figure 5.48** *Outline and pin notation common to the LM380 2W and LM384 5W audio power amplifier ICs*

148  Audio power amplifier circuits

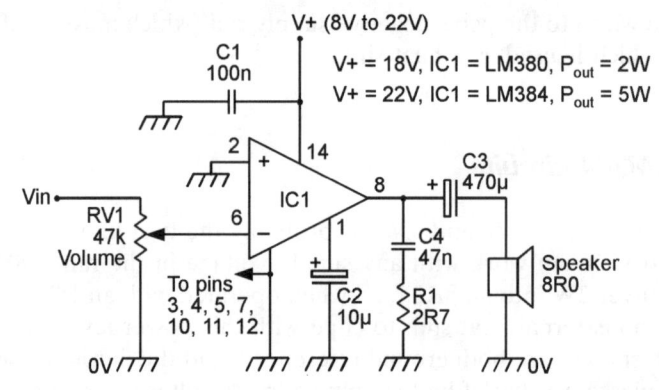

**Figure 5.49**  2W or 5W amplifier with simple volume control and ripple rejection

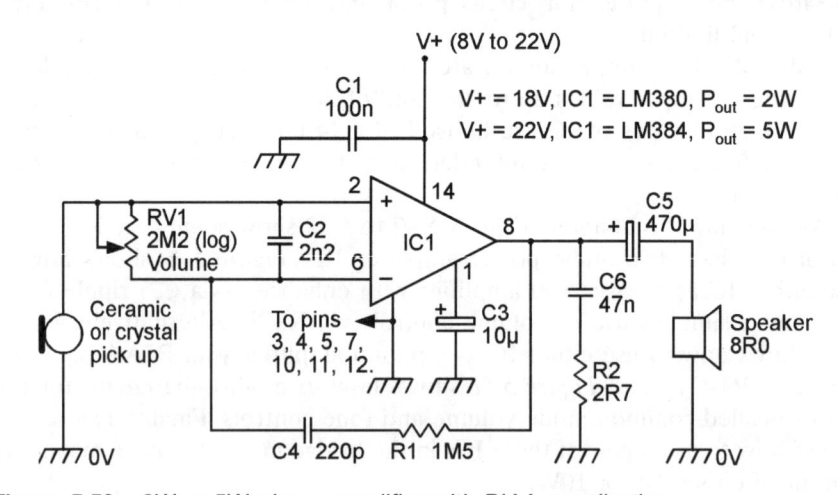

**Figure 5.50**  2W or 5W phono amplifier with RIAA equalization

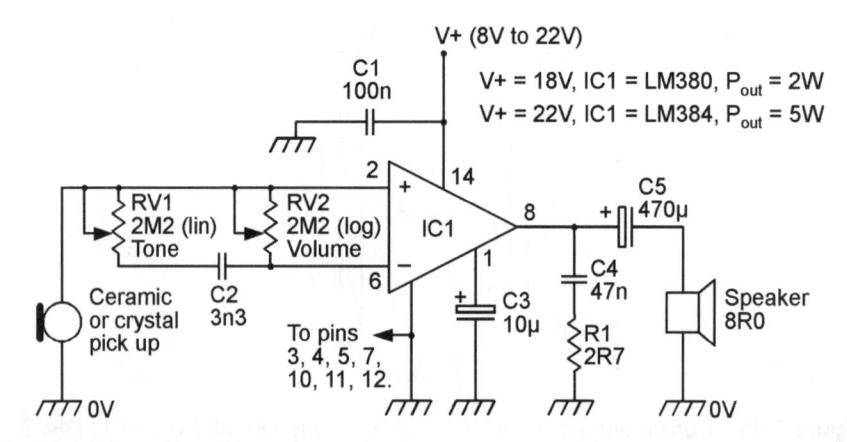

**Figure 5.51**  2W or 5W phono amp with common mode volume and tone controls

*Audio power amplifier circuits* 149

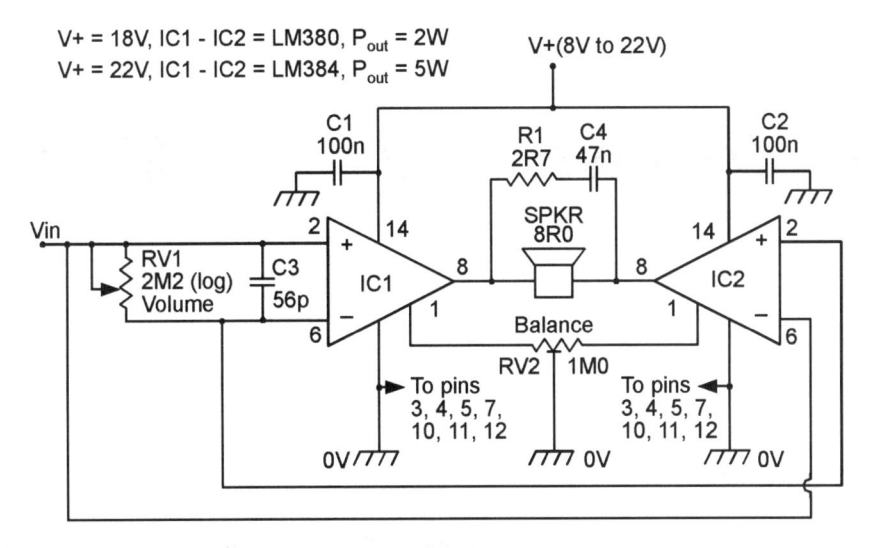

**Figure 5.52** *4W or 10W bridge-configured amplifier*

A further selection of applications of audio power amplifier ICs (including dual types), with maximum power ratings in the approximate range 6W to 68W, is given in the next chapter of this volume.

# 6
# High-power audio amplifiers

## Introduction

The previous chapter explained various audio power amplifier IC operating principles and presented a selection of practical application circuits based on popular audio power amplifier ICs with maximum output power ratings in the approximate range 325mW to 5.5W. The present chapter continues the 'audio power amplifier' theme by looking at a further selection of these ICs and their application circuits, but in this case deals with 'high-power' devices with maximum output power ratings in the approximate range 6W to 68W.

There are three very important practical points to keep in mind when attempting to build any of the high-power circuits described in this chapter. The first point is that, if the circuits are to work correctly, great care may have to be taken in designing the pcb of each circuit, to avoid harmful interaction between the output and input signal paths of each circuit (these interaction problems increase in direct proportion to the power rating of the circuit). General points to note are (1), that the IC's power-line connection terminals must be directly RF-decoupled to the common (ground) line via 100nF ceramic capacitors, (2), that the pcb tracking must be designed so that neither the IC's output-to-common signal (speaker) currents nor the IC's supply currents can flow in any part of the IC's input circuitry, and (3), that all tracking is adequately current rated and kept as short as possible (to minimize its inductance).

The second point to note is that the IC must (where appropriate) always be bolted to an adequate heatsink, using either a silicon grease or (preferably) a modern heat transfer compound to enhance the thermal union between the two components. The IC's case temperature, under sustained maximum power dissipation conditions, must be kept at low as possible, and must under no circumstances ever be allowed to exceed 100°C. An easy way to test the operating temperature is via the spit-and-touch method. Here, first simply use a finger to wipe a dollop of spit across the body of the IC when it is running

High-power audio amplifiers 151

at full power on its heatsink; if the spit rapidly evaporates, the IC temperature is above 100°C, and a much larger heatsink is needed. If the IC passes the spit test, try pressing a finger against the IC body; if you don't feel pain, the heatsink size is adequate. Most amateur circuit designers/buiders choose their heatsinks on a suck-it-and-see basis, starting off by fitting a fairly large heatsink and then keeping it if the IC passes the spit-and-touch tests, or enlarging it if the IC fails the tests. A rather more scientific method of determining the heatsink size is described in the final chapter (Chapter 10) of this book.

The third point to note about all of the practical circuits shown or mentioned in this chapter is that all of the power supply voltages quoted in the text and/or diagrams refer to the actual voltage values *under full power conditions*, rather than to those pertaining under quiescent conditions. Typically, when unregulated power supplies are used, the quiescent voltage is about 15% greater than the full power value (the actual percentage value is determined by the supply's so-called 'regulation factor'). Thus, a supply with a full-load value of (say) 30V may typically rise to 34.5V under quiescent conditions, and possible by a further 10% (to 37.95V) under high AC power-line conditions, and it is important to check that these latter values do not exceed the IC's maximum voltage rating; if they do, the circuit may have to be operated from a regulated power supply. Chapter 9 of this volume deals in depth with the subject of power supply circuit design, and parts of Chapter 10 deal with the matter of calculating the audio amplifier's power supply requirements and its speaker load impedance options.

Returning now to the present chapter, this chapter is roughly divided into two halves, with the first half dealing with ICs with maximum output power ratings in the range 6W to 12W, and the second half dealing with ICs with output power ratings ranging from about 18W to 68W. *Figure 6.1* gives basic details of the eight IC types that are dealt with in the first half of the chapter. Note that three of these ICs (the LM2877, LM2878, and LM2879 are 'dual' types which each house a pair of independently accessible amplifiers, and that four ICs (the TBA810P, TDA2003, LM383, and the TDA1020 are specifically designed for use in car (automobile) radio circuits.

Throughout the first half of this chapter all ICs are dealt with in the order in which they are listed in *Figure 6.1*. Practical application circuits are given for each IC type, but in some cases only very brief descriptions are given of individual IC circuit theory.

## LM2877/2878/2879 circuits

National Semiconductor have for many years produced well known ranges of 'dual' audio power amplifier ICs for use in stereo amplifiers and in bridge-configured mono amplifier applications. For most of that time the best known of these devices were their LM377 dual 2W, LM378 dual 4W, and

152   High-power audio amplifiers

| Device number | Amplifier type | Maximum output power | Supply voltage range | Distortion, into 8R0 | $Z_{IN}$ | $A_V$ | Bandwidth | Quiescent current |
|---|---|---|---|---|---|---|---|---|
| LM2877 | Dual | 4W per channel into 8R0 | 6 to 24V | 0.07%, Vs = 20V, Po = 2W/channel | 4M0 | 70dB | 65kHz | 25mA |
| LM2878 | Dual | 5W per channel into 8R0 | 6 to 32V | 0.14%, Vs = 22V, Po = 2W/channel | 4M0 | 70dB | 65kHz | 10mA |
| LM2879 | Dual | 8W per channel into 8R0 | 6 to 32V | 0.05%, Vs = 28V, Po = 1W/channel | 4M0 | 70dB | 65kHz | 12mA |
| TBA810P | Mono | 6W into 4R0 | 4 to 20V | 0.3%, Vs = 14.4V, Po = 2.5W | 5M0 | 37dB | 40Hz - 20kHz | 12mA |
| LM383 | Mono | 7W into 4R0 | 5 to 20V | 0.2%, Vs = 14.4V, Po = 4W | 150k | 40dB | 30kHz | 45mA |
| TDA2003 | Mono | 6W into 4R0 | 8 to 18V | 0.15%, Vs = 14.4V, Po = 1W | 150k | 40dB | 40Hz - 15kHz | 45mA |
| TDA1020 | Mono | 9.5W into 2R0 | 6 to 18V | — | 40k | 30dB | 15kHz | 30mA |
| TDA2006 | Mono | 12W into 4R0 | ±6 to ±15V | 0.1%, Vs = ±12V, Po = 4W | 5M0 | 30dB | 150kHz | 40mA |

**Figure 6.1**   Basic details of the eight ICs with output power ratings in the range 6W to 12W

LM379 dual 6W amplifier ICs, but a few years ago these famous old devices were withdrawn from production and replaced by improved designs with even better power ratings. The new devices are the LM2877 dual 4W, LM2878 dual 5W, and LM2879 dual 8W ICs.

The LM2877/2878/2879 ICs all use the same chip design and have the simplified equivalent circuit shown in basic form in *Figure 6.2*, i.e. each IC

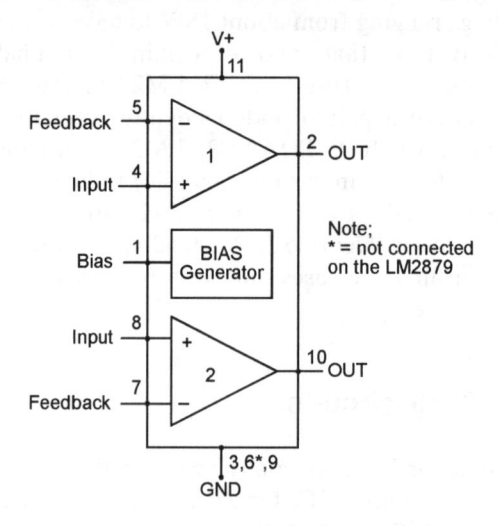

**Figure 6.2**   Simplified equivalent circuit common to the LM2877, LM2878 and LM2879 ICs

houses two identical power op-amps and a bias generator, but the ICs differ in their packaging styles and voltage/power ratings. All three ICs use an 11-pin single-in-line package with an attached heat sink, but the LM2877 and LM2878 use a package with the outline and pin notation shown in *Figure 6.3*, and the LM2879 uses a package with the outline and pin notation shown in *Figure 6.4*. Regarding the internal circuits of the three ICs, each amplifier actually consists of a differential input stage, a common-emitter amplifier 'driver' stage with a high-impedance constant-current collector load, and a quasi-complementary output stage that is biased via a fixed-gain 'amplified diode'. The input stages of the two amplifiers are powered via a common bias network that provides them with excellent supply-line ripple rejection, and that also makes a useful bias voltage externally available for biasing the amplifier outputs at a nominal 'half-supply' voltage value.

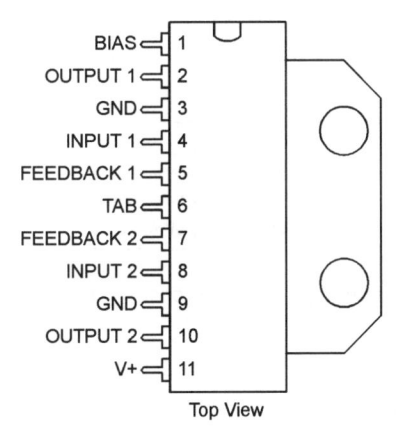

**Figure 6.3** *Outline and pin notation common to the LM2877 dual 4W and LM2878 dual 5W audio power amplifier ICs*

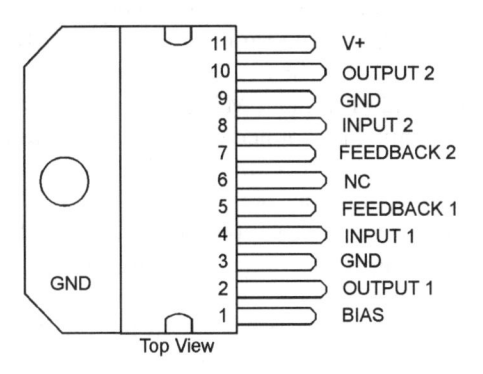

**Figure 6.4** *Outline and pin notation of the LM2879 dual 8W audio power amplifier IC*

154  *High-power audio amplifiers*

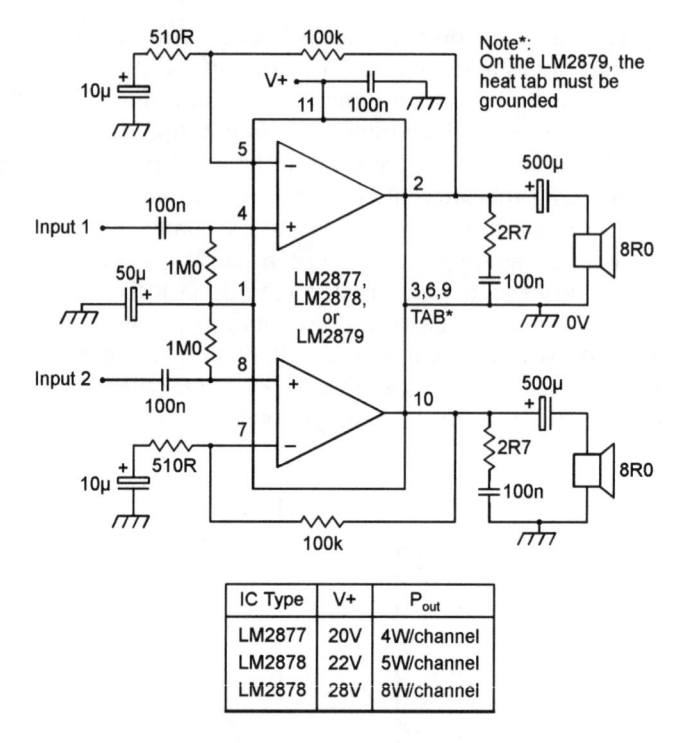

**Figure 6.5**  *Simple non-inverting stereo amplifier with a voltage gain of ×200*

| IC Type | V+ | $P_{out}$ |
|---|---|---|
| LM2877 | 20V | 4W/channel |
| LM2878 | 22V | 5W/channel |
| LM2878 | 28V | 8W/channel |

The three dual amplifier ICs are very easy to use. *Figure 6.5* shows the connections for making a simple non-inverting stereo amplifier that is operated from a single-ended power supply and can be used with any one of the three ICs, as shown in the table. In this circuit, the amplifier is biased by connecting each non-inverting input pin to the pin-1 BIAS terminal via a 1M0 isolating resistor, and the closed-loop voltage gain of each amplifier is set at approximately ×200 by the ratio of the 100k/510R feedback resistors. The circuit is fitted with 2R7–100n Zobel networks.

Note in *Figure 6.5* and all other LM2877/8/9 circuits shown in this section that the IC's heat tab must be connected to an adequate heat sink, that pins 3, 6, and 9 must be connected to the circuit's *negative* supply rail on the LM2877 and LM2878, and that pins 3 and 9 and the heat tab must be connected to the circuits negative supply rail on the LM2879. In circuits using a single-ended power supply, the 'negative' rail is, of course, the 0V line, but in split-supply circuits it is the V– line.

*Figure 6.6* shows the basic *Figure 6.5* circuit modified for operation from a split (dual) power supply, and with its voltage gain set at ×50 by the ratios of the 100k/2k0 feedback resistors and with its input signal levels made variable via stereo volume control pot RV1. Note that each amplifier's output

*High-power audio amplifiers* 155

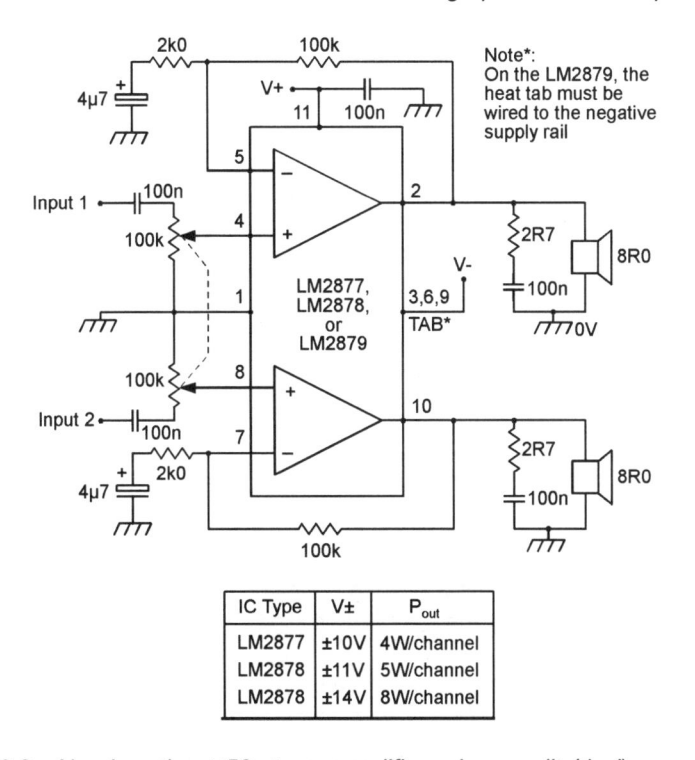

**Figure 6.6** *Non-inverting ×50 stereo amplifier using a split (dual) power supply*

| IC Type | V± | P<sub>out</sub> |
|---|---|---|
| LM2877 | ±10V | 4W/channel |
| LM2878 | ±11V | 5W/channel |
| LM2878 | ±14V | 8W/channel |

is direct coupled to its 8R0 speaker, and that the IC's pin-1 BIAS terminal is connected directly to the power supply 0V (common) line, which is grounded.

*Figure 6.7* shows how the basic circuit can be modified for use as an inverting stereo amplifier that uses a single-ended power supply. Here, the amplifier is biased by connecting each non-inverting input pin to the pin-1 BIAS terminal, which is ac-decoupled to ground, and the closed-loop voltage gain of each amplifier is set at approximately ×50 by the 1M0/22k resistance ratios. Each input channel has an input impedance of 22k (equal to the 22k resistor value).

The available output power of each channel of the LM2878 can easily be boosted to 15W (into a 4R0 speaker) via a high-power complementary emitter follower stage that is effectively wired in series with the amplifier output and incoporated in the circuit's negative feedback loop. *Figure 6.8* shows the circuit of one channel of a stereo amplifier of this type. This remarkably simple circuit is configured as a non-inverting amplifier with an overall ×50 voltage gain, and generates a typical THD of about 0.05% at a 10W output power level. At very low power levels, Q1 and Q2 are inoperative and power is fed directly to the speaker via the 4R7 series resistor. At higher power levels Q1 and Q2 act as a normal complementary emitter follower and provide most of the power drive to the speaker. The 4R7 series resistor and the

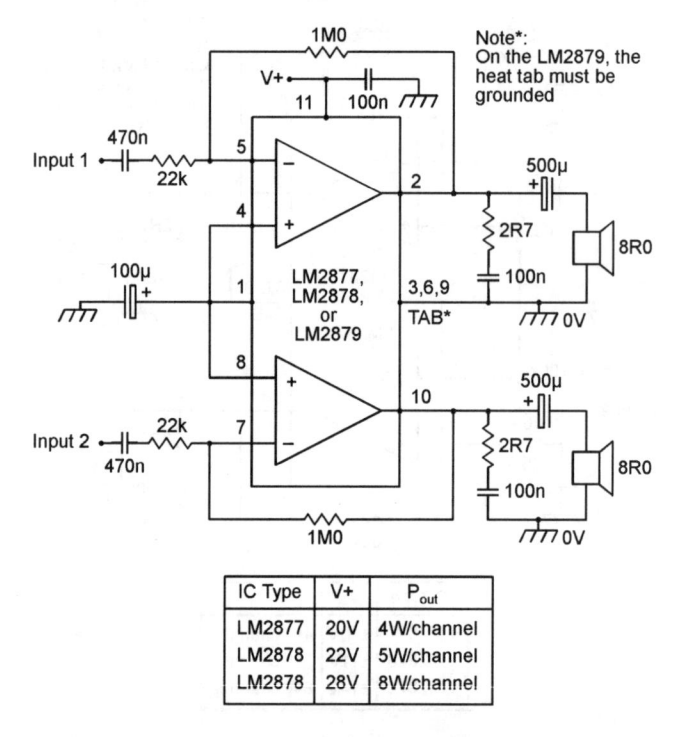

**Figure 6.7** Simple inverting stereo amplifier with ×50 voltage gain and a 22k input impedance

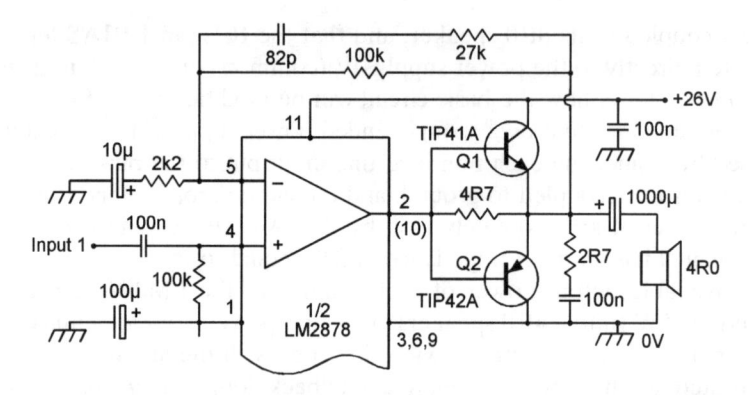

**Figure 6.8** One channel of a 15W per channel (into 4R0) stereo amplifier using a single-ended supply

base–emitter junctions of Q1 and Q2 are effectively wired into the negative feedback network of the circuit, thus minimizing signal cross-over distortion. The 82p–27k components wired across the 100k feedback resistor provide the circuit with high-frequency roll-off and help enhance circuit stability.

High-power audio amplifiers 157

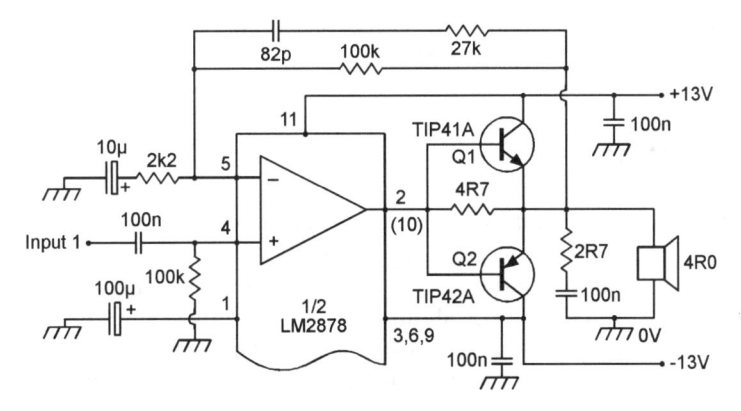

**Figure 6.9** *One channel of a 15W per channel (into 4R0) stereo amplifier using a split supply*

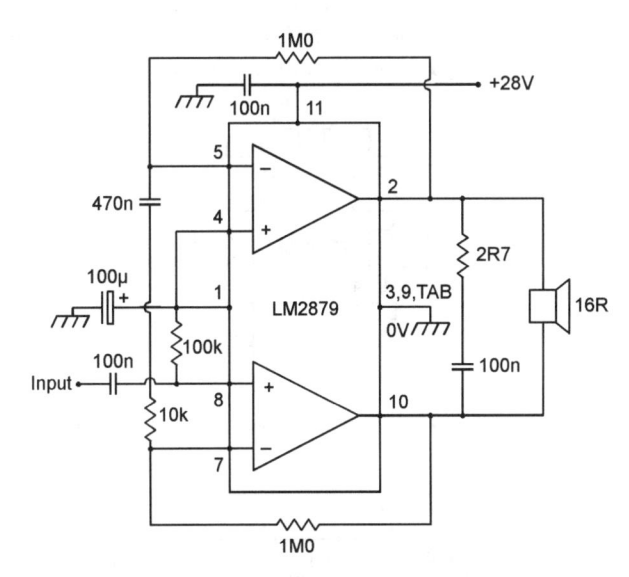

**Figure 6.10** *12W bridge amplifier circuit using a LM2879 IC*

*Figure 6.9* shows how the above circuit can be adapted for use with a split power supply. This circuit produces negligible output dc-offset, thus enabling the 4R0 speaker to be direct-coupled to the circuit's output.

The two built-in amplifiers of the LM2877/8/9 ICs can easily be used in the bridge-driving mode and used to feed relatively high power levels into direct-coupled speaker loads, but in such cases the available output power is severely limited by the power dissipation and limited current-driving capabilities of the individual IC. The most useful IC in this respect is the LM2879, and *Figure 6.10* shows how this specific IC can be used in this mode, in which

## The TBA810P

The TBA810P is a popular but rather ancient medium-power IC specifically designed for use in automobile applications, in which the '12V' supply voltage has an actual 'running' value of 14.4V nominal, under which condition the TBA810P can deliver about 6W of audio power into a 4R0 load, or 7W into a 2R0 load (using two 4R0 speakers wired in parallel). The IC is moderately sophisticated, and is internally protected against accidental supply polarity inversion and high supply-line transients, etc., and has a typical signal power bandwidth of 20kHz. Like most 'old' audio power amplifier ICs, it required the use of a fairly large number of external components.

*Figure 6.11* shows the outline of the TBA810P IC, and *Figure 6.12* shows a practical application circuit for the device. Here, the IC's voltage gain is

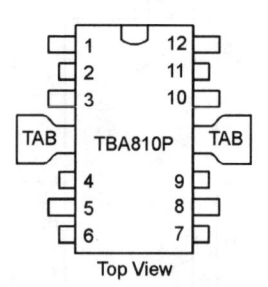

**Figure 6.11** *Outline of the TBA810P IC, which is designed for use as a 6W or 7W amplifier in automobiles*

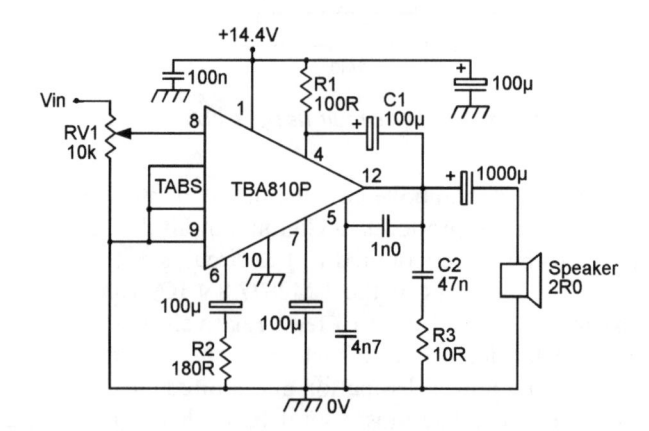

**Figure 6.12** *TBA810P 7W amplifier for use in automobiles*

*High-power audio amplifiers* 159

determined by R2 (minimum value is 68R, for maximum gain), R1 is an output biasing load resistor that is bootstrapped via C1, and R3–C2 is a Zobel network.

## LM383 (TDA2003) circuits

The LM383 and the TDA2003 are, internally and physically, almost identical devices and are usually described in manufacturer's literature as 8W audio power amplifier ICs. The two ICs do, however, differ somewhat in their actual electrical specifications, and the LM383 has a wider operating voltage (5V to 20V) and bandwidth (30kHz) than the TDA2003 (8V to 18V, and 15kHz). *Figure 6.13* shows the internal circuit that is common to both devices, and *Figure 6.14* shows the normal outline and pin notation of the ICs, which use a 5-pin TO220 plastic package. Both ICs are specifically designed for use in automobile applications, in which – at a normal 'running' supply value of 14.4V – they can typically deliver 5.5W into a 4R0 load or

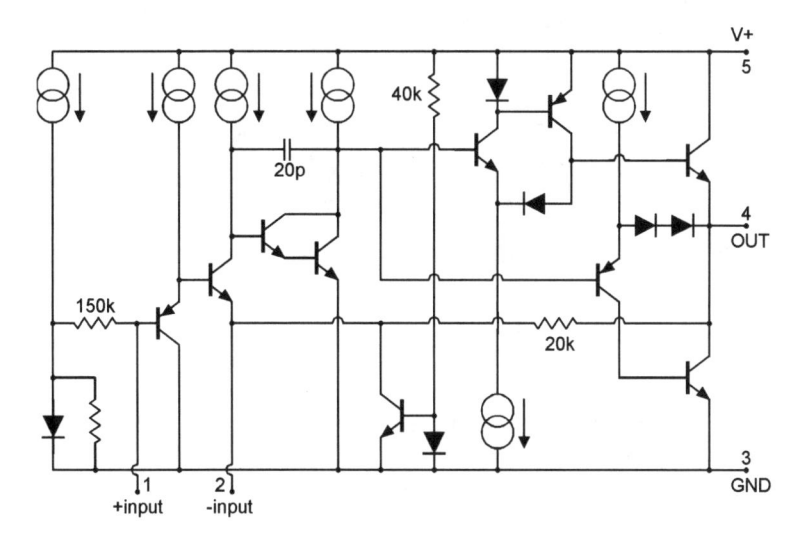

**Figure 6.13** *Internal circuit of the LM383 (or TDA2003) 8W audio power amplifier IC*

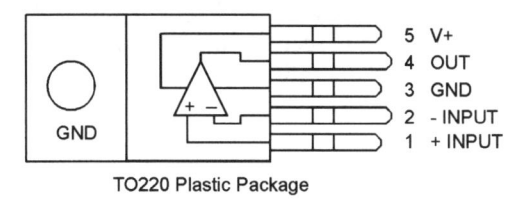

**Figure 6.14** *Outline and pin notation of the LM383 or TDA2003 IC*

160  High-power audio amplifiers

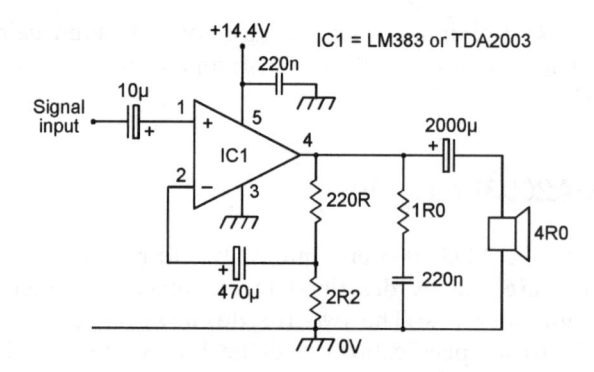

**Figure 6.15** *LM383 (or TDA2003) 5.5W amplifier for use in automobiles*

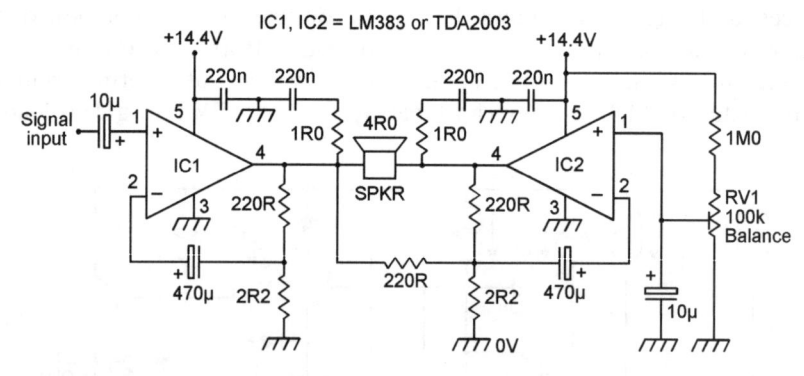

**Figure 6.16** *LM383 (or TDA2003) 16W bridge-type amplifier for use in automobiles*

8.6W into a 2R0 load. The LM383 can supply peak output currents of 3.5A, and both ICs have a current-limited and thermally protected output stage.

The LM383 (or TDA2003) is a very easy device to use. *Figure 6.15* shows a practical application circuit (complete with a Zobel network) for using the IC as a simple 5.5W audio amplifier in automobiles. Here, the closed-loop voltage gain is set at ×100 via the 220R/2R2 feedback network, and the IC is operated in the non-inverting mode by simply feeding the input signal to pin 1 via a 10μF electrolytic capacitor.

*Figure 6.16* shows another automobile application circuit, in which pair of LM383 or TDA2003 ICs are used as a 16W bridge amplifier. Pre-set pot RV1 is used here to balance the quiescent output voltages of the two ICs and to thus minimize the quiescent operating current of the circuit.

## The TDA1020

The TDA1020 is manufactured by Philips and is yet another audio power amplifier IC that is specifically designed for use in automobile applications,

*High-power audio amplifiers* 161

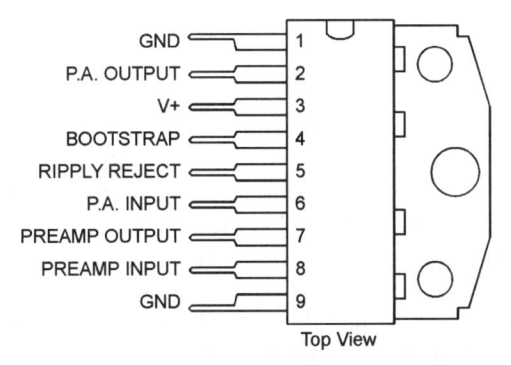

**Figure 6.17** *Outline and pin notation of the TDA1020 9.5W audio power amplifier IC*

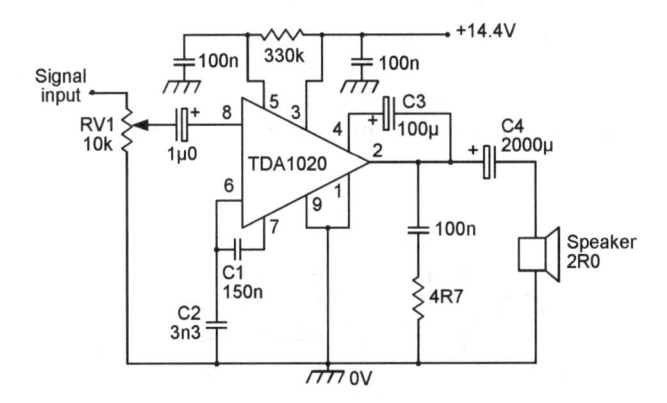

**Figure 6.18** *TDA1020 9.5W amplifier for use in automobiles*

in which it can deliver 6W into a 4R0 speaker load or 9.5W into a 2R0 load at 1% THD. The IC is housed in a 9-pin single-in-line plastic package, using the outline and pin notation shown in *Figure 6.17*. The IC incorporates separately accessible pre-amplifier and power amplifier circuits, which when used together give an overall voltage gain of 47dB.

*Figure 6.18* shows a practical application circuit for the TDA1020, as an audio power amplifier that can pump 9.5W into a 2R0 load (two 4R0 speakers wired in parallel) at 1% THD, or 12W at 10% THD. Alternatively, the circuit can be used to pump 6W into a 4R0 load, in which case the C4 value can be reduced to 1000µF. In *Figure 6.18*, the input signal is applied (via RV1) to the pin-8 input terminal of the pre-amp stage, then removed from the pre-amp at pin 7 and fed (via C1) into the pin-6 input terminal of the power amplifier circuit, which has its internal 'driver' load bootstrapped via C3. The power amplifier's output signal is applied to the speaker load from pin 2.

## The TDA2006

The TDA2006 is a high-quality general-purpose audio amplifier IC that can be used with either split or single-ended power supplies and can deliver up to 12W into a 4R0 load, and which typically generates less than 0.1% distortion when feeding 8W into a 4R0 speaker. The IC is housed in a 5-pin TO220 package (see *Figure 6.19*) that has a heat tab which can be bolted directly to an external heat sink (without need of an insulating washer) in single-ended supply circuits.

*Figure 6.20* shows how to use the TDA2006 with a single-ended supply. The non-inverting input pin is biased at half-supply volts via R3 and the R1–R2 potential divider, and the voltage gain is set at ×22 via R5/R4. D1 and D2 protect the output of the IC against damage from back EMF voltages from the speaker, and R6–C6 form a Zobel network.

Finally, to complete this section of this chapter, *Figure 6.21* shows how to modify the *Figure 6.20* circuit for use with split power supplies. In this case the IC's non-inverting input is tied to ground via R1. This circuit also shows how high-frequency roll-off can be applied to the amplifier via C5–R4.

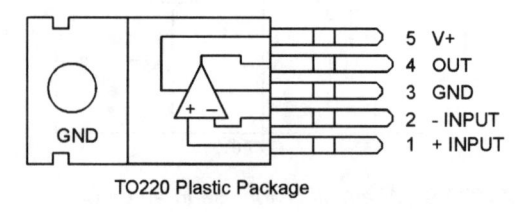

**Figure 6.19** Outline and pin notation of the TDA2006 12W audio power amplifier IC

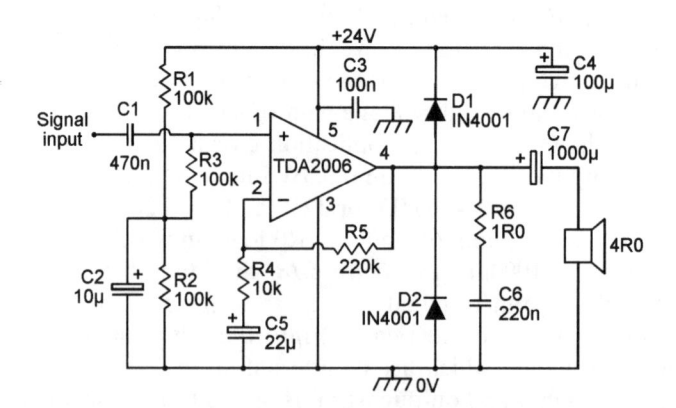

**Figure 6.20** TDA2006 8W amplifier with a single-ended supply

*High-power audio amplifiers* 163

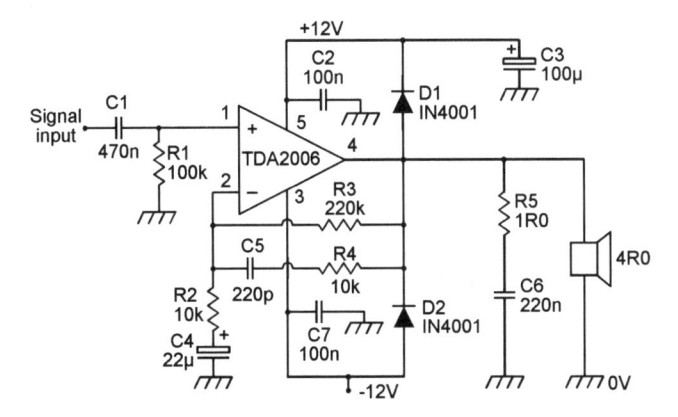

**Figure 6.21** *TDA2006 8W amplifier with a split power supply*

## High-power (18W to 68W) ICs

The remaining half of this chapter deals with audio amplifier ICs with output power ratings ranging from about 18W to 68W. *Figure 6.22* gives basic details of the nine IC types that are dealt with in this half of the chapter. Note that all but one of these ICs are mono types, that a pair of these mono ICs are

| Device number | Amplifier type | Maximum output power | Supply voltage range | Distortion, into 8R0 | $Z_{IN}$ | $A_V$ | Bandwidth | Quiescent current |
|---|---|---|---|---|---|---|---|---|
| TDA2030 | Mono | 18W into 4R0 | ±6 to ±18V | 0.1%, Vs = ±18V, Po = 4W | 5M0 | 30dB | 150kHz | 40mA |
| TDA2005M | Dual | 20W into 2R0 | 6 to 18V | 0.25%, Vs = 14.4V, Po = 16W (2R0) | 100k | 50dB | 40Hz - 20kHz | 75mA |
| TDA2040 | Mono | 22W into 4R0 | ±3 to ±20V | 0.08%, Vs = ±16V, Po = 10W into 4R0 | 5M0 | 30dB | 100kHz | 30mA |
| LM1875 | Mono | 25W into 4R0 | 20 to 60V | 0.015%, Vs = 50V, Po = 20W | 1M0 | 26dB | 70kHz | 70mA |
| TDA2050 | Mono | 32W into 4R0 | 9 to 50V | 0.05%, Vs = ±19V, Po = 15W | 500k | 30dB | 20Hz - 20kHz | 55mA |
| TDA1514A | Mono | 40W into 4R0 | ±7.5 to ±30V | 0.003%, Vs = ±28V, Po = 32W | 1M0 | 30dB | 25kHz | 60mA |
| LM3875 | Mono | 56W into 8R0 | 20 to 84V | 0.06%, Vs = ±35V, Po = 40W | 150k | 30dB | 80kHz | 30mA |
| LM3876 | Mono | 56W into 8R0 | 24 to 84V | 0.06%, Vs = ±35V, Po = 40W | 150k | 30dB | 80kHz | 30mA |
| LM3886 | Mono | 68W into 4R0 | 20 to 84V | 0.03%, Vs = ±28V, Po = 60W (4R0) | 150k | 30dB | 80kHz | 50mA |

**Figure 6.22** *Basic details of the nine ICs with output power ratings in the range 18W to 68W*

thus needed to make a stereo system, and that such an amplifier has a total output power equal to double the per-channel value. The only 'dual' IC in the list is the TDA2005M, which actually houses a pair of independently accessible power amplifiers that in most practical applications are connected in the bridge configuration to provide 20W of drive into a 2R0 mono load in automobile applications. Throughout the remainder of this chapter the nine listed ICs are dealt with in the order in which they appear in *Figure 6.22*. Practical application circuits are given for each IC type, but in some cases only very brief descriptions are given of individual IC circuit theory.

## The TDA2030

The TDA2030 is a very popular high-quality audio amplifier IC that can be regarded as an uprated version of the TDA2006, and is housed in a similar 5-pin TO220 package with built-in heat tab, as shown in *Figure 6.23*. The IC can operate with single-ended supplies of up to 36V, or with balanced split supplies of up to ±18V, and can deliver up to 18W into a 4R0 load. When used with a +28V single-ended supply it gives a guaranteed output of 12W into 4R0 or 8W into 8R0. Typical THD is 0.05% at 1kHz at 7W output, rising to less than 0.1% at 8W.

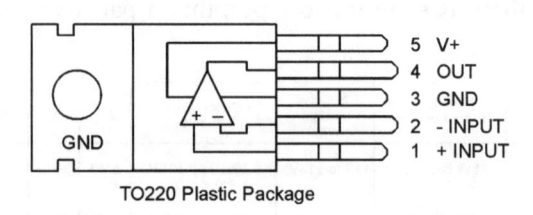

**Figure 6.23** *Outline and pin notation of the TDA2030 18W power amplifier*

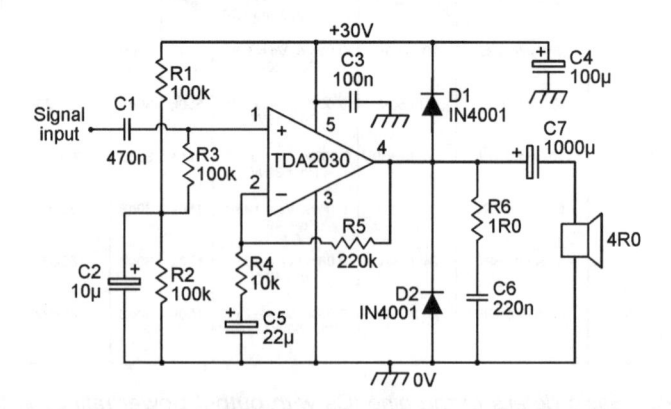

**Figure 6.24** *TDA2030 15W amplifier with a single-ended supply*

High-power audio amplifiers 165

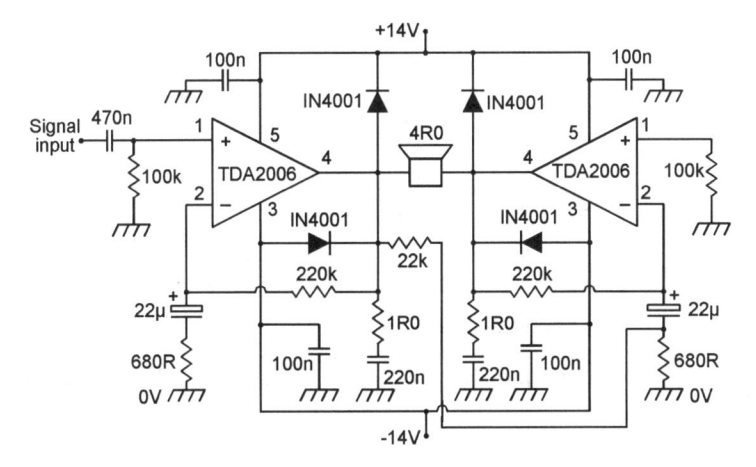

**Figure 6.25** *TDA2030 24W bridge amplifier with a split power supply*

The TDA2030 can be used in the same basic audio amplifier circuits as the TDA2006, but with suitable increases in the circuit supply voltages. *Figure 6.24*, for example, shows how to connect the TDA2030 as a 15W amplifier using a single-ended +30V supply and a 4R0 speaker load and which gives a voltage gain of 30dB. Alternatively, *Figure 6.25* shows how to wire a pair of TDA2030 ICs as a split-supply bridge amplifier that can deliver 24W into a direct-coupled 4R0 speaker load while generating typical total harmonic distortion of less than 0.5%.

## The TDA2005M

The TDA2005M is a 20W audio power booster IC specifically designed for use in automobiles, and is fully protected against output short-circuits, etc. The IC actually houses a pair of independently accessible power amplifiers that can each pump about 4W into a 4R0 speaker load, but in most practical applications are connected in the bridge configuration to provide 20W of drive into a 2R0 mono load when the IC is operated from the 14.4V (nominal) power supply of an automobile. The IC is housed in an 11-pin package, as shown in *Figure 6.26*. *Figure 6.27* shows a practical applications circuit that can deliver 20W to a 2R0 speaker load; note that all capacitors must be rated at 25V minimum.

## The TDA2040

The TDA2040 is a high-quality power amplifier IC intended for use in hi-fi applications, and is designed to operate from split power supplies. The device

166  High-power audio amplifiers

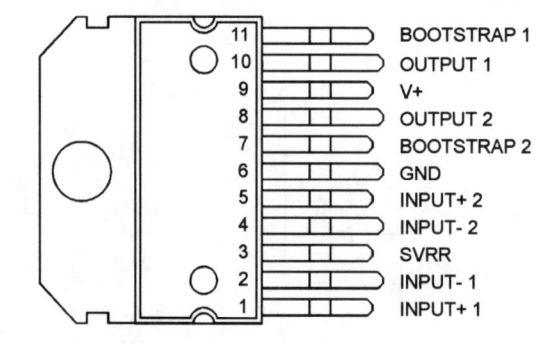

**Figure 6.26** *Outline and pin notation of the TDA2005M 20W bridge-connected amplifier IC*

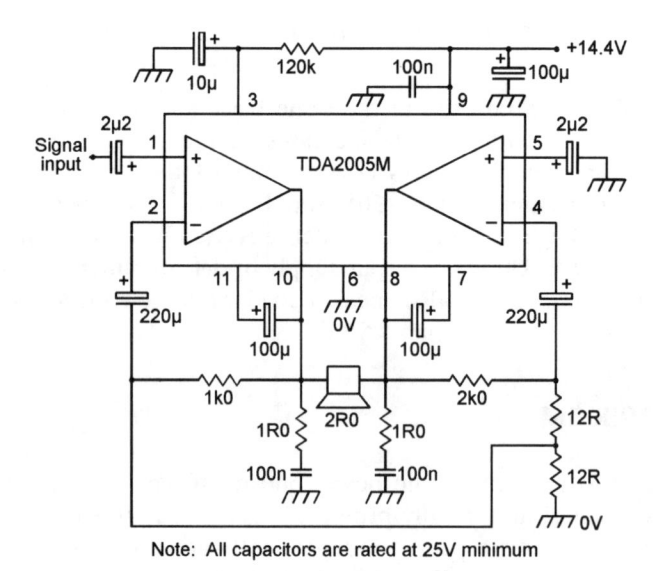

**Figure 6.27** *TDA2005M 20W power booster for use in automobiles*

typically generates up to 22W of audio power (at 0.5% THD) in a 4R0 speaker when powered from a split ±16V supply. The IC is internally protected against temporary overloads or output short-circuits, and incorporates automatic thermal shutdown circuitry. The IC is housed in a 5-pin package designed for vertical mounting on the pcb and has the outline and pin notation shown in *Figure 6.28*. *Figure 6.29* shows the IC's basic application circuit, as a 22W amplifier powered from split (dual) supplies. With the component values shown, the circuit gives a ×32 voltage gain and has an input impedance of 22k.

*High-power audio amplifiers* 167

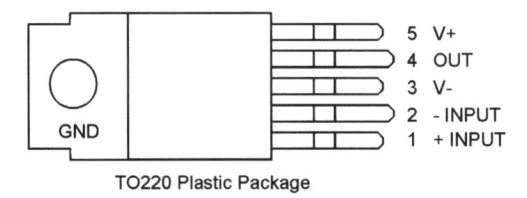

**Figure 6.28** *Outline and pin notation of the TDA2040 22W audio IC*

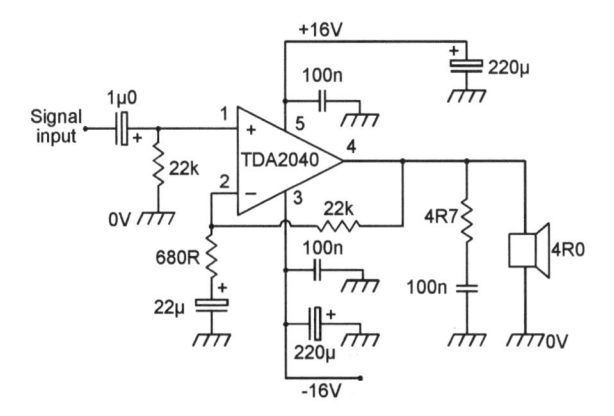

**Figure 6.29** *TDA2040 22W audio amplifier using split supplies*

## The LM1875

The LM1875 is a very popular very high quality audio amplifier that can deliver a maximum of 25W into a 4R0 load; it will deliver 20W into a 4R0 load when using a 50V supply and generating a mere 0.015% of THD. The IC is housed in a 5-pin TO220 package that does not require the use of an insulating washer between its metal tab and an external heat sink in single-ended supply applications; note, however, that an insulating washer must be used if the device is powered from dual (split) supplies.

*Figure 6.30* shows the outline and pin notation of the LM1875, which, like most modern audio high-power amplifier ICs, is very easy to use but requires

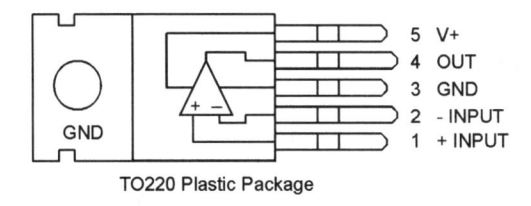

**Figure 6.30** *Outline and pin notation of the LM1875 25W amplifier*

168  High-power audio amplifiers

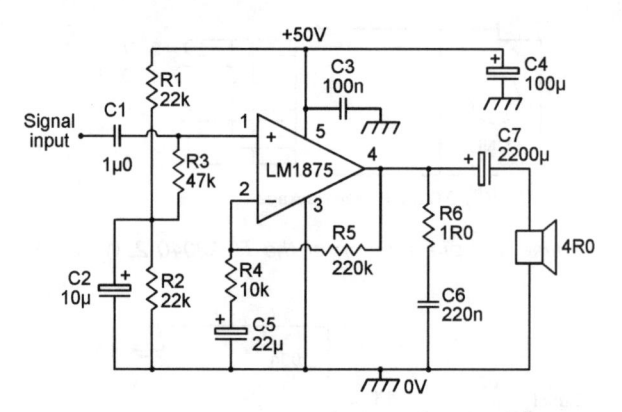

**Figure 6.31**  LM1875 20W amplifier using a single-ended power supply

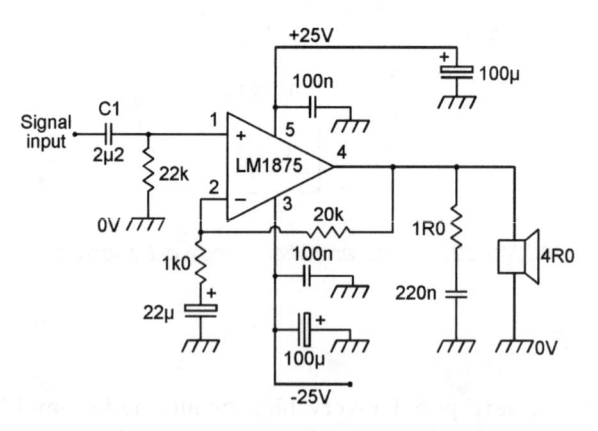

**Figure 6.32**  LM1875 20W amplifier using a dual (split) power supply

some care in the design of its pcb. *Figures 6.31* and *6.32* show practical ways of using the IC in audio application circuits using single and dual power supplies, respectively. Note in both of these circuits that input capacitor C1 is a non-polarized electrolytic type, and that the closed-loop voltage gain (×22 in *Figure 6.31*, ×20 in *Figure 6.32*) is set by the ratios of the feedback resistors (R5/R4 in *Figure 6.31*).

## The TDA2050

The TDA2050 is another popular high-quality audio amplifier IC that is housed in a 5-pin TO220 package that does not require the use of an insulating washer between its metal tab and an external heat sink in single-ended supply applications. The TDA2050 can deliver a maximum of 32W into a

High-power audio amplifiers 169

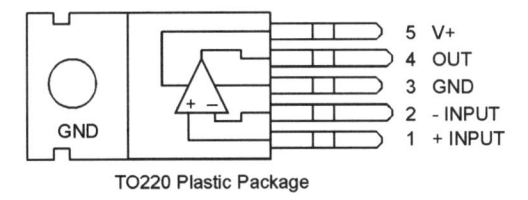

**Figure 6.33** *Outline and pin notation of the TDA2050 32W amplifier*

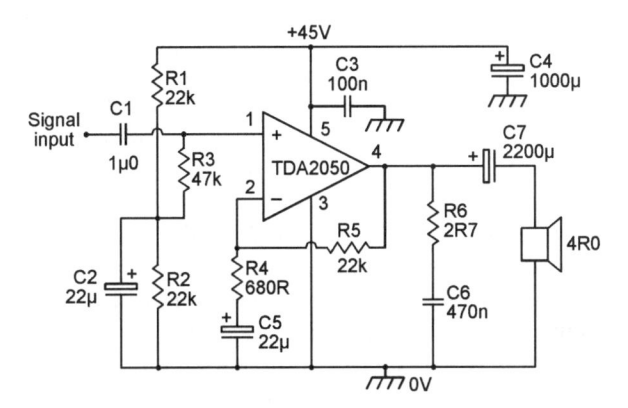

**Figure 6.34** *TDA2050 32W amplifier using a single-ended power supply*

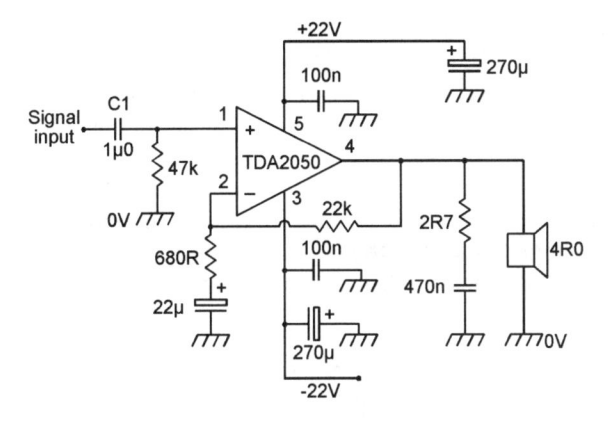

**Figure 6.35** *TDA2050 32W amplifier using a dual (split) power supply*

4R0 load or 25W into an 8R0 load when powered from a 45V (single ended) supply; when delivering 24W into a 4R0 load it typically generates a modest 0.03% of THD.

*Figure 6.33* shows the outline and pin notation of the TDA2050, and *Figures 6.34* and *6.35* show practical audio amplifier application circuits that use single-ended or dual power supplies and can generate up to 32W in a

170  High-power audio amplifiers

4R0 speaker load. The circuits are very similar to that of the LM1875 (see *Figures 6.31* and *6.32*); note in both circuits that input capacitor C1 is a non-polarized electrolytic type, that the input impedance is set at about 47k by the input resistor (R3 in *Figure 6.34*), and that the closed-loop voltage gain is set at ×32 by the ratios of the feedback resistors (R5/R4 in *Figure 6.34*).

## The TDA1514A

The TDA1514A is a very high quality 'super-fi' audio amplifier IC that is designed for use with split power supplies and can deliver a maximum of 40W into a 4R0 load when using a ±21V supply or 40W into an 8R0 load when using a ±27.5V supply; it typically generates a mere 0.0032% of THD when delivering a 32W output. The IC is housed in a 9-pin flat package with an integral heatsink that is internally connected to the IC's negative supply pin (pin 4); the heatsink must be insulated from ground in all split-supply applications. The IC is quite sophisticated, and incorporates output mute circuitry that eliminates speaker 'thumps' at power switch-on and switch-off, plus other circuitry that prevents damage from output short-circuits or overloads, and from thermal runaway problems.

*Figure 6.36* shows the outline and pin notation of the TDA1514A. Note that this is an underside view, seen from the metal mounting base side of the IC. Also note that the pin-2 'SOAR' title refers to the IC's *safe operating area region* thermal protection system.

*Figure 6.37* shows a basic application circuit for the IC, as a super-fi audio amplifier that can generate up to 40W in a 4R0 speaker. This same circuit can be used to generate 40W in an 8R0 load by simply using supply voltages of ±27.5V, or 25W in an 8RO load by using a ±22V supply. Note in *Figure 6.37* that the circuit's stability is enhanced with the aid of a 220p capacitor wired between input pin 1 and ground, that pin 7 is bootstrapped from the pin-5 output terminal, and that the IC's closed-loop voltage gain is set at ×32 by the 22k/680R feedback resistors. In practical versions of this basic design,

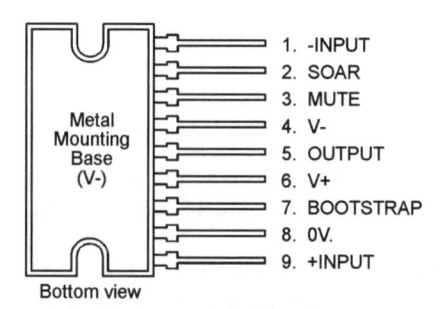

**Figure 6.36**  *Outline and pin notation of the TDA1514A 40W amplifier*

High-power audio amplifiers 171

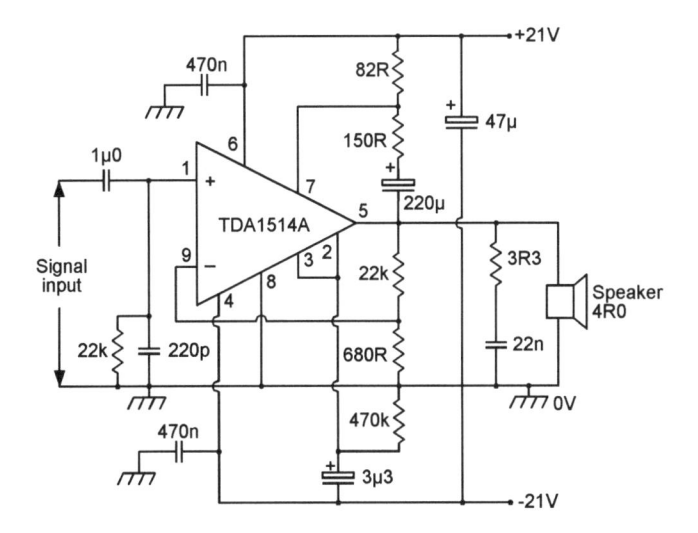

**Figure 6.37** *Basic application circuit for the TDA1514A, as a super-fi 40W audio amplifier*

the circuit's high-frequency performance may be enhanced by shunting all signal-carrying electrolytics with 220n ceramic capacitors.

## LM3875/76/86 basics

National Semiconductor produce a range of three super-fi audio amplifier ICs with maximum output power ratings that cover the range 56W (into 8R0) to 68W (into 4R0). The three ICs are the LM3875, the LM3876, and the LM3886, and they all use the same basic chip design. This design consists of a high-impedance differential input stage that has its output direct-coupled to the input of a common-emitter output-driver that uses a constant-current generator as its high-impedance collector load. This driver transistor drives a high-power 5-transistor quasi-complementary output stage via a 3-diode auto-bias network. The overall basic design is quite sophisticated, and incorporates automatic current sensing circuitry that protects the IC against output shorts or overloads, thermal sensing circuitry that gives protection against overheating problems, and voltage-sensing circuitry that protects the IC against damage from load-induced output transient over-voltages or from thump-generating switch-on or switch-off conditions in the supply lines. All three ICs can operate from maximum supply voltages of 84V (or ±42V in split supply circuits).

The design and state-of-the-art manufacturing techniques used in the construction of the basic semiconductor chips are such that all three ICs offer

172  High-power audio amplifiers

exceptionally good operating characteristics. They all have signal-to-noise ratios better than 95dB, have typical power supply rejection ratios of 120dB (enabling the ICs to use unregulated supplies), have typical open-loop gains of 120dB, and have typical gain–bandwidth product values of 8MHz. All three ICs use an 11-pin single-in-line TO-220 plastic package with an integral heat tab, but use their own unique pin notation. In the standard versions of these ICs (which carry a 'T' suffix at the end of their part number) the heat tab is internally connected to the IC's pin-4 supply-negative terminal, and the IC must thus be bolted to an external heatsink via an insulating washer in dual-supply applications. Special versions of the ICs are available in packages with electrically isolated heat tabs that can be bolted directly to external heatsinks; these ICs carry a 'TF' suffix. In all cases, pin 1 of the IC is indicated by a small round indent in the moulded package. Brief descriptions of the three individual ICs are as follows.

## The LM3875

The LM3875 is the most basic of the three high-power super-fi amplifier ICs. *Figure 6.38* shows the outline and pin notation of this particular device. Note that only five of the IC's eleven pins are internally connected, and the IC thus functions as a 5-pin (V+, V–, Input–, Input+, and Output) device that can be regarded as a high-quality high-power op-amp. *Figure 6.39* shows a practical LM3875 application circuit as an amplifier that (when the IC is bolted to an adequate heat sink) can generate up to 56W of audio power into an 8R0 load when the IC is operated from a split ±35V supply.

In *Figure 6.38*, RV1 acts as a simple volume control, and R1 is an input protection resistor. Feedback components R3/R2 set the circuit's ac (signal) voltage gain at ×20, and C1 ensures that the loop provides unity dc voltage

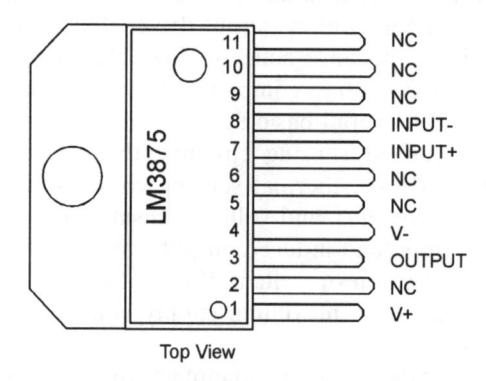

**Figure 6.38**  *Outline and pin notation of the LM3875 56W amplifier*

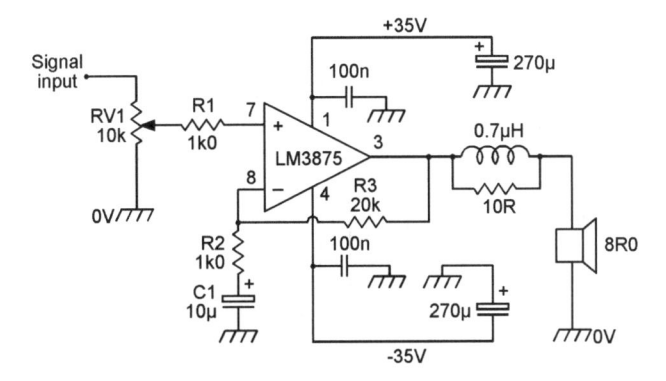

**Figure 6.39** *LM3875 56W amplifier using a dual (split) power supply*

gain. The parallel-connected heavy-duty 0.7µH inductor (which can be made from 20 turns of 18swg enamelled wire close wound on an 8mm diameter former) and 10R resistor wired in series with the IC's output are used to prevent instability when the IC is feeding the speaker load via long (and significantly capacitive) connecting leads. In practical versions of this circuit, the high-frequency stability can sometimes be enhanced by wiring a 220pF capacitor between the IC's inverting and non-inverting input pins.

## The LM3876

The LM3876 can be regarded as a simple variant of the LM3875, with a highly effective built-in input muting facility. *Figure 6.40* shows the outline and pin notation of this particular device. Note that only seven of the IC's

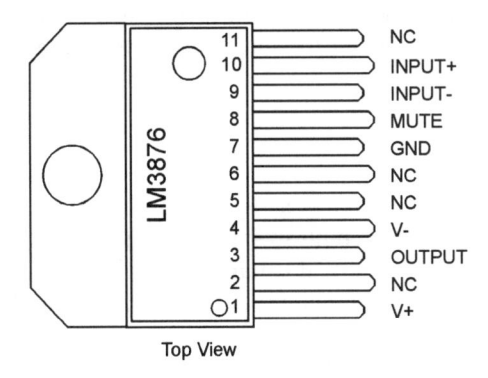

**Figure 6.40** *Outline and pin notation of the LM3876 56W amplifier with input muting facility*

174  High-power audio amplifiers

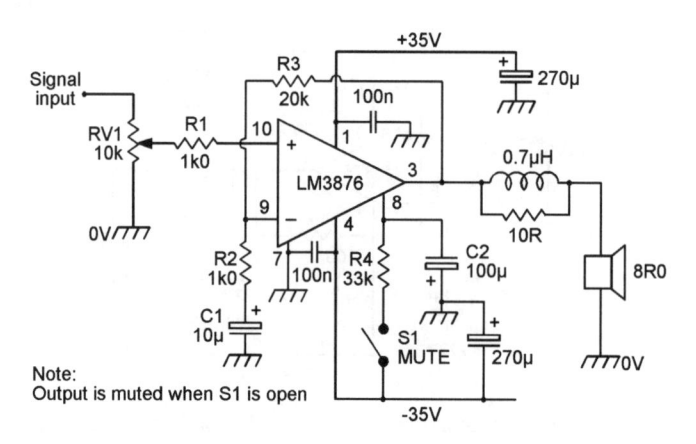

**Figure 6.41** *LM3876 56W dual-supply amplifier, with muting facility*

eleven pins are internally connected. The IC can be regarded as a special-purpose high-quality high-power op-amp. *Figure 6.41* shows the practical circuit of an LM3876 mutable amplifier that (when the IC is bolted to an adequate heat sink) can generate up to 56W of audio power into an 8R0 load when the IC is operated from a split ±35V supply.

The *Figure 6.41* circuit is similar to that of *Figure 6.39*, with its signal gain set at ×20 via feedback components R3/R2, etc., but uses different pin numbering and is provided with a switch-controlled (via S1) muting facility. The mute circuit actually controls the supply current feed to the IC's internal input and driver stages. The IC's action is such that the input is fully muted (the IC gives zero audio output) if pin 8 (the MUTE terminal) is open, and only turns on (to give a normal audio output) if a current of 0.5mA or greater flows out of pin 8 towards the circuit's negative supply rail. Thus, in the diagram, the circuit is muted when S1 is open, but gives normal amplifier operation when S1 is closed. In the mute switching network, R4 controls the muting current, and C2 adds a time constant that smooths out the mute-switching action.

## The LM3886

The LM3886 can be regarded as a slightly modified and uprated version of the LM3876, with the same built-in input muting facility. *Figure 6.42* shows the outline and pin notation of this particular device. Note that eight of this IC's eleven pins are internally connected, with pins 1 and 5 both serving as V+ connecting points. *Figure 6.43* shows a practical LM3886 application circuit as a mutable amplifier that (when the IC is bolted to an adequate heat sink) can generate up to 68W of audio power into a 4R0 load when the IC is operated from a split ±28V supply.

High-power audio amplifiers 175

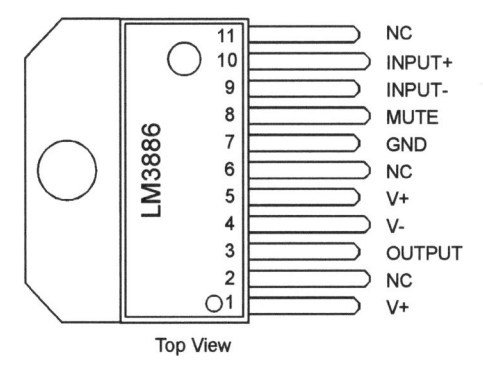

**Figure 6.42** *Outline and pin notation of the LM3886 68W amplifier with input muting facility*

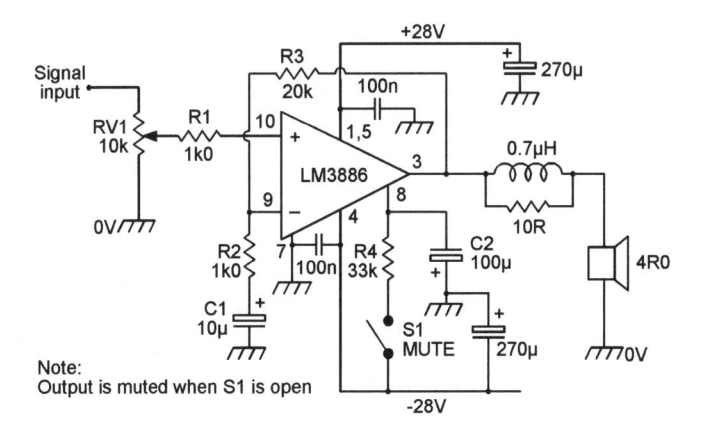

**Figure 6.43** *LM3886 68W (into 4R0) dual-supply amplifier, with muting facility*

The *Figure 6.43* circuit is very similar to that of *Figure 6.39*, with its signal gain set at ×20 via feedback components R3/R2, etc., but uses slightly different pin numbering, plus different supply rail and speaker impedance values. If this circuit is used with an 8R0 speaker it will generate ouput powers up to 38W when using ±28V supply rails, or up to 50W when using ±35V supply rails.

## Biamplification

The weakest link in most audio power amplifier systems is the loudspeaker. It is almost impossible to design a self-contained loudspeaker that will – at a reasonable cost – linearly span the full audio frequency range. In most

176  High-power audio amplifiers

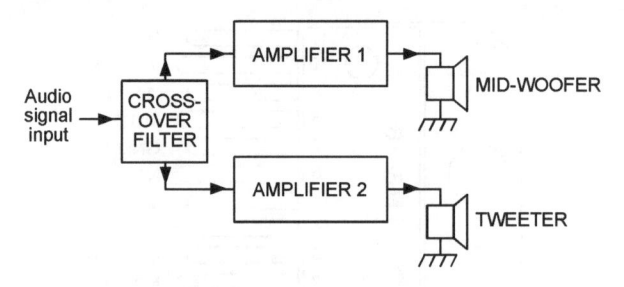

**Figure 6.44** *Diagram illustrating the basic principle of the so-called 'biamplification' technique*

modern hi-fi systems this problem is overcome by feeding each channel's output to a speaker unit that contains a passive cross-over filter and two (or more) speakers. One of the speakers (sometimes called a mid-woofer) is designed to efficiently span the low-to-middle frequency range (typically 40Hz to 10kHz), and the other (called a tweeter) to span the mid-to-high frequency range (typically 2kHz to 20kHz). The cross-over filter feeds the power amplifier's bass signals to the mid-woofer, the treble signals to tweeter, and mid-band signals to both speakers, and is a vital part of this system.

The above type of sound distribution system is reasonably efficient at input power levels up to about 25W, but beyond that level its efficiency falls off significantly due to power lossed in the passive filter system. An increasingly popular solution to this particular problem is – in high-power audio systems – to use the so-called 'biamplification' technique illustrated in *Figure 6.44*. This system uses two power amplifiers, with one directly driving the mid-woofer speaker and the other directly driving the tweeter, but with the system's cross-over filter placed ahead of the two amplifiers, where it is used to split audio input signals into two paths. Usually, the filter's cross-over frequency is set somewhere between 500Hz and 1.6kHz.

A major feature of the biamplification system is that its available output power is double that of a single amplifier. Thus, pairs of the various power amplifier IC circuits shown in this chapter can, when they are used in the biamplification mode, be used to generate very high per-channel output powers (up to a maximum limit of 112W when used with 8R0 speakers).

# 7
# LED bar-graph displays

LED (light emitting diode) bar-graph displays are widely used to replace moving-coil meters in modern audio equipment, and give fast-acting visual indications of parameters such as audio-amplifier output levels and various ac signal and DC power supply values. Several manufacturers produce special types of bar-graph driver ICs, and this chapter describes the two most popular families of these devices, the U237 family from AEG and the LM3914 family from National Semiconductors.

## LED bar-graph basics

LED bar-graph displays can be regarded as modern all-electronic moving-light replacements for the old fashioned moving coil-and-pointer types of analogue indicating meter, and are widely used in modern domestic equipment such as audio amplifiers and associated equipment. In a bar-graph display a line of ordinary LEDs are used to give an analogue representation of a scale length, and in use several adjacent LEDs may be illuminated simultaneously to form a light bar that gives an analogue indication of the value of a measured parameter.

*Figure 7.1* illustrates the bar-graph indicating principle, and shows a line of ten LEDs used to represent a linear-scale 0V-to-10V meter. *Figure 7.1(a)* shows the display indicating an input of 7V, with seven LEDs on and three LEDs off. *Figure 7.1(b)* shows the display indicating an input of 4V, with four LEDs on and six LEDs off.

A number of special bar-graph driver ICs are available for activating LED bar-graph displays. The two most useful types are the U237 (etc.) family from AEG, and the LM3914 (etc.) family from National Semiconductors. The U237 family are simple 'dedicated' devices which can usefully be cascaded to drive a maximum of 10 LEDs in bar mode only. The LM3914 family are more-complex and highly versatile devices, which can usefully be cascaded

178  LED bar-graph displays

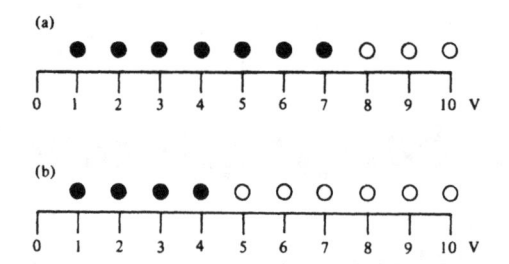

**Figure 7.1**  (a) Bar-graph indication of 7V on a 10V 10-LED scale, and (b) bar-graph indication of 4V on a 10V 10-LED scale

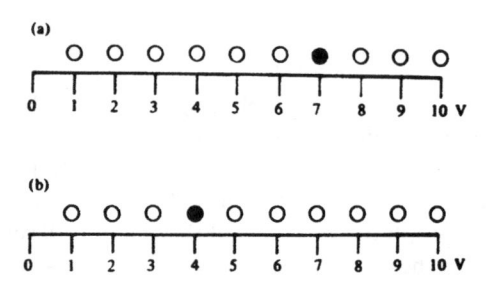

**Figure 7.2**  (a) Dot indication of 7V on a 10V 10-LED scale, and (b) dot indication of 4V on a 10V 10-LED scale

to drive as many as 100 LEDs, and can drive them in either bar or 'dot' mode. *Figure 7.2(a)* shows a 10-LED 10V meter indicating 7V in the 'dot' mode, and *Figure 7.2(b)* shows the same meter indicating an input of 4V. Note in each case that only a single LED is illuminated, and that its scale position indicates the analogue value of the input voltage.

Bar-graph displays (with suitable drivers) act as inexpensive and superior alternatives to analogue-indicating moving-coil meters. They are immune to inertia and 'sticking' problems, so are fast acting and are unaffected by vibration or attitude. Their scales can easily be given any desired shape (a vertical or horizontal straight line, an arc or circle, etc.). In a given display, individual LED colours can be mixed to emphasize particular sections of the display. Electronic over-range detectors can easily be activated from the driver ICs and used to sound an alarm and/or flash the entire display under the over-range condition.

Bar-graph displays have far better linearity than conventional moving-coil meters, typical linear accuracy being within 0.5%. Their scale definition depends on the number of LEDs used; a 10-LED display gives adequate resolution for most audio signal-level indicating purposes.

Let's now move on and take detailed looks at the two main families of bar-graph driver ICs, starting off with the U237 (etc.) types.

## U237 bar-graph driver ICs

### Basic principles

AEG's U237 family of bar-graph driver ICs are simple, dedicated devices, housed in 8-pin DIL packages and each capable of directly driving up to five LEDs. The family comprises four individual devices. The U237B and U247B produce a linear-scaled display and are intended to be used as a pair, driving a total of ten LEDs. The U257B and U267B produce a log-scaled display and are also intended to be used as a pair driving a total of ten LEDs.

All ICs of the U237 family use the same basic internal circuitry, which is shown in block diagram form (together with external connections) in *Figure 7.3*.

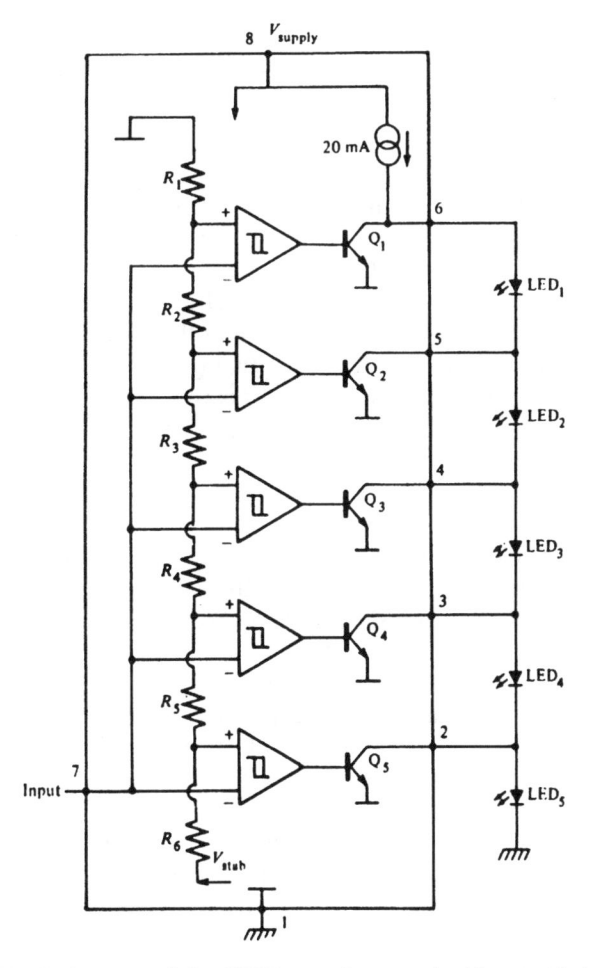

**Figure 7.3** Block diagram of the U237-type bar-graph driver, with basic external connections

180  LED bar-graph displays

| $V_{in}$ | $Q_1$ | $Q_2$ | $Q_3$ | $Q_4$ | $Q_5$ |
|---|---|---|---|---|---|
| 1.0 V | Off | Off | Off | Off | Off |
| 0.8 V | Off | Off | Off | Off | On |
| 0.6 V | Off | Off | Off | On | On |
| 0.4 V | Off | Off | On | On | On |
| 0.2 V | Off | On | On | On | On |
| 0 V | On | On | On | On | On |

**Figure 7.4**  States of the U237B internal transistors at various input voltages

| Device | Step 1 | Step 2 | Step 3 | Step 4 | Step 5 |
|---|---|---|---|---|---|
| U237 B | 200 mV | 400 mV | 600 mV | 800 mV | 1.00 V |
| U247 B | 100 mV | 300 mV | 500 mV | 700 mV | 900 mV |
| U257 B | 0.18 V/−15 dB | 0.5 V/−6 dB | 0.84 V/−1.5 dB | 1.19 V/+1.5 dB | 2.0 V/+6 dB |
| U267 B | 0.1 V/−20 dB | 0.32 V/−10 dB | 0.71 V/−3 dB | 1.0 V/0 dB | 1.41 V/+3 dB |

**Figure 7.5**  Step voltage values of the U237 family of bar-graph driver ICs

The IC houses five sets of Schmitt voltage comparators and transistor switches, each of which has its threshold switching or 'step' voltage individually determined by a tapping point on the R1 to R6 voltage divider, which is powered from a built-in voltage regulator: the input of each comparator is connected to the pin 7 input terminal of the IC. The IC also houses a constant-current generator (20mA nominal), and the five external LEDs are wired in series between this generator and ground (pin 1), as shown in the diagram.

The basic action of the circuit is such that groups of LEDs are turned on or off by activating individual switching transistors within the IC. Thus, if Q3 is turned on it sinks the 20mA constant-current via LEDs 1 and 2, so LEDs 1 and 2 turn on and LEDs 3 to 5 turn off.

The U237B has step voltages spaced at 200mV intervals, and *Figure 7.4* shows the states of its five internal transistors at various values of input voltage. Thus, at 0V input, all five transistors are switched on, so Q1 sinks the full 20mA of constant current, and all five LEDs are off. At 200mV input, Q1 turns off but all other transistors are on, so Q2 sinks the 20mA via LED1, driving LED1 on and causing all other LEDs to turn off, and so on. Eventually, at 1V input, all transistors are off and the 20mA flows to ground via all LEDs, so all five LEDs are on. Note that the circuit's operating current is virtually independent of the number of LEDs turned on, so the IC generates negligible RFI as it switches transistors/LEDs.

The four ICs in the U237 family differ only in their values of step voltages, which are determined by the R1 to R6 potential divider values. *Figure 7.5* shows the step values of the four ICs. Note that the U237B and U247B are linearly scaled, and can be coupled together to make a 10-LED linear meter with a basic full-scale value of 1V. The U257B and U267B are log scaled, and can be coupled together to make a 10-LED meter that spans 100mV (−20dB) to 2V (+6dB) in ten steps.

*LED bar-graph displays* 181

| Parameter | Minimum | Typical | Maximum |
|---|---|---|---|
| Supply voltage (see text) | 8 V | 12 V | 25 V |
| Input voltage | | | 5 V |
| Input current | | | 0.5 mA |
| Maximum supply current | | 25 mA | 30 mA |
| Power dissipation (at 60°C) | | | 690 mW |
| Step tolerance | −30 mV | | +30 mV |
| Step hysteresis | | 5 mV | 10 mV |
| Input resistance | | 100 k | |
| Output saturation voltage | | | 1 V |

**Figure 7.6** *General specification of the U237 family of ICs*

## What supply voltage?

*Figure 7.6* shows the basic specification of the U237 family of ICs. Note that the supply voltage range is specified as '8V to 25V', but in practice the minimum supply voltage is one of the few design points that must be considered when using these devices, and must be at least equal to the sum of the ON voltages of the five LEDs, plus a couple of volts to allow correct operation of the internal constant-current generator. Thus, when driving five red LEDs, each with a forward volt drop of 2V, the supply value must be at least 12V. Different coloured LEDs, with different forward volt drop values, can be used together in the circuit, provided that the supply voltage is adequate.

The only other 'usage' point concerns the input impedance of the IC. Although the input impedance is high (typically greater than 100k), the IC in fact tends to become unstable if fed from a source impedance in excess of 20k or so. Ideally, the signal feeding the input should have a source impedance less than 10k. If the source impedance is greater than 10k, stability can be enhanced by wiring a 10n capacitor between pins 7 and 1.

## Practical U237 circuits

*Figures 7.7* to *7.14* show some practical ways of using the U237 family of devices. In all of these diagrams the supply voltages are shown as '+12 to 25V', but the reader should keep in mind the constraints already mentioned.

*Figure 7.7* shows the practical connections for making a 0-to-1V 5-LED linear-scaled meter, using a single U237B IC, and *Figure 7.8* shows how a U237B/U247B pair of ICs can be coupled together to make a 0-to-1V 10-LED linear-scaled meter. Note in the latter case that the two ICs are operated as individual '*Figure 7.7*' circuits (needing only a 5-LED supply voltage), but have their input terminals tied together and have their LEDs physically alternated, to give a 10-LED display.

182  LED bar-graph displays

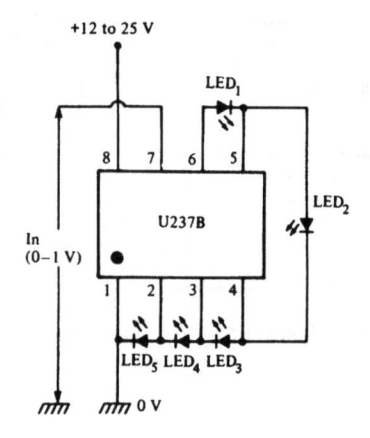

**Figure 7.7**  Practical connections for making a 0–1V 5-LED linear-scaled meter

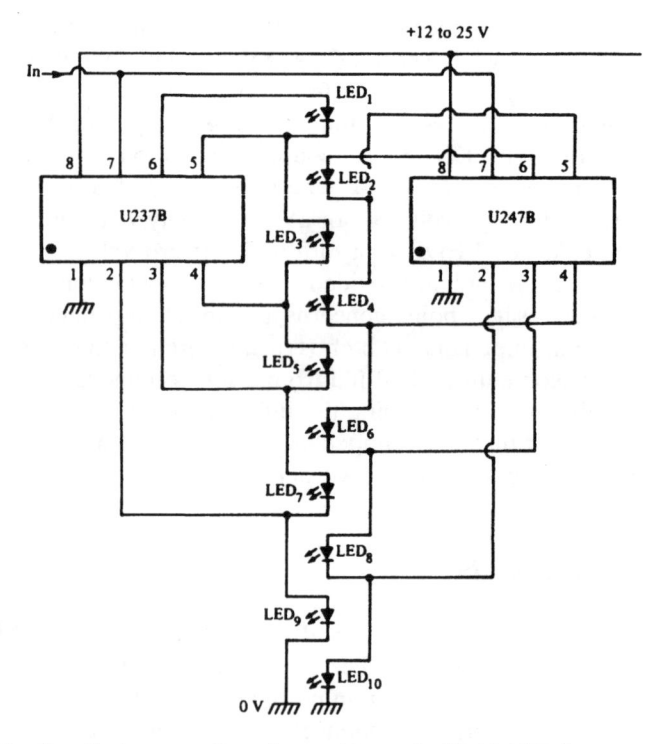

**Figure 7.8**  Practical connections for making a 0–1V 10-LED linear-scaled meter

*Figures 7.9* and *7.10* show how the full-scale sensitivity of the basic circuit can be altered, via external input circuitry, to suit particular applications. In *Figure 7.9*, the sensitivity is reduced by the R1–R2–RV1 input attenuator, which has a 15:1 ratio and thus gives an effective full-scale sensitivity of 15V.

*LED bar-graph displays* 183

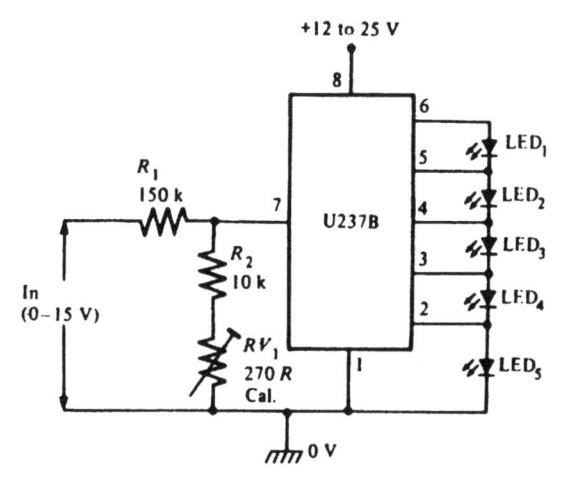

**Figure 7.9** *Method of reducing the sensitivity of the Figure 7.7 circuit, via an input potential divider, to make a 0–15V 5-LED meter*

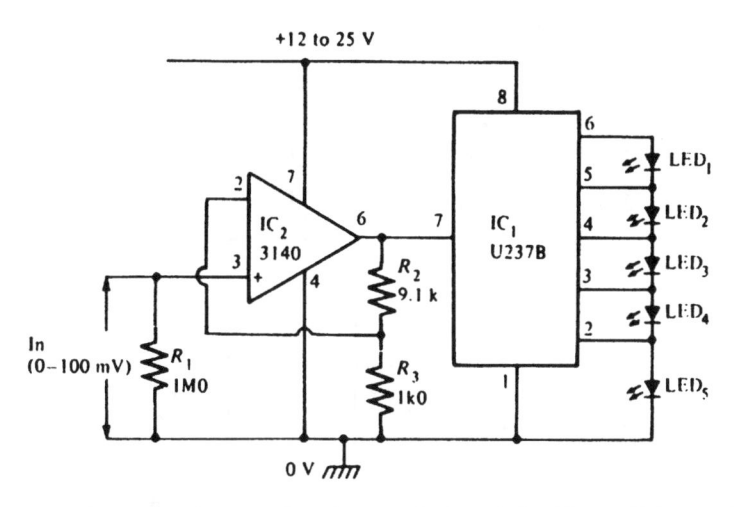

**Figure 7.10** *Method of increasing the sensitivity of the Figure 7.7 circuit, via a ×10 buffer amplifier, to make a 0–100mV 5-LED meter*

In *Figure 7.10*, the sensitivity is increased by a factor of ten, to 100mV full-scale, via non-inverting ×10 amplifier IC2, which also raises the input impedance to 1M0 (determined by R1).

*Figures 7.11* and *7.12* show how the basic *Figure 7.7* circuit can be used to indicate the value of a physical parameter, such as light, heat, liquid level, etc., that can be represented by an analogue resistive value in a transducer ($R_T$). In both of these circuits the transducer is simply fed from a

184  LED bar-graph displays

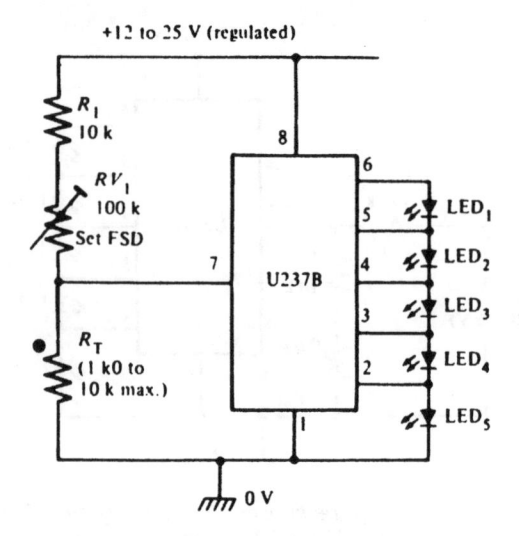

**Figure 7.11** Simple method of using a transducer sensor to indicate the value of a physical quantity

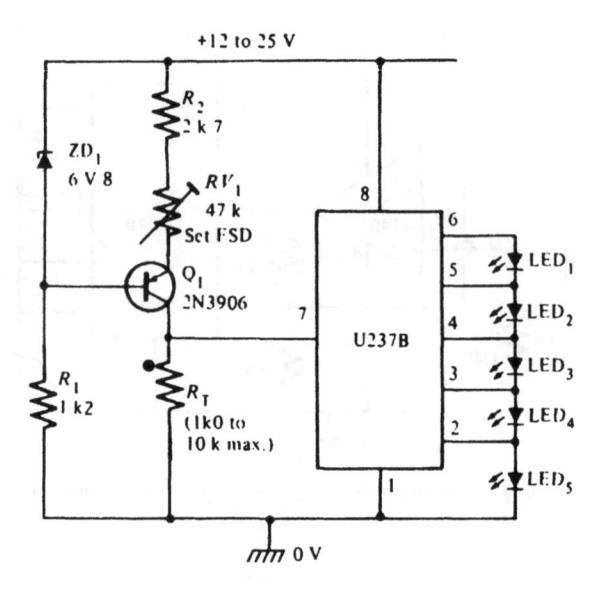

**Figure 7.12** Alternative method of using a transducer sensor to indicate the value of a physical quantity

constant-current generator, so that the input voltage reaching the IC is directly proportional to the transducer resistance.

In the *Figure 7.11* circuit, the constant current is derived from the regulated supply line via R1–RV1, and current constancy relies on the fact

*LED bar-graph displays* 185

that the supply voltage is large relative to the 1V full-scale value of the meter. Thus, if the supply value is 20V, the transducer current varies by only 5% when the transducer resistance varies between the zero volts and full-scale volts values. *Figure 7.12* shows how the linearity can be further improved, without the need for a regulated supply, by feeding the transducer from the output of the Q1 constant-current generator.

## *Over-range alarms, etc.*

*Figure 7.13* shows how the basic U237B circuit can be fitted with an audio-visual over-range alarm, which generates a pulsed tone and flashes the entire display at a rate of 2 flashes/second when f.s.d. (full-scale deflection) is reached or exceeded. The circuit theory is fairly simple, as follows.

The current of LED5 (the f.s.d. LED) flows to ground via R1 and the base–emitter junction of Q1, so Q1 turns on and pulls pin 1 of IC2a low whenever LED5 turns on. IC2a–IC2b are wired as a gated semi-latching 0.5Hz astable, which is gated on via a low input on pin 1. The pin-3 output of this astable is normally low, and the pin-4 output is normally high, but both pins give an astable output when the circuit is active. The pin-3 output feeds the base of Q2, and the pin-4 output feeds gated tone generator IC2c–IC2d, which feeds piezo acoustic (sounder) transducer Tx.

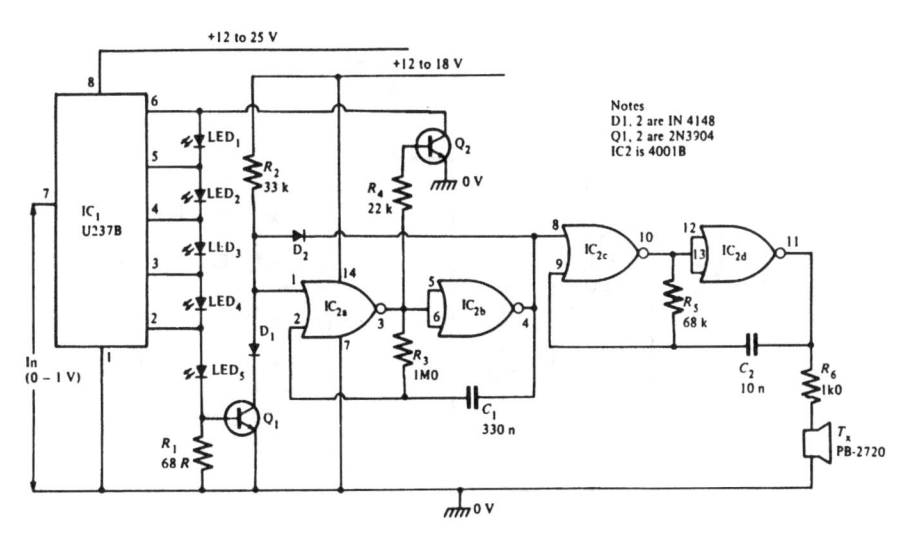

**Figure 7.13** *Method of fitting an audio-visual over-range alarm to the basic* Figure 7.7 *circuit; the entire display flashes (at a rate of 2 flashes/second) when f.s.d. is reached*

186  LED bar-graph displays

Thus, when IC1 output is below f.s.d., Q1 is off and Q2 and the tone generator are inactive. When f.s.d. is reached, LED5 and Q1 turn on, and pin 1 of IC2a is pulled low via D1, and IC2a–IC2b enters an astable mode in which pin 4 switches low and activates the IC2c–IC2d acoustic alarm and also pulls pin 1 low via D2; simultaneously, pin 3 switches high and turns Q2 on, thereby sinking the entire 20mA constant current of IC1 and blanking all LEDs for half the duration of the astable cycle. Note that Q1 turns off as soon as Q1 turns on, but that the input (pin 1) of the astable is maintained low at this stage via D2. At the end of the half-cycle, Q2 turns off and restores the display and the IC2c–IC2d acoustic alarm turns off. As the display is restored, Q1 turns back on (if LED5 is still active), but the IC2a–IC2b astable cannot re-trigger until it completes the second 'free-wheeling' half of its cycle.

Thus, the *Figure 7.13* circuit flashes the entire display and generates a pulsed-tone alarm under the f.s.d. or over-range condition. Note that if the IC1 supply rail is restricted to the 12V to 18V range, it can be made common with the supply of the alarm circuitry. If the acoustic alarm is not required, simply omit R5–C2–R6–Tx and connect the inputs of IC2c and IC2d to ground. The circuit can be made to flash only part of the display, if required, by simply connecting Q2 collector to the appropriate one of the pin 2–6 terminals of IC1; connection to pin 3, for example, causes only LEDs 4 and 5 to flash. If the over-range alarm circuit is to be used with the *Figure 7.8* 10-LED circuit, the feed to R1 and the base of Q1 should be taken from LED10.

Finally, to complete this look at the U237 range of ICs, *Figure 7.14* shows how the U267B 'log' IC can be used to make a 5-LED audio-level meter. A

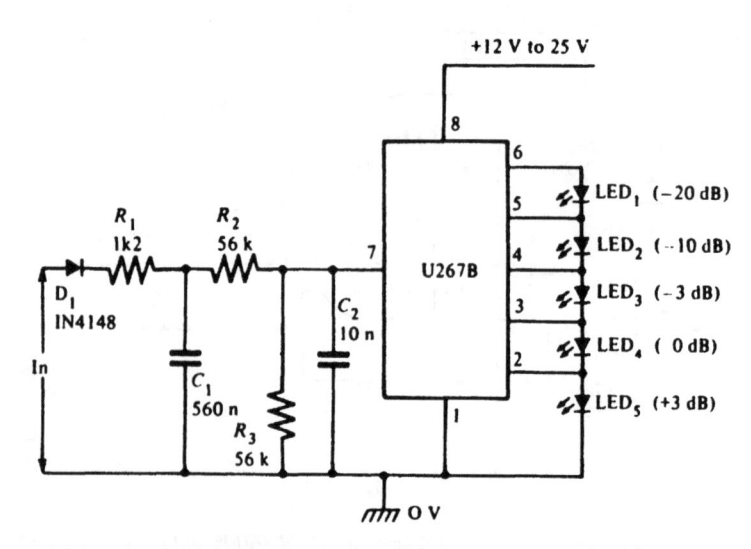

**Figure 7.14** 5-LED AF-level meter. A 10-LED version can be made by using a U257B/U267B pair of ICs

10-LED meter can be made by connecting the R1–R2–R3–C1–D1 input circuit to the input of a U257B/U267B pair of ICs connected in the same configuration as shown in *Figure 7.8*.

The basic *Figure 7.14* circuit has a discharge time constant of about 70ms; the sensitivity is determined by the R2–R3 ratio and, with the values shown, indicates 0dB with an input of 3V. The circuit requires a low-impedance drive, such as can be obtained (directly or via a potential divider) from a loudspeaker, etc.

## LM3914 dot/bar-graph driver ICs

### Basic principles

National Semiconductor's LM3914 family of ICs are fairly complex and highly versatile devices, housed in 18-pin DIL packages and each capable of directly driving up to ten LEDs in either dot or bar mode. The family comprises three devices, these being the LM3914, the LM3915 and the LM3916; they all use the same *basic* internal circuitry (see *Figure 7.15*), but differ in the style of scaling of the LED-driving circuitry, as shown in *Figure 7.16*.

Thus, the LM3914 is a linear-scaled unit, specifically meant for use in LED voltmeter applications in which the number of illuminated LEDs gives a direct indication of the value of input volts. The LM3915, on the other hand, has a log-scaled output designed to span the range 0dB to −27dB in ten −3dB steps, and is specifically meant for use in power meter applications, etc. Finally, the LM3916 has a semi-log scale, and is specifically designed for use in VU meter applications.

All three devices in the LM3914 family use the same basic internal circuitry, and *Figure 7.15* shows the specific internal circuit of the linear-scaled LM3914, together with the connections for making it act as a simple 10-LED 0-to-1.2V meter. The IC contains ten voltage comparators, each with its non-inverting terminal taken to a specific tap on a floating precision multi-stage potential divider and with all inverting terminals wired in parallel and taken to input pin 5 via a unity-gain buffer amplifier. The output of each comparator is externally available, and can sink up to 30mA; the sink currents are internally limited, and can be externally pre-set via a single resistor (R1).

The IC also contains a floating 1.2V reference source between pins 7 and 8. In *Figure 7.15* the reference is shown externally connected to potential divider pins 6 and 4, with pins 8 and 4 grounded, so in this case the bottom of the divider is at zero volts and the top is at 1.2V. The IC also contains a logic network that can be externally set to give either a dot of a bar display from the outputs of the ten comparators. The IC operates as follows.

188  LED bar-graph displays

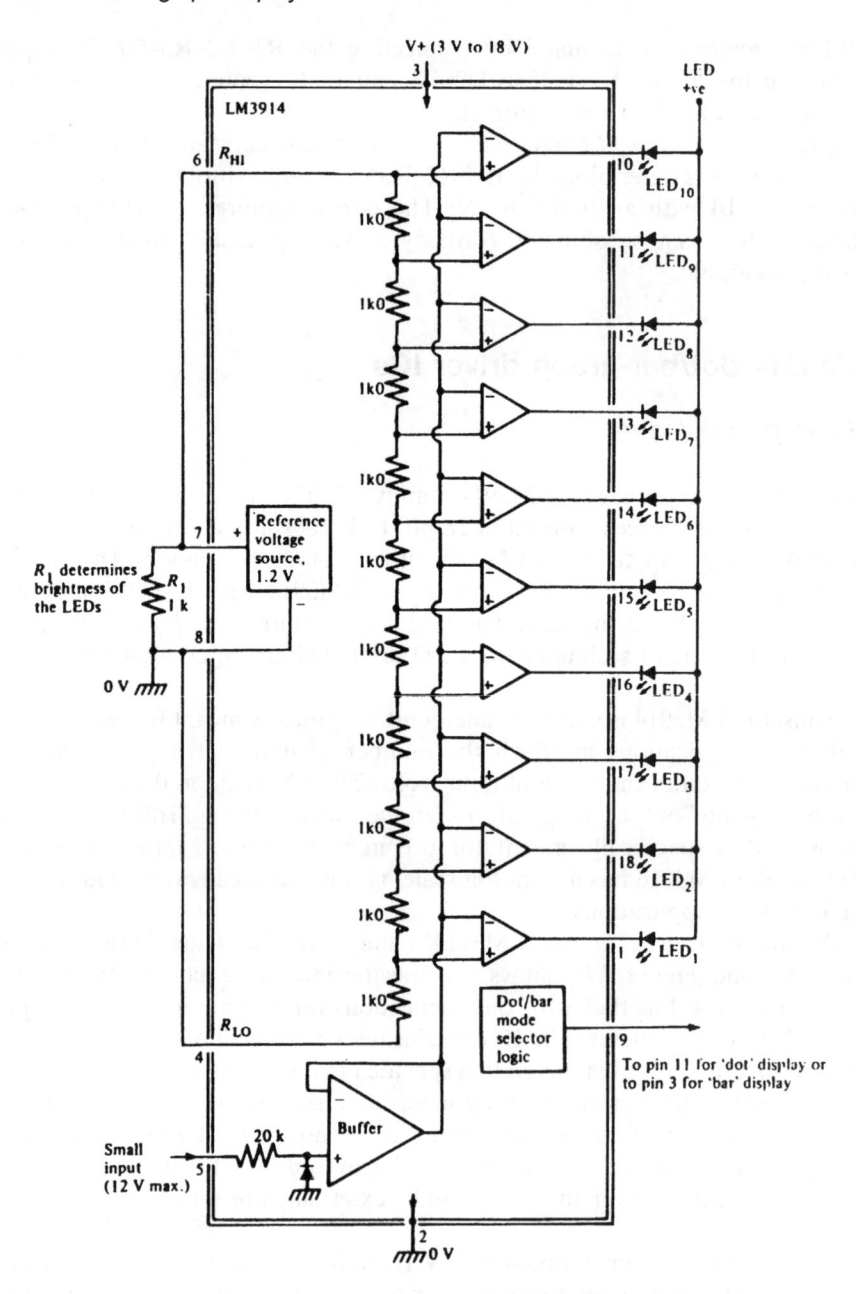

**Figure 7.15** Internal circuit of the LM3914, with connections for making a 10-LED 0–1.2V linear meter with dot or bar display

*LED bar-graph displays* 189

| LED number | TYPICAL THRESHOLD-POINT VALUE, at 10V f.s.d |||||| 
|---|---|---|---|---|---|---|
| | LM3914 | LM3915 || LM3916 |||
| | V | V | dB | V | dB | VU |
| 1 | 1.00 | 0.447 | -27dB | 0.708 | -23dB | -20 |
| 2 | 2.00 | 0.631 | -24dB | 2.239 | -13dB | -10 |
| 3 | 3.00 | 0.891 | -21dB | 3.162 | -10dB | -7 |
| 4 | 4.00 | 1.259 | -18dB | 3.981 | -8dB | -5 |
| 5 | 5.00 | 1.778 | -15dB | 5.012 | -6dB | -3 |
| 6 | 6.00 | 2.512 | -12dB | 6.310 | -4dB | -1 |
| 7 | 7.00 | 3.548 | -9dB | 7.079 | -3dB | 0 |
| 8 | 8.00 | 5.012 | -6dB | 7.943 | -2dB | +1 |
| 9 | 9.00 | 7.079 | -3dB | 8.913 | -1dB | +2 |
| 10 | 10.00 | 10.000 | 0dB | 10.000 | 0dB | +3 |

**Figure 7.16** *Threshold-point values of the LM3914/15/16 range of ICs, when designed to drive ten LEDs at a full-scale sensitivity of 10V*

Assume that the *Figure 7.15* circuit's logic is set for bar-mode operation and that, as already shown, the 1.2V reference is applied across the internal 10-stage divider. Thus, 0.12V is applied to the inverting or reference input of the lower comparator, 0.24V to the next, 0.36V to the next, and so on. If a slowly rising input voltage is now applied to pin 5 of the IC, the following sequence of actions takes place.

When the input voltage is zero, the outputs of all ten comparators are disabled and all LEDs are off. When the input voltage reaches the 0.12V reference value of the first comparator, its output conducts and turns LED1 on. When the input reaches the 0.24V reference value of the second comparator, its output also conducts and turns on LED2, so at this stage LED1 and LED2 are both on. As the input voltage is further increased, progressively more and more comparators and LEDs are turned on, until eventually, when the input rises to 1.2V, the last comparator and LED10 turn on, at which point all ten LEDs are illuminated.

A similar kind of action is obtained when the LM3914 logic is set for dot mode operation, except that only one LED turns on at any given time. At zero volts, no LEDs are on, and at above 1.2V only LED10 is on.

## Some finer details

In *Figure 7.15*, R1 is shown connected between pins 7 and 8 (the output of the 1.2V reference), and determines the ON currents of the LEDs. The ON current of each LED is roughly ten times the output current of the 1.2V source, which can supply up to 3mA and thus enables LED currents of up to 30mA to be set via R1. If, for example, a total resistance of 1k2 (equal to

190  LED bar-graph displays

the paralleled values of R1 and the 10k of the IC's internal potential divider) is placed across pins 7 and 8, the 1.2V source will pass 1mA and each LED will pass 10mA in the ON mode.

Note from the above that the IC can pass total currents up to 300mA when used in the bar mode with all ten LEDs on. The IC has a maximum power rating of only 660mW, so there is a danger of exceeding this rating when the IC is used in the bar mode. In practice, the IC can be powered from DC supplies in the range 3V to 25V, and the LEDs can use the same supply as the IC or can be independently powered; this latter option can be used to keep the power dissipation of the IC at minimal level.

The internal 10-stage potential divider of the IC is floating, with both ends externally available for maximum versatility, and can be powered from either the internal reference or from an external source or sources. If, for example, the top of the chain is connected to a 10V source, the IC will function as a 0-to-10V meter if the low end of the chain is grounded, or as a restricted-range 5V-to-10V meter if the low end of the chain is connected to a 5V source. The only constraint on using the divider is that its voltage must not be greater than 2V less than the IC's supply voltage (which is limited to 25V maximum). The input (pin 5) to the IC is fully protected against overload voltages up to plus or minus 35V.

The IC's internal voltage reference produces a nominal output of 1.28V (limits are 1.2V to 1.32V), but can be externally programmed to produce effective reference values up to 12V (as shown later in this chapter). The IC can be made to give a dot-mode display by wiring pin 9 to pin 11, or a bar display by wiring pin 9 to positive supply pin 3.

Finally, note that the major difference between the three members of the LM3914 family of ICs lays in the values of resistance used in the internal 10-stage potential divider. In the LM3914, all resistors in the chain have equal values, and thus produce a linear display of ten equal steps. In the LM3915, the resistors are logarithmically weighted, and thus produce a log display that spans 30dB in ten 3dB steps. In the LM3916, the resistors are weighted in a semi-log fashion and produce a display that is specifically suited to VU-meter applications. With all of these points in mind, let us now move on and look at some practical applications of this series of devices, paying particular attention to the linear LM3914 IC.

## Practical LM3914 circuits

### *Dot-mode voltmeters*

*Figures 7.17* to *7.20* show various ways of using the LM3914 IC to make 10-LED dot-mode voltmeters. Note in all of these circuits that pin 9 is wired to

*LED bar-graph displays* 191

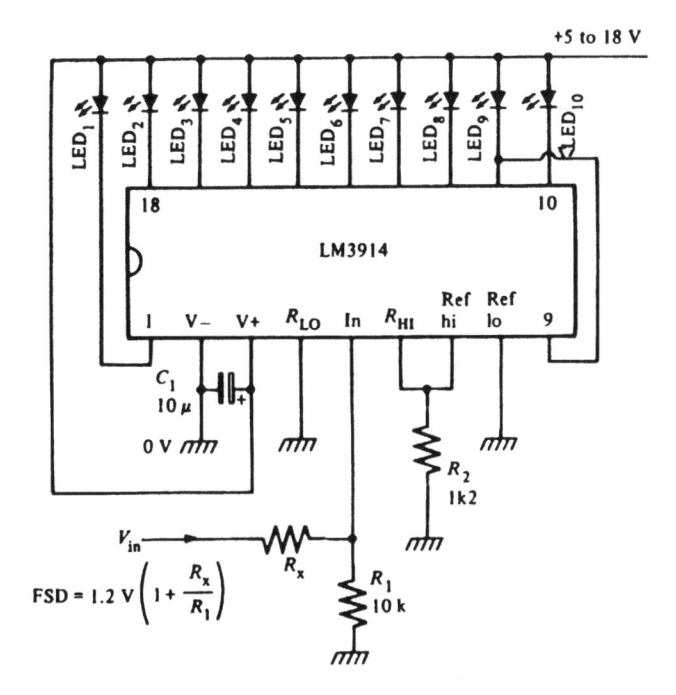

**Figure 7.17** *1.2V to 1000V f.s.d. dot-mode voltmeter*

pin 11, to give dot-mode operation, and that a 10μ capacitor is wired directly between pins 2 and 3 to enhance circuit stability.

*Figure 7.17* shows the connections for making a variable-range (1.2V to 1000V f.s.d.) voltmeter. The low ends of the internal reference and divider are grounded and their top ends are joined together, so the meter has a basic full-scale sensitivity of 1.2V, but variable ranging is provided by the Rx–R1 potential divider at the input of the circuit. Thus, when Rx is zero, f.s.d. is 1.2V, but when Rx is 90k the f.s.d. is 12V. Resistor R2 is wired across the internal reference and sets the ON current of all LEDs at about 10mA.

*Figure 7.18* shows how to make a fixed-range 0-to-10V meter, using an external 10V zener diode (connected to the top of the internal divider) to provide a reference voltage. The supply voltage to this circuit must be at least two volts greater than the zener reference voltage.

*Figure 7.19* shows how the internal reference of the IC can be made to effectively provide a variable voltage, enabling the meter f.s.d. value to be set anywhere in the range 1.2V to 10V. In this case the 1mA current (determined by R1) of the floating 1.2V internal reference flows to ground via RV1, and the resulting RV1-voltage raises the reference pins (7 and 8) above zero. If, for example, RV1 is set to 2k4, pin 8 will be at 2.4V and pin 7 at 3.6V. RV1 thus enables the pin-7 voltage, which is connected to the top of

192  LED bar-graph displays

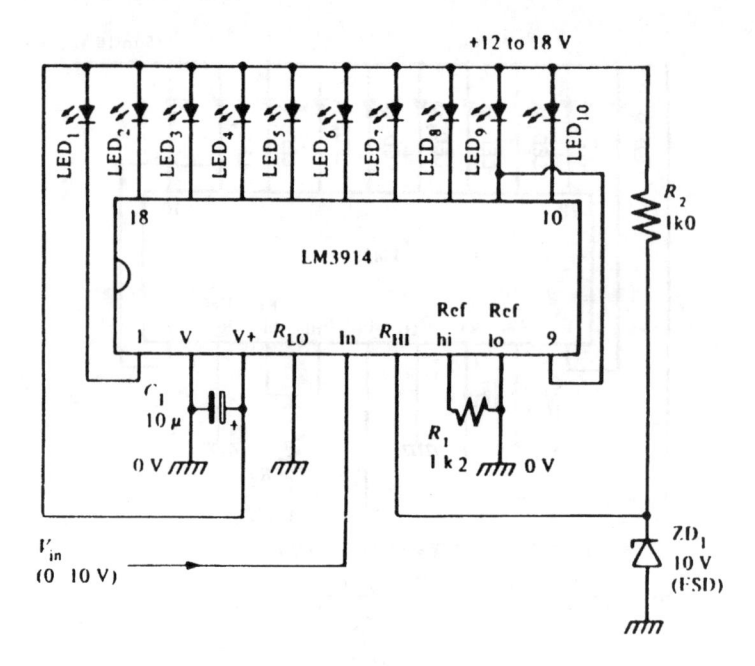

**Figure 7.18**  *10V f.s.d. meter using an external reference*

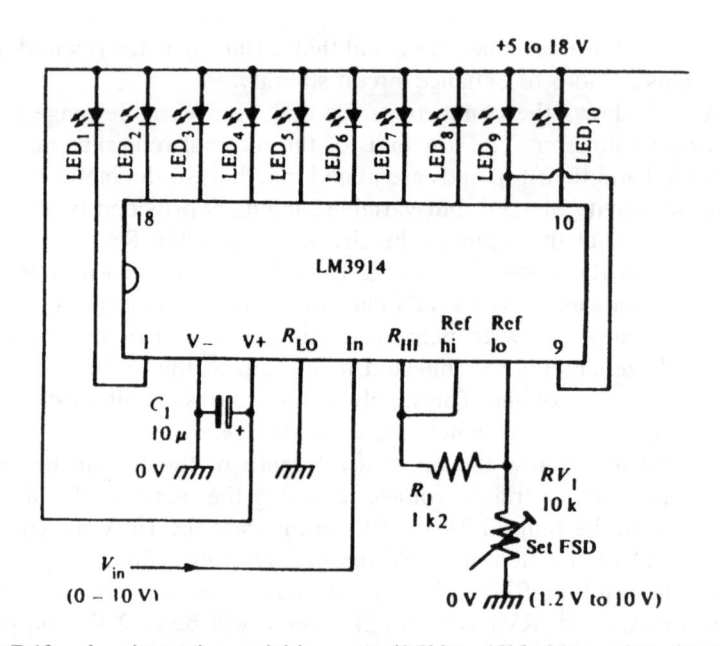

**Figure 7.19**  *An alternative variable-range (1.2V to 10V) dot-mode voltmeter*

*LED bar-graph displays* 193

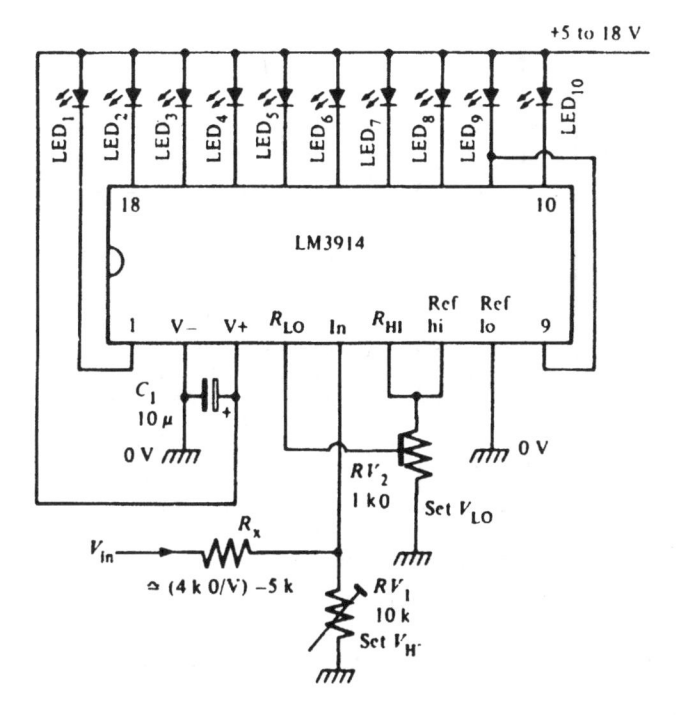

**Figure 7.20** *Expanded-scale (10V–15V) dot-mode voltmeter*

the internal divider, to be varied from 1.2V to approximately 10V, and thus sets the f.s.d. value of the meter within these values.

Finally, *Figure 7.20* shows the connections for making an expanded-scale meter that, for example, reads voltages in the range 10V to 15V. RV2 sets the LED current at about 12mA, but also enables a reference value in the range 0V to 1.2V to be set on the low end of the internal divider. Thus, if RV2 is set to apply 0.8V to pin 4, the basic meter will read voltages in the range 0.8V to 1.2V only. By fitting potential divider Rx–RV1 to the input of the circuit, this range can be 'amplified' to, say, 10V to 15V, or whatever range is desired.

## Bar-mode voltmeters

The dot-mode circuits of *Figures 7.17* to *7.20* can be made to give bar-mode operation by simply connecting pin 9 to pin 3, rather than to pin 11. When using the bar mode, however, it must be remembered that the IC's power rating must not be exceeded by allowing excessive output-terminal voltages to be developed when all ten LEDs are on. LEDs 'drop' about 2V when they

194  LED bar-graph displays

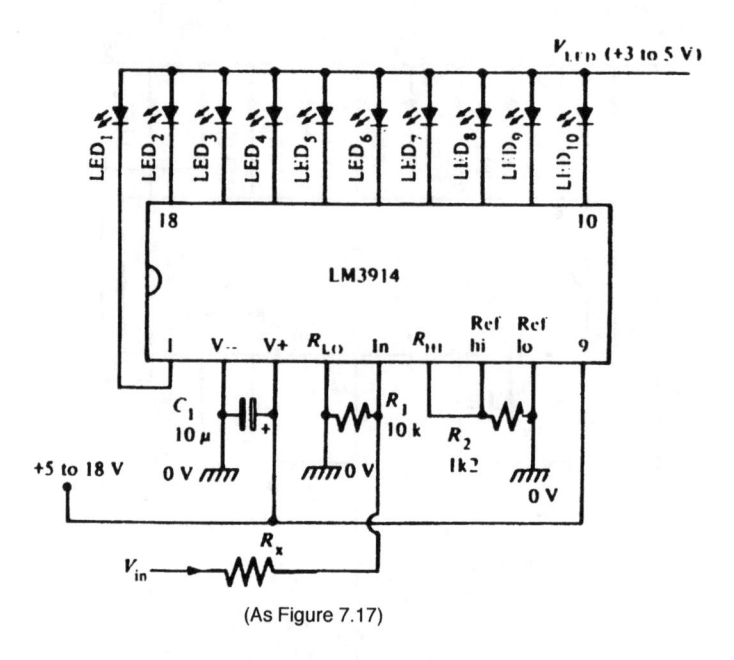

**Figure 7.21**  Bar-display voltmeter with separate LED supply

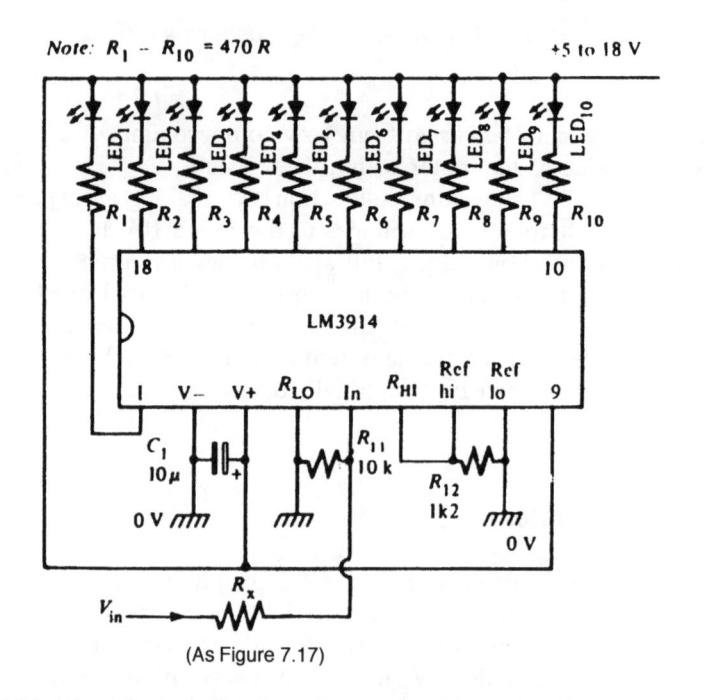

**Figure 7.22**  Bar-display voltmeter with common LED supply

are conducting, so one way around this problem is to power the LEDs from their own low-voltage (3 to 5V) supply, as shown in *Figure 7.21*.

An alternative solution is to power the IC and the LEDs from the same supply, but to wire a current-limiting resistor in series with each LED, as shown in *Figure 7.22*, so that the IC's output terminals saturate when the LEDs are on.

An alternative way of obtaining a bar display without excessive power dissipation is shown in *Figure 7.23*. Here, the LEDs are all wired in series, but with each one connected to an individual output of the IC, and the IC is wired for dot-mode operation. Thus, when (for example) LED5 is driven on it draws its current via LEDs 1 to 4, so all five LEDs are on. In this case, however, the total LED current is equal to that of a single LED, so power dissipation is quite low: in this mode the LM3914 in fact operates in a similar fashion to the U237-type of IC. The LED supply to this circuit must be greater than the sum of all LED volt drops when all LEDs are on, but within the voltage limits of the IC; a regulated 24 volt supply is thus needed.

*Figure 7.24* shows a modification of the above circuit, which enables it to be powered from an unregulated 12V to 18V supply. In this case the LEDs are split into two chains, and the transistors are used to switch the lower (LEDs 1 to 5) chain on when the upper chain is active; the maximum total LED current is equal to twice the current of a single LED.

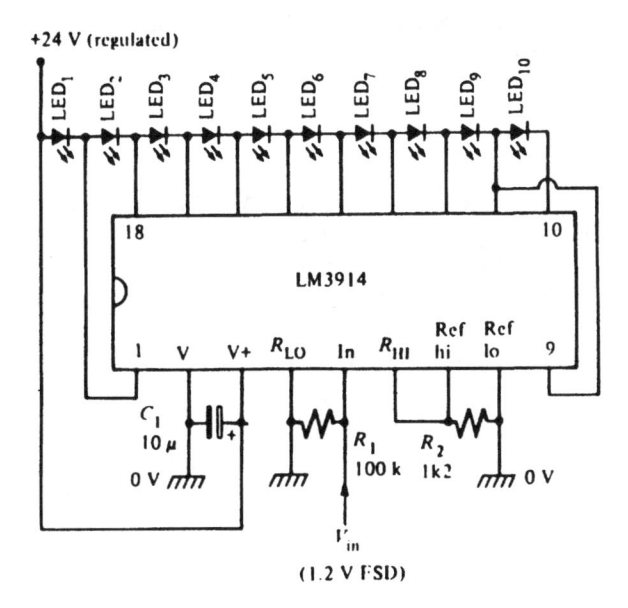

**Figure 7.23** Method of obtaining a bar display with dot-mode operation and minimal current consumption

196  LED bar-graph displays

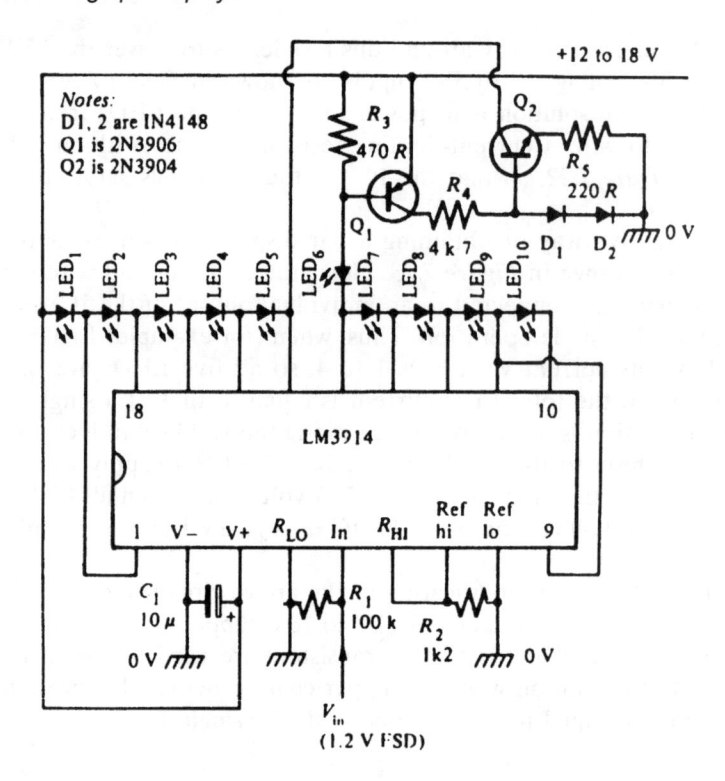

**Figure 7.24** Modification of the Figure 7.23 circuit, for operation from unregulated 12V to 18V supplies

### 20-LED voltmeters

*Figures 7.25* and *7.26* show how pairs of LM3914s can be interconnected to make 20-LED 0-to-2.4V voltmeters. Here, the input terminals of the two ICs are wired in parallel, but IC1 is configured so that it reads 0-to-1.2V and IC2 is configured so that it reads 1.2V-to-2.4V. In the latter case, the low end of the IC2 potential divider is coupled to the 1.2V reference of IC1 and the top end of the divider is taken to the top of the 1.2V reference of IC2, which is raised 1.2V above that of IC1.

The *Figure 7.25* circuit is wired for dot-mode operation. Note in this case that pin 9 of IC1 is wired to pin 1 of IC2, and pin 9 of IC2 is wired to pin 11 of IC2. Also note that a 22k resistor is wired in parallel with LED9 of IC1.

The *Figure 7.26* circuit is wired for bar-mode operation. The connections are similar to those above, except that pin 9 is taken to pin 3 of each IC, and a 470R current-limiting resistor is wired in series with each LED to reduce power dissipation of the ICs.

*LED bar-graph displays* 197

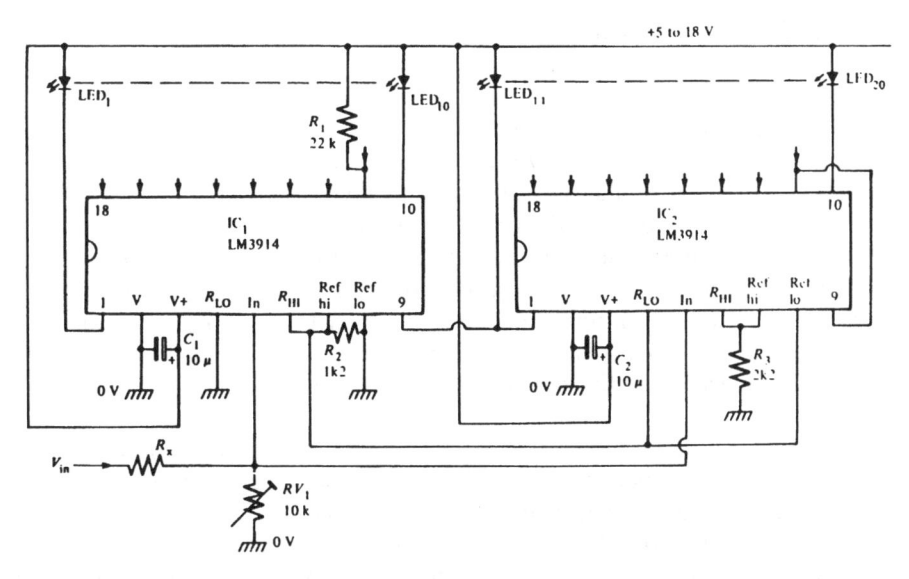

**Figure 7.25** *Dot-mode 20-LED voltmeter (f.s.d. = 2.4V when Rx = 0)*

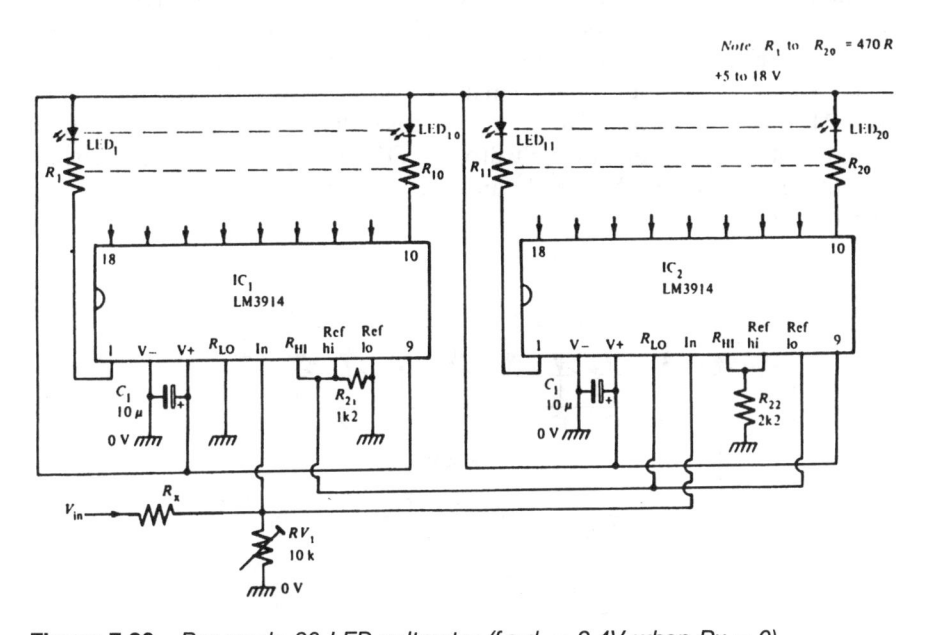

**Figure 7.26** *Bar-mode 20-LED voltmeter (f.s.d. = 2.4V when Rx = 0)*

In both cases, the voltage-ranging of the 20-LED *Figures 7.25* and *7.26* circuits can easily be altered by using techniques that have already been described in this chapter.

198  LED bar-graph displays

## LM3915/LM3916 circuits

The LM3915 log and LM3916 semilog ICs operate in the same basic way as the LM3914, and can in fact be directly substituted in most of the circuits shown in *Figures 7.17* to *7.26*. In most practical applications, however, these particular ICs are used to give a meter indication of the value of an ac input signal, and the simplest way of achieving such a display is to connect the ac signal directly to the pin-5 input terminal, as shown in *Figure 7.27*. The IC responds only to the positive halves of such input signals, and the number of illuminated LEDs is thus proportional to the instantaneous peak value of the input signal. In such circuits, the IC should be operated in the dot mode and set to give about 30mA of LED drive current.

The *Figure 7.27* circuit is that of a simple LM3915-based audio power meter. Pin 9 is left open-circuit to ensure dot-mode operation, and R1 has a value of 390R to give a LED current of about 30mA. The meter gives audio power indications over the range 200mW to 100W.

A more sophisticated way of using these ICs to show the value of an ac input signal is to use a half-wave converter to change the ac signal into dc that is then fed to the input of the IC. *Figures 7.28* and *7.29* show practical LM3916-based VU-meter (volume unit meter) circuits of this type.

In *Figure 7.28*, the input signal is converted to dc via the simple D1–R1–R2–C1 network. Note in this case that rectifier D2 is used to compensate for the forward volt drop of D1. Also note that this particular circuit operates in the bar mode and uses separate supplies for the IC and the LED display.

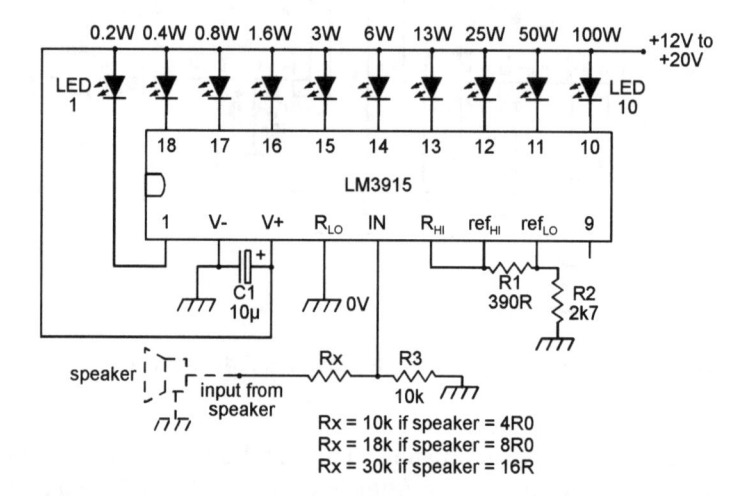

**Figure 7.27** Simple audio power meter

LED bar-graph displays 199

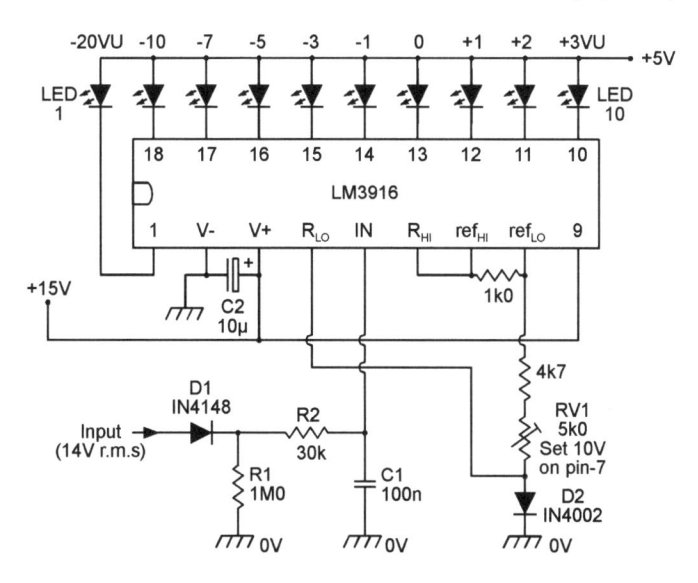

**Figure 7.28** *Simple VU-meter*

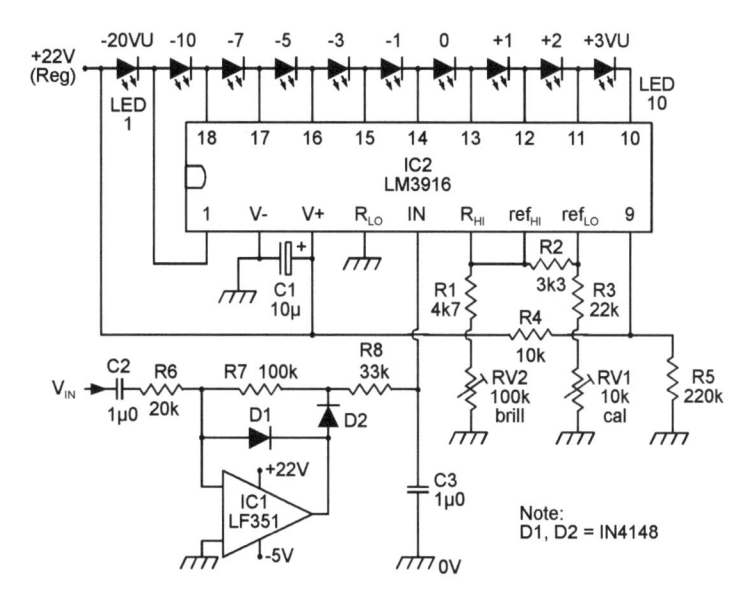

**Figure 7.29** *Precision VU-meter, with low current drain*

Finally, to complete this look at the LM3914 range of devices, *Figure 7.29* shows how the LM3916 can be used as a precision VU-meter by using a precision half-wave rectifier (IC1) to give ac/dc conversion. Note in this circuit that the LEDs are wired in series and IC2 is wired in the dot mode,

## 200 LED bar-graph displays

to give a low-consumption bar display of the type shown in *Figure 7.23*. To set up this circuit, simply adjust RV1 to set 10V on pin 7; RV2 can then be used as a LED brightness control.

# 8

# Audio delay-line systems and circuits

Solid-state audio-frequency delay lines are widely used in modern music and audio systems. They can be used to produce popular sound effects such as echo, reverb, chorus or phasing in music and karaoke systems, or rare effects such as ambience synthesis or pseudo 'room expansion' in expensive hi-fi systems, or to give predictive/anticipatory effects such as automatic pre-switching in cassette tape recorders or click/scratch elimination in record (disc) players, or to enhance message intelligibility in multi-output public address systems by inserting correctly balanced acoustic delays between individual loudspeakers.

Modern solid-state audio delay-line systems can be built in either analogue or digital form. In the former case they are built around one or more dedicated analogue delay-line ICs, and in the latter case are usually built around collections of ICs such as D-to-A and A-to-D converters, RAMs, and a variety of less exotic digital ICs. In both cases, the delay-line systems are driven via 'clock' waveform generators, and the delay period is controlled by the clock frequency. This chapter explains the operating principles of both systems, and shows how dedicated analogue and digital delay-line ICs can be used in practical delay-line systems.

## Audio delay-line basics

An audio delay-line is a unit in which an audio (usually speech or music) signal is applied to the line's input, and an identical copy of the signal appears at the line's output but is time-delayed by a pre-set amount (usually in the range 1 to 500ms), as shown in *Figure 8.1*. *Figures 8.2* to *8.4* show some simple delay-line applications.

*Figure 8.2* shows a delay line used to give a double-tracking effect, in which the delay is in the range 10 to 25ms and the delayed and direct signals are added in an audio mixer to give a composite 'two signals' output. If a solo

## 202 Audio delay-line systems and circuits

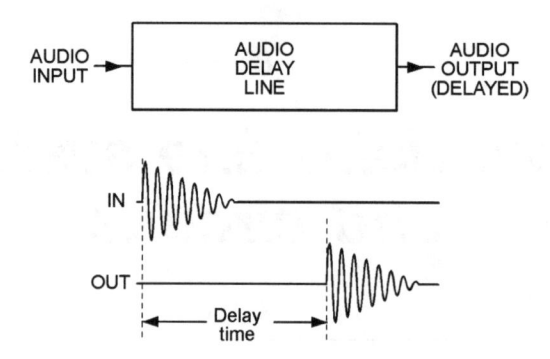

**Figure 8.1** Basic audio delay-line system

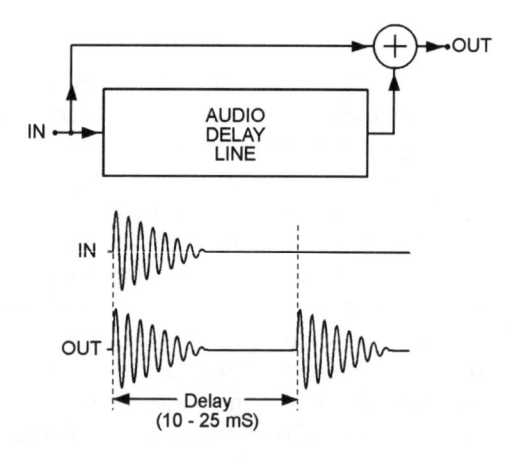

**Figure 8.2** Basic double-tracking circuit

singer's voice is played through this unit, the output sounds like that of a pair of singers in very close harmony. Alternative names for this type of circuit are mini-echo and mini-chorus.

*Figure 8.3* shows the above circuit modified to act as an 'echo' unit. In this case the delay line's output is attenuated before being added to that of the direct input, and the 'delay' time is in the range 10 to 250ms. At sea level, at a temperature of 20°C, sound travels through the air at a speed of 0.343 metres per ms. Thus, if this unit gives a delay of (for example) 20ms, its output will sound like the original signal accompanied by a single echo that has bounced off a surface that is 3.43 metres from the original sound source. Note that this circuit produces only a single echo, but that the echo volume is variable.

*Figure 8.4* shows the above circuit modified to act as an echo/reverb unit that produces multiple echoes (reverberation). This circuit uses two mixers, one ahead of the delay line and the other at the output. Part of the delay

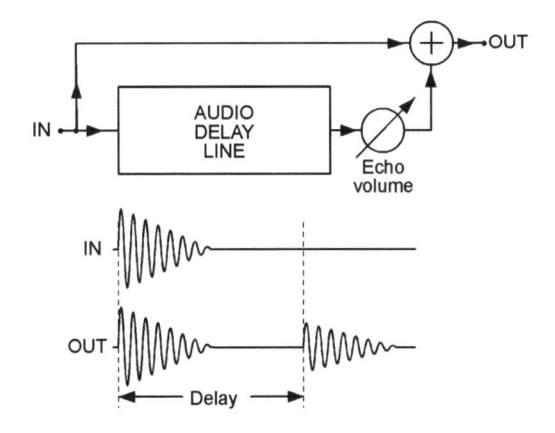

**Figure 8.3** Basic 'echo' unit

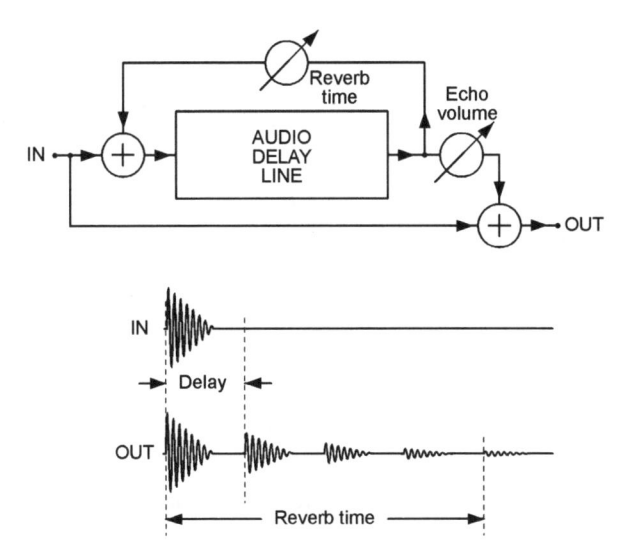

**Figure 8.4** Basic 'echo-reverb' circuit

output is fed back to the input mixer, so that the circuit gives echoes of echoes of echoes, etc. The reverb time is defined as the time taken for the repeating echoes to fall by 60dB relative to the original input signal, and depends on the delay time and the overall attenuation of the feedback signals. Echo delay time, echo volume and reverb time are all independently variable.

Prior to the end of the 1960s, circuits of the above types were usually built around electromechanical devices such as coil-spring delay lines or continuous-loop tape recorders. Coil-spring units rely on the facts that acoustic

204  *Audio delay-line systems and circuits*

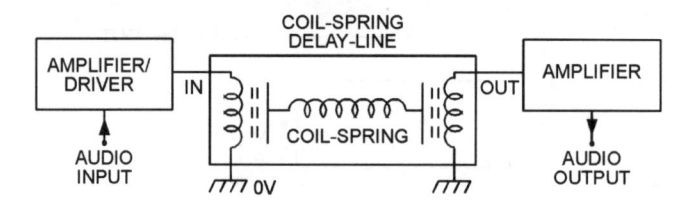

**Figure 8.5**  *Basic coil-spring delay-line usage circuit*

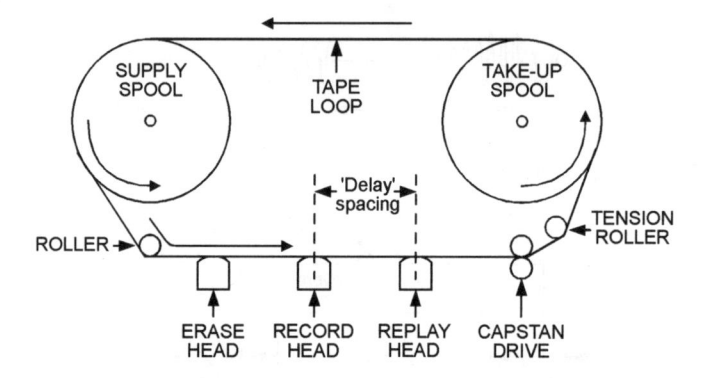

**Figure 8.6**  *Basic 'continuous-loop' tape recorder delay-line system*

vibrations travel quite efficiently along steel wire that is under tension, and that considerable lengths of such wire can be housed in a relatively small area by winding it in the shape of a coil-spring. A 100-turn coil-spring that is 5cm in diameter and 25cm in length may, for example, contain about 15.5 metres of steel wire, and an acoustic signal takes about 45ms to travel such a length. Practical coil-spring delay lines have an electromagnetic (coil-and-magnet) driver at the 'input' end of the spring, and a magnetoelectric (magnet-and-coil) pickup at the 'output' end. *Figure 8.5* shows the basic way of using such a unit.

Continuous-loop tape recorder delay lines operate in the basic way illustrated in *Figure 8.6*. The unit has separate erase, record and replay heads; the input is applied to the record head, and the time-delayed output is taken from the replay head. The delay period (in seconds) equals the record-to-replay head spacing (in cm) divided by the tape speed in cm/s. Thus, at the normal tape speed of 4.75cm/s, the delay equals 300ms at a head spacing of 1.425cm. In practice, the tape speed is typically variable from (about) 50% to 200% of nominal, enabling the delay period to be varied between 150ms and 600ms in the above example.

'Spring' and 'tape' delay lines are delicate devices with very limited delay-period range control. In the early 1970s both types of device were made obsolete by the arrival of a new type of integrated circuit that is variously

*Audio delay-line systems and circuits* 205

known as a charge-coupled device (CCD) or bucket brigade delay line (BBD). In essence, these ICs house a chain of hundreds of IGFET-plus-capacitor sample-and-hold analogue storage elements that act like charge-storage 'buckets' (hence the bucket brigade title) that are interconnected via IGFET electronic switches that are clocked via an external 2-phase generator so that the IC acts like an analogue shift register; analogue samples of the input signals are clocked in at the front of the bucket chain, and are clocked out in reconstructed time-delayed form at the chain's end. The IC's delay time is proportional to its number of 'bucket' stages and to its clock-frequency period, and can easily be varied (via the clock generator) over a very wide range. Such devices are thus robust and versatile and require very little external support circuitry, other than a clock generator and a couple of simple active filters.

## BBD principles

The first primitive BBD IC was developed by Philips Laboratory in 1968, and in use was connected into the basic circuit shown in *Figure 8.7*, in which the audio input signal is ac-coupled to the IC's IN terminal, which is biased into its linear operating range via voltage divider R1–R2, and the IC's bucket stages are clocked by a pair of antiphase squarewave inputs at a frequency that is high relative to that of the audio input signal. The IC has two outputs (OUT A and OUT B), and these are effectively shorted together to generate the final reconstructed and delayed audio output.

The basic operating principle of this type of BBD can be understood with the help of *Figure 8.8*, which depicts the first six stages of the IC as indivual 'buckets', shown operating through the first six half-cycles of the clock phase-1 waveform (annotated +1 to –3 in the diagram). Switch S1 represents an IGFET switch that is connected between the IC's IN terminal and the first bucket stage, and the IC's basic action is such that S1 is open and all odd-numbered buckets transfer their charges into the even-numbered buckets to

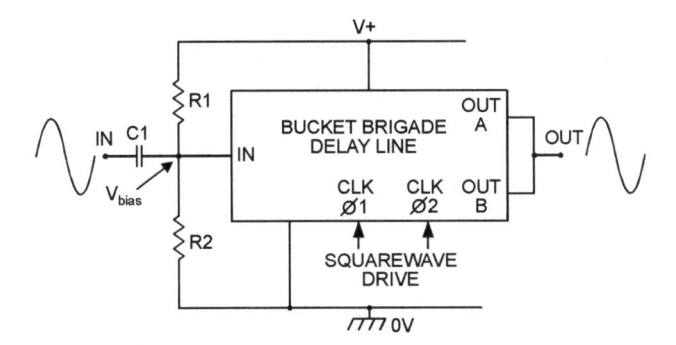

**Figure 8.7** *Basic usage circuit of the original (1968) type of BBD IC*

206  Audio delay-line systems and circuits

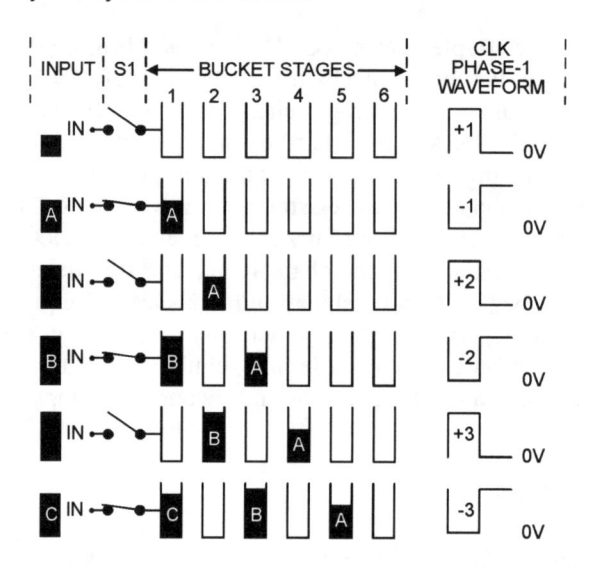

**Figure 8.8** *Diagram illustrating the basic operating principle of the original (1968) type of bucket-brigade delay line*

their right when the phase-1 waveform is high (+), and S1 is closed and all even-numbered buckets transfer their charges into the odd-numbered buckets to their right when the phase-1 waveform is low (−).

Thus, in the first (+1) clock period, S1 is open and all buckets are empty, but in the second (−1) period S1 is closed and the 'A' input is connected to the 1st sample-and-hold bucket stage. In the +2 period, S1 opens again and charge 'A' is moved into bucket 2, leaving bucket 1 empty. In the −2 period S1 is closed and charge 'B' is loaded into bucket 1, and simultaneously charge 'A' is transferred into bucket 3. This basic process repeats on each successive clock cycle, with all bucket charges shifting one step to the right on each half-cycle, and with a fresh input sample being loaded into bucket 1 each time S1 closes.

Note in *Figure 8.8* that each charge steps two places to the right in each complete clock cycle, and that there is always an empty bucket between each charge-carrying pair of buckets. Consequently, the output waveform from the final bucket stage is too rough to be of direct value, and to overcome this problem the IC incorporates voltage-follower buffers on the final two bucket stages, making both of these outputs externally accessible as shown in *Figure 8.9(a)*, and enabling a useful 'composite' output waveform to be obtained by shorting the two outputs together (either directly or via a balance pot) as shown in *Figure 8.9(b)*, thus effectively adding the two output together and generating a gap-free 'composite' output as shown in *Figure 8.9(c)*. This final output is thus a quantized but time-delayed replica of the original input signal.

*Audio delay-line systems and circuits* 207

**Figure 8.9** *Diagram showing how a composite output waveform is derived from the last two bucket stages in the delay-line chain*

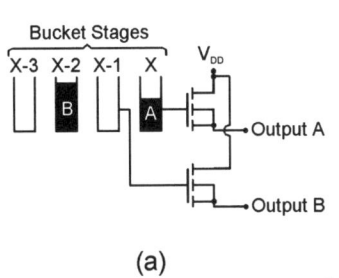

(a) The final two bucket stages are accessible via voltage-follower buffer stages.

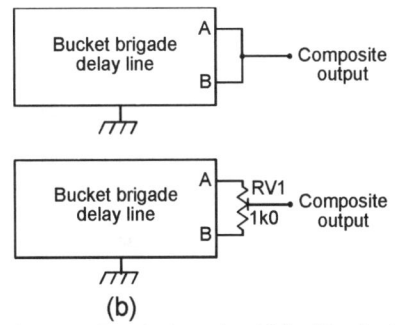

(b) A composite output can be obtained by shorting the A and B outputs together, either directly or via a balance pot.

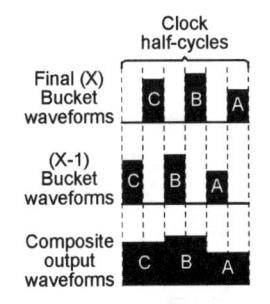

(c) The composite output waveforms are generated in the way shown here.

That, then, is how the original (1968) Philips bucket-brigade delay line worked, and is, in essence, how all modern BBD ICs operate. The original device did, however, suffer from several major weaknesses due to the near-impossibility of fully discharging each of its several-hundred 'sample-and-hold' storage capacitors at the end of each transfer cycle, and this caused the IC to give poor transfer efficiency and to offer a very poor dynamic operating range (typically less than 60dB). Throughout the early 1970s, companies such as Philips, Bell, and Reticon devoted much effort to improving the basic design, and this resulted in the modified concept illustrated in *Figure 8.10*.

The major feature of the *Figure 8.10* bucket brigade system is that all 'buckets' start off fully charged via an external bias generator, and that S1 opens and each odd-numbered bucket fully tops-up the even-numbered buckets to its left when the phase-1 waveform is high (+), and S1 closes and each even-numbered bucket fully tops-up the odd-numbered bucket to its left when the phase-1 waveform is low (−).

Thus, when S1 first closes on the second (−1) clock period, the 'A' input is connected to the 1st sample-and-hold bucket stage, and when S1 opens

208  *Audio delay-line systems and circuits*

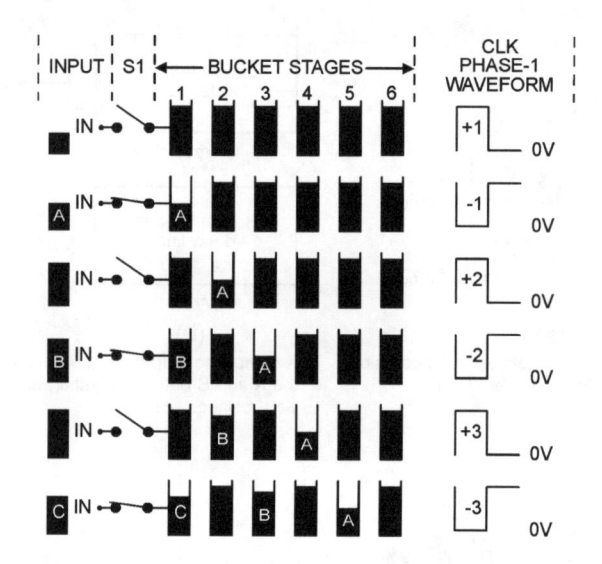

**Figure 8.10**  *Diagram illustrating the basic operating principle of the modern type of bucket-brigade delay line*

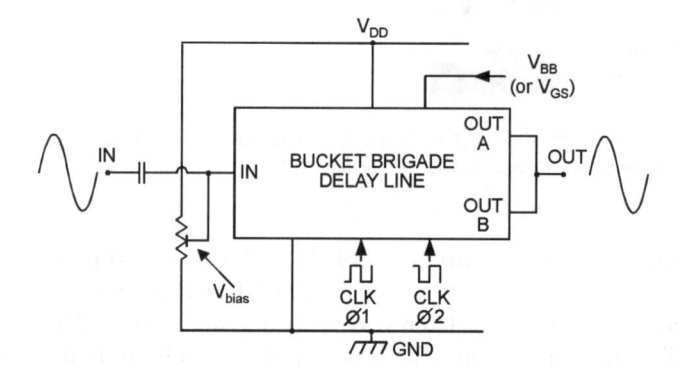

**Figure 8.11**  *Basic 'usage' connections of a modern BBD IC*

again in the +2 period the '2' bucket tops-up bucket 1, so at the end of this period the residual charge in bucket '2' precisely equals the original 'A' charge, i.e. charge 'A' is effectively transferred into bucket 2. In the −2 period, S1 closes and charge 'B' is loaded into bucket 1, and simultaneously bucket 3 tops-up bucket 2, effectively transferring charge 'A' into bucket 3. This basic process repeats on each successive clock cycle, with all input charges effectively shifting one step to the right on each half-cycle, and with a fresh input sample being loaded into bucket 1 each time S1 closes.

Thus, this modified form of bucket-brigade operation achieves the same end results as the basic *Figure 8.8* design, but does so very efficiently, since it only partially discharges its internal capacitors during each clocked operating cycle. This modified system is used in all modern BBDs, and offers very high transfer efficiency and has an excellent dynamic operating range (usually better than 80dB).

*Figure 8.11* shows the basic 'usage' connections of a modern BBD IC, which is powered via $V_{DD}$ and ground but also uses another supply, annotated $V_{BB}$ or $V_{GG}$, to correctly bias its sample-and-hold IGFETS. The IC's input terminal must be biased into the linear mode by voltage $V_{bias}$, and the IC's two outputs must be shorted together as already described; direct-shorting is shown used in the diagram, and in some ICs this is provided internally and only a single 'composite' output is available from the IC. Finally, the IC needs a 2-phase clock signal, normally consisting of a pair of anti-phase squarewaves that switch fully between the $V_{DD}$ and GND potentials.

## How much delay?

Each of a BBD's bucket charges are shifted through two bucket stages during each complete clock cycle, and the maximum number of samples held by a BBD is equal to half the number of bucket stages. The actual time-delay available from a BBD is thus given by:

Time Delay = $S.p/2$ or $S/2.f$

where $S$ is the number of bucket stages, $p$ is the clock period, and $f$ is the clock frequency.

Practical analogue delay line ICs usually have 512, 1024, 1536, 3328 or 4096 stages. Thus, a 1024-stage line using a 10kHz (100µs) clock gives a delay of 51.2ms. A 4096-stage line gives a 204.8ms delay at the same clock frequency. In practice, the maximum useful signal frequency of a delay line is equal to one third of the clock frequency, so a delay line clocked at 10kHz actually has a useful bandwidth of only 3.3kHz.

*Figure 8.12* shows the block diagram of a practical BBD analogue delay line system. The input signal is fed to the BBD line via a low-pass filter with a cut-off frequency that is one third (or less) of the clock value, and is vital to overcome 'aliasing' or inter-modulation problems. The BBD's output is passed through a second low-pass filter, which also has a cut-off frequency one third (or less) of that of the clock and which serves the dual purposes of rejecting clock break-through signals and integrating the delay-line output pulses, so that the final output signal is a faithful (but time-delayed) copy of the original.

A closer look at practical 'clock' and filter circuits is taken later in this chapter. In the meantime, it is wise to look at some more popular delay-line

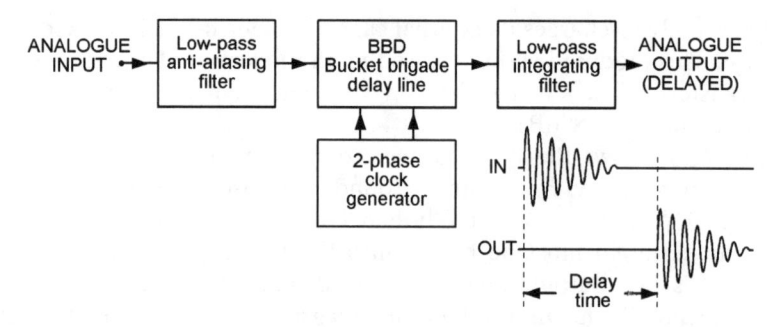

**Figure 8.12** *Block diagram of a basic BBD analogue delay-line system*

applications, and in order to fully understand these it is necessary to first briefly study the subject known as psycho-acoustics.

## Psycho-acoustics

Many of the special effects obtainable with delay lines rely on the human brain's idiosyncratic behaviour when interpreting sounds. Basically, the brain does not always perceive sounds as they truly are, but simply 'interprets' them so that they conform to a preconceived pattern; it can easily be fooled into misinterpreting the nature of sound. The study of this particular subject is known as psycho-acoustics. Here are some relevant psycho-acoustic 'laws' that are worth knowing:

(1) If a human's ears receive two sounds that are identical in form but are time-displaced by less than 10ms, the brain integrates them and perceives them as a single (undisplaced) sound.

(2) If a human's ears receive two sounds that are identical in form but are time-displaced by 10–50ms, the brain perceives them as two independent sounds but integrates their information content into a single easily recognizable pattern, with no loss of information fidelity.

(3) If a human's ears receive two signals that are identical in form but time-displaced by greater than 50ms, the brain perceives them as two independent sounds but may not be able to integrate them into a recognizable pattern.

(4) If a human's ears receive two sounds that are identical in basic form but not in magnitude, and which are time-displaced by more than 10ms, the brain interprets them as two sound sources (primary and secondary) and draws conclusions concerning (a) the *location* of the primary sound source and (b) the relative *distances apart* of the two sources.

Regarding 'location' identification, the brain identifies the first perceived signal as the prime sound source, even if its magnitude is substantially lower than that of the second perceived signal (this is known as the Hass effect). Delay lines can thus be used to trick the brain into wrongly identifying the location of a sound source.

Regarding 'distance' identification, the brain correlates distance and time-delay in terms of approximately 0.34 metres (13.4 inches) per millisecond of delay. Delay lines can thus be used to trick the brain about 'distance' information.

(5) The brain uses echo and reverberation (repeating echoes of diminishing amplitude) information to construct an image of environmental conditions, e.g. if echo times are 50ms but reverb time is 2 seconds, the brain may interpret the environment as being a 17 metre (56 ft) cave or similar hard-faced structure, but if the reverb time is only 150ms it may interpret the environment as being a 17 metre wide softly-furnished room. Delay lines can thus be used to trick the brain into drawing false conclusions concerning its environment, as in ambience synthesizers or acoustic 'room expanders'.

(6) The brain is very sensitive to brief transient increases in volume, such as caused by clicks and scratches on records (discs), but is blind to transient drops in volume. Delay lines can be used to take advantage of this effect in record players, where they can be used (in conjunction with other circuitry) to effectively predict the arrival of a noisy click or scratch and replace it with an 'unseen' neutral or negative transient.

## Practical delay-line applications

### Simple musical effects

*Figures 8.13* to *8.22* show a variety of fairly basic analogue delay-line applications. For the sake of simplicity, these diagrams do not show the input/output low-pass filters that are used in most practical circuits. The first three diagrams show simple musical effects circuits.

*Figure 8.13* shows how the delay line can be used to apply vibrato (frequency modulation) to any input signal. The low-frequency sinewave generator modulates the voltage-controlled oscillator (VCO) clock generator frequency and thus causes the output signals to be similarly time-delay or 'frequency' modulated; this circuit is widely used in karaoke units.

*Figure 8.2* has already shown how a delay line can be used to give a double-tracking effect, and *Figure 8.14* shows how this basic circuit can be modified to act as an auto-double-tracking (ADT) mini-chorus unit. Clock signals come from a VCO that is modulated by a slow oscillator, so that the delay times slowly vary. The effect is that, when a solo singer's voice is played through the unit, the output sounds like a pair of singers in loose or natural harmony.

212   Audio delay-line systems and circuits

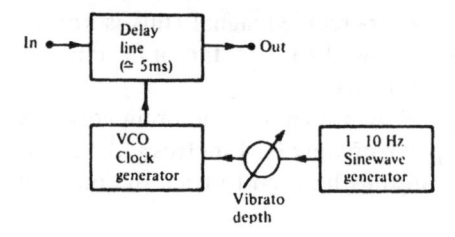

**Figure 8.13**   True vibrato circuit, which applies slow frequency modulation to all input signals

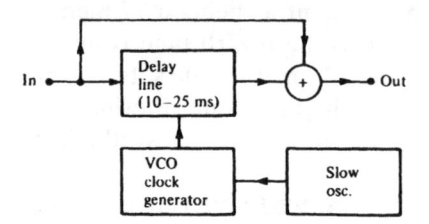

**Figure 8.14**   Auto-double-tracking (ADT) or mini-chorus circuit

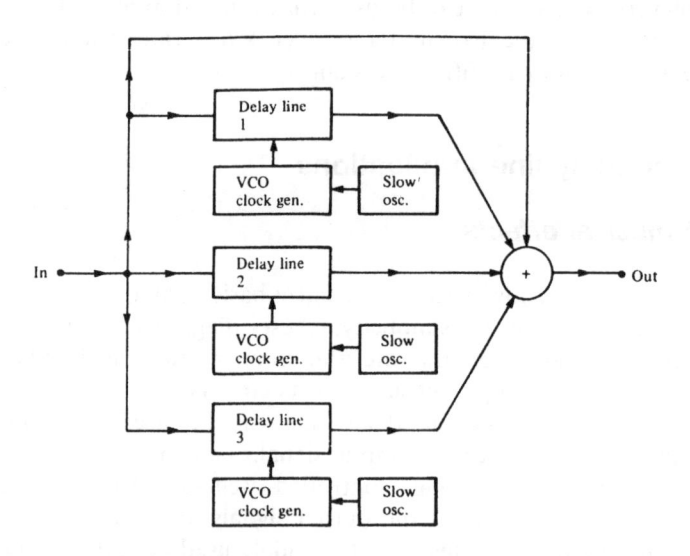

**Figure 8.15**   'Chorus' generator

*Figure 8.15* shows how three ADT circuits can be wired together to make a chorus machine. All three delay lines have slightly different delay times. The original input and the three delay signals are added together, the net effect being that a solo singer sounds like a quartet, or a duet sounds like an octet, etc.

*Audio delay-line systems and circuits* 213

## Comb filter circuits

*Figure 8.16* shows a delay line used to make a 'comb' filter. Here, the direct and delayed signals are added together; signal components that are in-phase when added give an increased output signal amplitude, and those that are in anti-phase tend to self-cancel and give a reduced output level. Consequently, the frequency response shows a series of notches, the notch spacing being the reciprocal of the line delay time (1kHz spacing at 1ms delay, 250Hz spacing at 4ms delay). These phase-induced notches are typically only 20–30dB deep.

The two most popular musical applications of the comb filter are in phasers and flangers. In the phaser (*Figure 8.17*) the notches are simply swept slowly up and down the audio band via a slow-scan oscillator, introducing a pleasant acoustic effect on music signals.

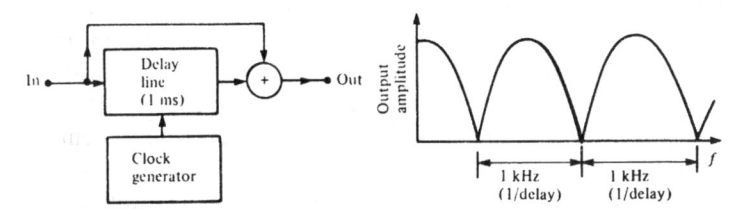

**Figure 8.16** *CCD comb filter. Notches are about 20 to 30 dB deep, 1kHz apart*

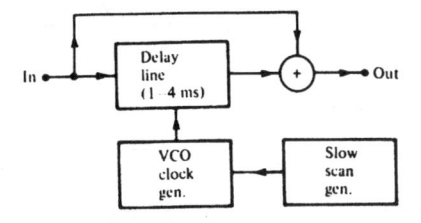

**Figure 8.17** *A phaser is a variable comb filter, in which the notches are slowly swept up and down the audio band*

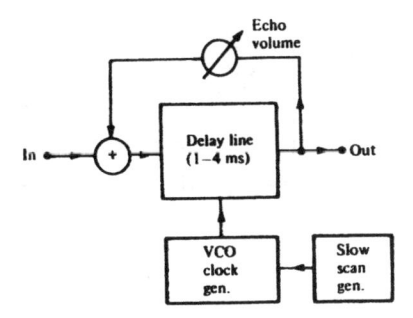

**Figure 8.18** *A flanger is a phaser with accentuated and variable notch depth*

214  *Audio delay-line systems and circuits*

The *Figure 8.18* flanger circuit differs from the phaser in that the mixer is placed ahead of the delay line and part of the delayed signal is fed back to one input of the mixer, so that in-phase signals add together regeneratively. Amplitudes of the peaks depend on the degree of feedback, and can be made very steep. These phase-induced peaks introduce very powerful acoustic effects as they are swept up and down through music signals via the slow-scan oscillator.

## *Echo/reverb circuits*

Basic echo and reverb circuits have already been shown in *Figures 8.3* and *8.4*, and *Figure 8.19* shows how several reverb circuits can be coupled together to make an ambience synthesizer or acoustic room expander. Here, the outputs of a conventional stereo hi-fi system are summed to give a mono image and the resulting signal is then passed to a pair of semi-independent reverb units (which produce repeating echoes but not the original signal). The reverb outputs are then summed and passed to a mono PA system and speaker, which is usually placed behind the listener. The system effectively synthesizes the echo and reverb characteristics of a chamber of any desired size, so that the listener can be given the impression of sitting in a cathedral,

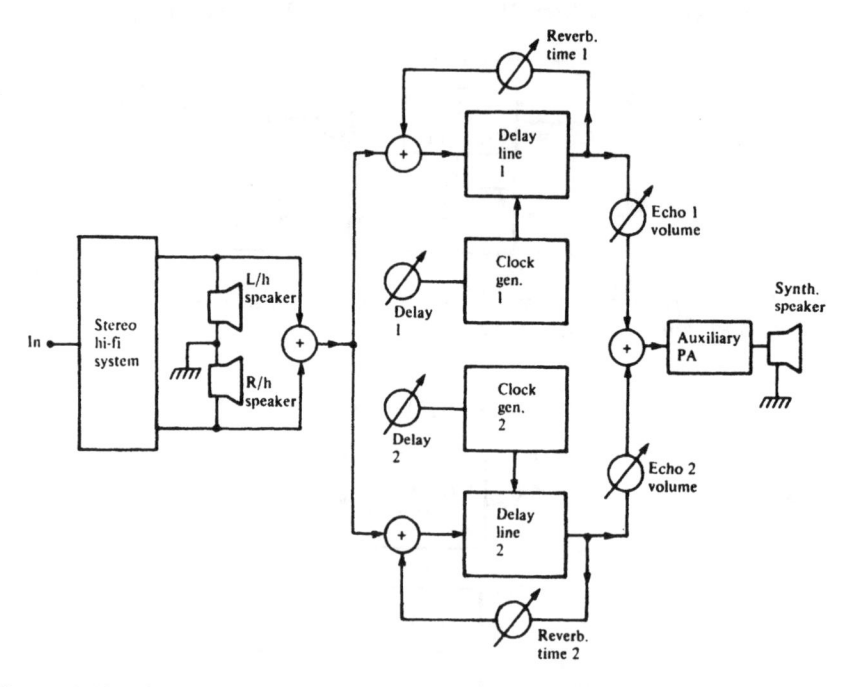

**Figure 8.19**  *Ambience synthesizer or acoustic 'room expander'*

concert hall, or small club house, etc., while in fact sitting in his own living room. Such units produce very impressive results.

There are lots of possible variations on the basic *Figure 8.19* circuit. In some cases the mono signal is derived by differencing (rather than summing) the stereo signals, thereby cancelling centre-stage signals and overcoming a rather disconcerting 'announcer-in-a-cave' effect that occurs in 'summing' systems. The number of delay (reverb) stages may vary from one in the cheapest units to four in the more expensive.

## *Predictive switching circuits*

Delay lines are particularly useful in helping to solve predictive or anticipatory switching problems, in which a switching action is required to occur slightly *before* some random event occurs. Suppose, for example, that you need to make recordings of random or intermittent sounds (thunder, speech, etc.). To have the recorder running continuously would be inefficient and expensive, and it would not be practical to simply try activating the recorder automatically via a sound switch, because part of the sound will already have occurred by the time the recorder turns on.

*Figure 8.20* shows the solution to this problem. The sound input activates a sound switch, which (because of mechanical inertia) turns the recorder's motor on within 20ms or so. In the meantime, the sound travels through the 50ms delay line towards the recorder's audio input terminal, so that the recorder has already been turned on for 30ms by the time the first part of the sound reaches it. When the original sound ceases, the sound switch turns off, but the switch extender maintains the motor drive for another 100ms or so, enabling the entire 'delayed' signal to be recorded.

*Figure 8.21* shows how predictive switching techniques can also be used to help eliminate the sounds of clicks and scratches from a record player. Such sounds can easily be detected by using stereo phase-comparison methods.

In the diagram, the record (disc) signals are fed to the audio amplifier via a 3ms delay line, a bilateral switch, and a track-and-hold circuit. Normally, the bilateral switch is closed, and the signal reaching the audio amplifier is

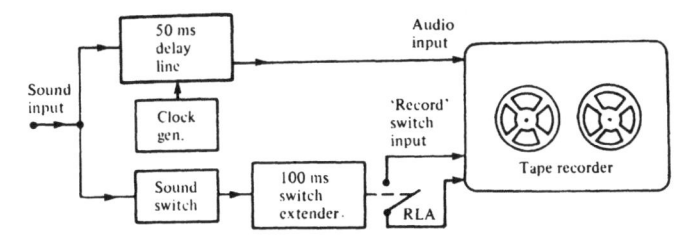

**Figure 8.20** *Automatic tape recorder with predictive switching*

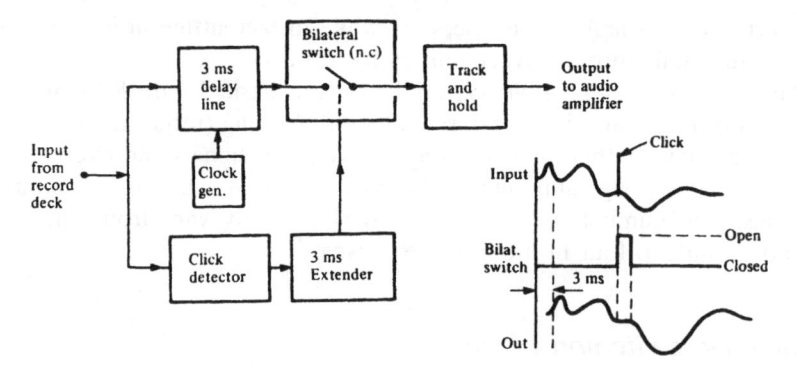

**Figure 8.21** *Record (disc) 'click' eliminator*

a delayed but otherwise unmodified replica of the disc signal. When a click or scratch occurs on the record, the detector/expander circuit opens the bilateral switch for a minimum of 3ms, briefly blanking the audio feed to the amplifier. Because of the presence of the delay line, the blanking period effectively straddles the 'click' period, enabling its sound effects to be completely eliminated from the system (see 'Psycho-acoustics').

## *Delay equalization*

One of the most important applications of delay line techniques is in sound-delay equalization of public address systems in theatres and at open-air venues such as air shows.

Sound travels through air at a rate of about 0.34 metres per millisecond. In simple multiple-speaker public address systems, in which all loudspeakers are fed with the same signal at the same moment, this factor inevitably creates multiple sound delays which can make voice signals unintelligible to the listener. This problem can be eliminated by delaying the signal feed to successive speakers (which are each driven by their own PA units) by progressive amounts (by 1ms per 0.34 metres of spacing from the original sound source), as shown in *Figure 8.22*. Ideally, the speakers should be spaced at 6 metre intervals; the spacing intervals should not exceed 60 metres.

## Practical BBD ICs

A major feature of the average BBD analogue delay line IC is that it inherently offers a very good 'medium-fi' performance; typically, it generates total signal distortion of less than one percent, has a signal-to-noise ratio

*Audio delay-line systems and circuits* 217

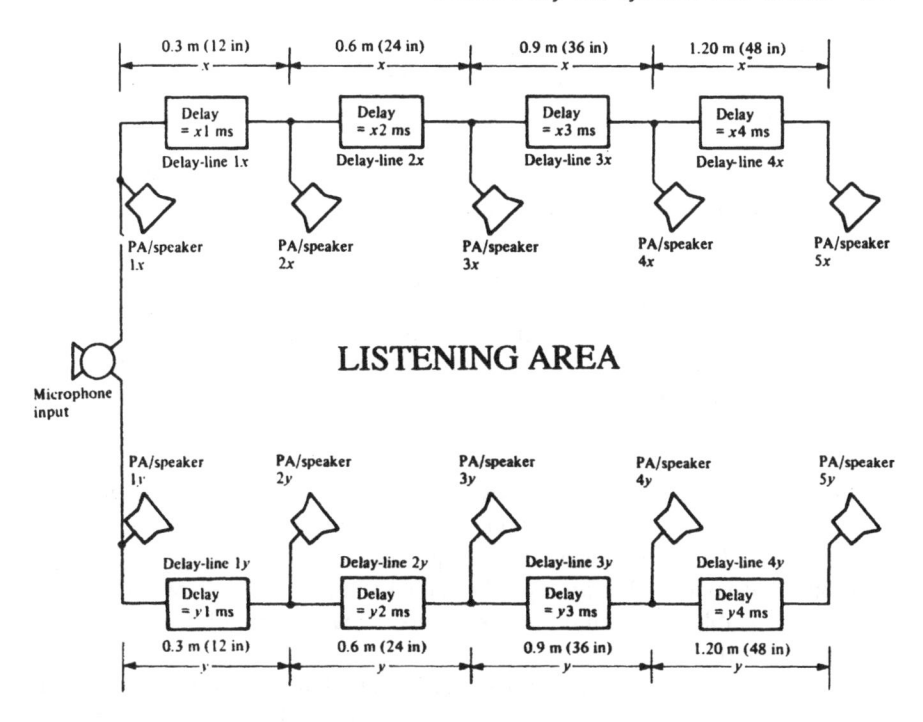

**Figure 8.22** *Public address speech-delay equalization system*

(effective dynamic range) of about 80dB at maximum output, and suffers only a few dB of insertion loss (signal attenuation) between the IC's input and output terminals. This performance greatly exceeds that required in many modern 'low-fi' applications, such as in simple speech-signal processors and karaoke systems, etc.

Throughout the 1970s and 1980s, BBD ICs were the most widely used analogue delay line device, and were produced by several manufacturers. In the early 1990s, however, simple *digital* delay line systems became popular in the types of 'low-fi' application mentioned above, and BBD sales began to decline. The current (late 1996) situation is that, in terms of cost-effectiveness, BBD analogue delay line systems still outperform digital systems in most medium-fi applications requiring delays of less than 300ms, but that digital systems are more cost-effective in most lo-fi applications and in all medium-fi applications requiring delays far greater than 300ms.

Currently, only one major producer (Panasonic) still manufactures a significant range of BBD ICs, but many older devices are still stocked by some large retail suppliers. *Figure 8.23* lists basic details of the eight best known BBD ICs that are still currently available, and *Figures 8.24* to *8.26* show the major parameters, IC outlines and pin notation of these devices. General details of the eight ICs are as follows.

218  *Audio delay-line systems and circuits*

The TDA1022 is a very popular general-purpose 512-stage delay line that can give a wide range of delays (from 0.85ms to 51.2ms). It needs a 2-phase clock drive, and gives 12.8ms delay at 7kHz bandwidth when clocked at 20kHz.

The SAD512D is a 512-stage delay line that gives a performance similar to that of the TDA1022. It is a modified version of an earlier device known as the SAD512, but has a built-in clock divider and line drivers that enable it to be driven by a single-phase clock input. The SAD512D offers a wide range of delay times, and can give a signal bandwidth of up to 200kHz.

The MN3004 is a modern (current-production) high-performance 512-stage device that typically generates only 0.4% total harmonic distortion and has a typical signal-to-noise ratio of 85dB. Its delay periods are variable from 2.56ms to 25.6ms, and its signal bandwidth is limited to 33kHz.

The SAD1024A is a dual version of the original SAD512, with each half needing a 2-phase clock drive. Its two halves can be used independently or can be wired in series to give a delay of 25.6ms at 7kHz bandwidth.

The MN3207 is a modern (current-production) low-voltage 1024-stage delay line that is housed in an 8-pin DIL package and is specifically designed for use in portable radios and karaoke units. It can operate from supplies in the range 4V to 10V.

The TDA1097 is a general-purpose 1536-stage delay line that is housed in an 8-pin DIL package. It needs a 2-phase clock drive, and can give a maximum delay of 153.6ms or a maximum signal bandwidth of 25kHz.

The MN3011 is a modern (current-production) 3328-stage delay line with six output taps (at stages 396, 662, 1194, 1726, 2790 and 3328) which each offer a composite output from one particular point in the delay line chain. When these outputs are mixed together they can be used to generate natural reverberation effects in ambience synthesizers, etc. The MN3011 is a high-performance device, with near-zero insertion loss and a typical distortion figure of only 0.4%. The IC needs a low-impedance clock drive, since its clock terminal input impedance is about 2000pF.

The SAD4096 is a 4096-stage delay line that gives a performance equal to eight SAD512s in series; it gives a delay of 102.4ms at 7kHz bandwidth, or 250ms at 3kHz bandwidth. The IC needs a low-impedance 2-phase clock drive, since its clock terminal input capacitance is about 1000pF.

## Practical circuits

### *Delay-line circuits*

The eight delay-line devices detailed in *Figures 8.23* to *8.26* are quite easy to use. All but the MN3207 are designed to operate from a supply with a nominal value of 15V, but some ICs are designed around n-channel IGFETs

*Audio delay-line systems and circuits* 219

| DEVICE NUMBER | STAGES | SAMPLES | DELAY TIME (mS) VS CLOCK FREQUENCY | DELAY AT 7kHz BANDWIDTH | NOTES |
|---|---|---|---|---|---|
| TDA1022 | 512 | 256 | 256/f | 12.8mS | Popular low-cost device. |
| SAD512D | 512 | 256 | 256/2f | 12.8mS | Has built-in clock divider; uses single-phase clock input |
| MN3004 | 512 | 256 | 256/f | 12.8mS | Modern high-performance device |
| SAD1024A | 1024 | 512 | 2 x 256/f | 25.6mS | Dual SAD512 device |
| MN3207 | 1024 | 512 | 512/f | 25.6mS | Modern low-voltage unit |
| TDA1097 | 1536 | 768 | 768/f | 38.4mS | General purpose unit |
| MN3011 | 3328 | 1664 | 1664/f | 83.2mS | Modern long-delay unit with six output taps (at stages 396, 662, 1194, 1726, 2790 and 3328 |
| SAD4096 | 4096 | 2048 | 8 x 256/f | 102.4mS | 4096-stage delay line. The clock terminal input capacitance = 1n0 |

**Figure 8.23** *Basic details of eight popular BBD delay-line ICs*

|  | TDA1022 | SAD512D | MN3004 | SAD1024A | MN3207 | TDA1097 | MN3011 | SAD4096 |
|---|---|---|---|---|---|---|---|---|
| Stages | 512 | 512 | 512 | 2 x 512 | 1024 | 1536 | 3328 | 4096 |
| $V_{SUPPLY}$ range | -12V to -16V | +10V to +17V | -14 to -16V | +10 to +17V | +4 to +10V | -12 to -16V | -14 to -16V | +8 to +18V |
| Clock f range | 5 to 3000 kHz | 1 to 1000 kHz | 10 to 100 kHz | 1 to 1000 kHz | 10 to 200 kHz | 5 to 100 kHz | 10 to 100 kHz | 8 to 1000 kHz |
| Delay range | 0.85 to 51.2 mS | 0.26 to 200 mS | 2.56 to 25.6 mS | 0.26 to 100 mS | 2.56 to 51.2 mS | 7.7 to 153.6 mS | 16.6 to 166.4 mS | 2.0 to 250 mS |
| Signal f range | dc to 45kHz | dc to 200kHz | dc to 33kHz | dc to 200kHz | dc to 50kHz | dc to 25kHz | dc to 20kHz | dc to 40kHz |
| Maximum signal $V_{IN}$ | 2V rms | 2V p-p | 1.8V rms | 2V p-p | 0.36V rms | 1.5V rms | 1V rms | 2V p-p |
| Insertion loss | 3.5dB | 2dB | 1.5dB | 0dB | 0dB | 4dB | 0dB | 2dB |
| S/N-ratio at max. output | 74dB | 70dB | 85dB | 70dB | 73dB | 77dB | 76dB | 70dB |
| IC package | 16-pin DIL | 8-pin DIL | 14-pin DIL | 16-pin DIL | 8-pin DIL | 8-pin DIL | 12-pin DIL | 16-pin DIL |

**Figure 8.24** *Major parameters of eight popular delay-line ICs*

220  Audio delay-line systems and circuits

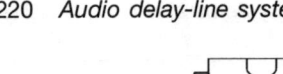

**Figure 8.25**  Outlines and pin notation of five popular but hard-to-find BBD delay-line ICs; these ICs are no longer manufactured

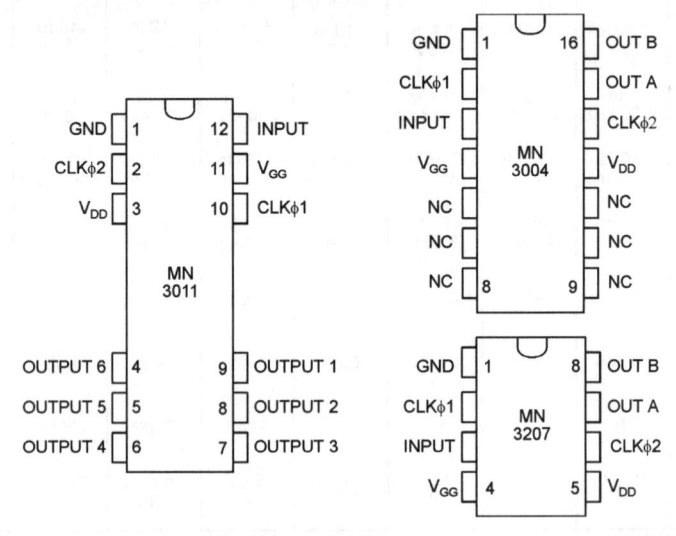

**Figure 8.26**  Outlines and pin notation of three current-production BBD delay-line ICs produced by Panasonic

*Audio delay-line systems and circuits* 221

and use a positive supply on the $V_{DD}$ terminal, while others are designed around p-channel IGFETS and use a negative supply. In all cases, the $V_{BB}$ (or $V_{GG}$) terminal is operated at about 14/15 of the $V_{DD}$ voltage (at 14V with a 15V supply, 8.4V with a 9V supply), and the input terminal is biased at about half-supply volts (the precise value is adjusted to give minimum output signal distortion). All ICs except the SAD512D need a symmetrical 2-phase clock drive, which must switch fully between the ground and supply rail values; the SAD512D accepts a simple single-phase clock drive.

*Figures 8.27* to *8.34* show how each of the above ICs can be wired as a simple delay line circuit that uses a +15V or +9V supply rail voltage and a grounded 'common' terminal. Note in the case of all ICs designed to operate with a negative $V_{DD}$ voltage (the TDA1022, MN3004, TDA1097 and

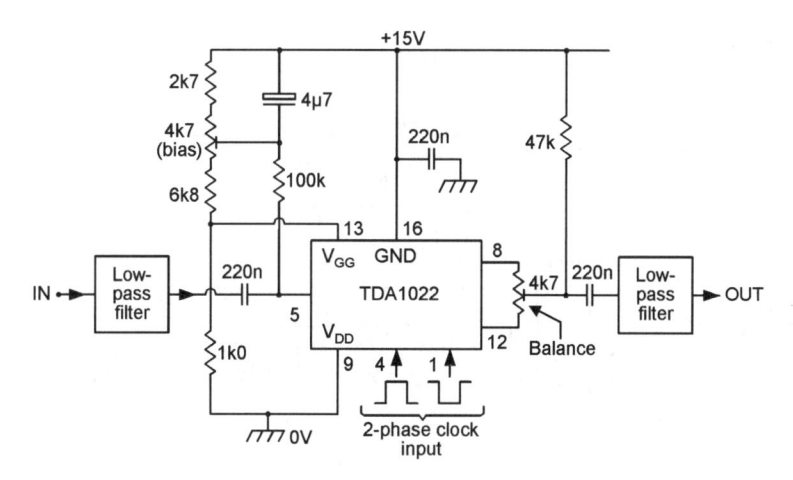

**Figure 8.27** *Delay line using the TDA1022*

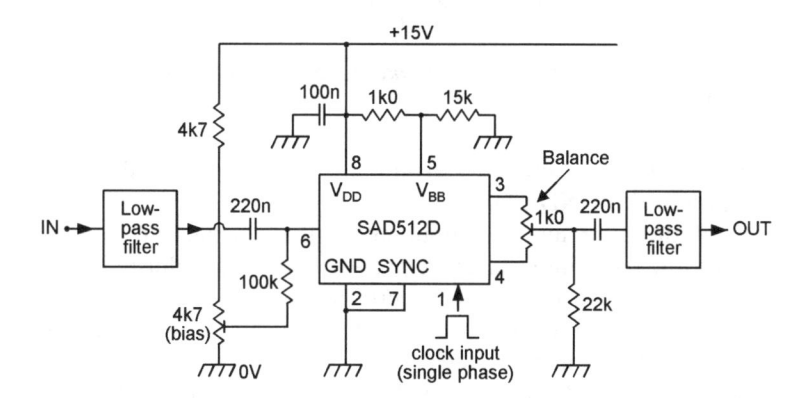

**Figure 8.28** *Delay line using the SAD512D*

222  Audio delay-line systems and circuits

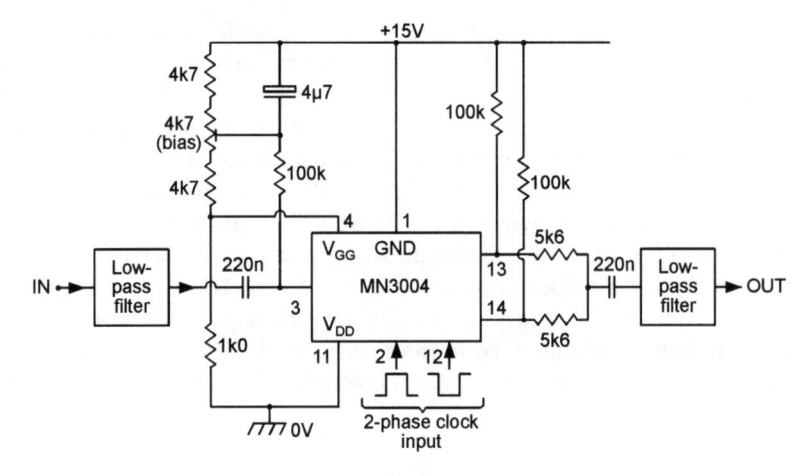

**Figure 8.29** Delay line using the MN3004

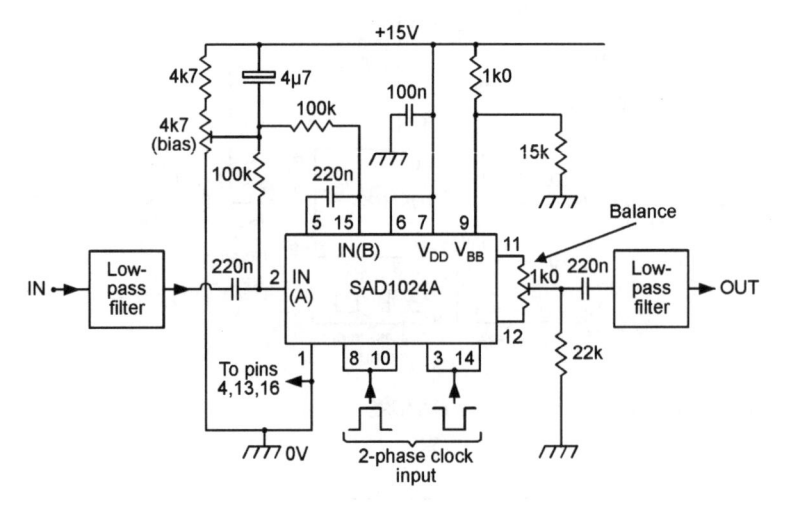

**Figure 8.30** Delay line using the two halves of the SAD1024A in the series-connected mode

MN3011) that this is achieved by grounding the $V_{DD}$ terminal and feeding the +15V line to the GND pin.

In each of these circuits the input and output signals are applied and removed via low-pass filter stages; a pre-set is used to adjust the input dc bias so that symmetrical clipping occurs under overdrive conditions, and (except in the case of the MN3011) another pre-set is used to balance the IC's two outputs for minimum clock break-through.

Note in the *Figure 8.30* circuit that the two halves of the SAD1024A chip are wired in series, with the pin-5 output of delay-line A feeding to the pin-15

Audio delay-line systems and circuits 223

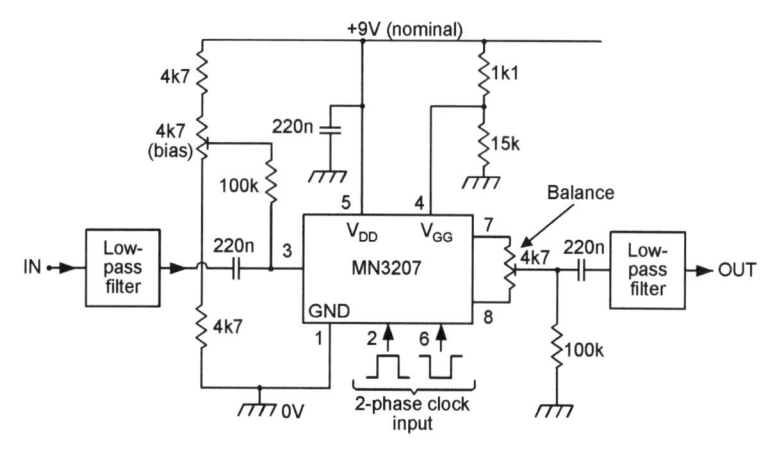

**Figure 8.31** *Delay line using the MN3207*

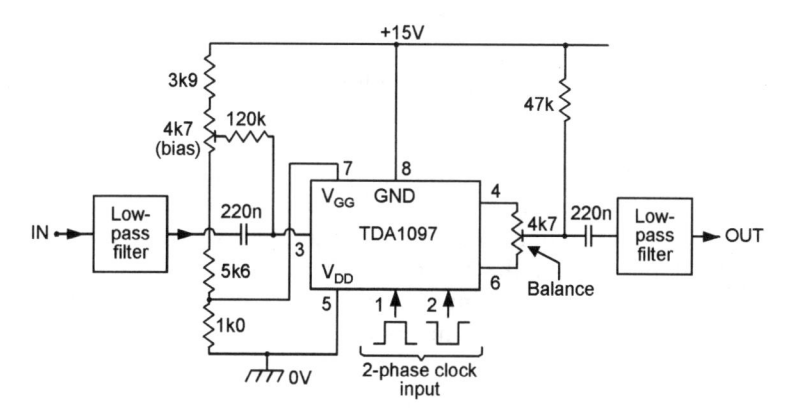

**Figure 8.32** *Delay line using the TDA1097*

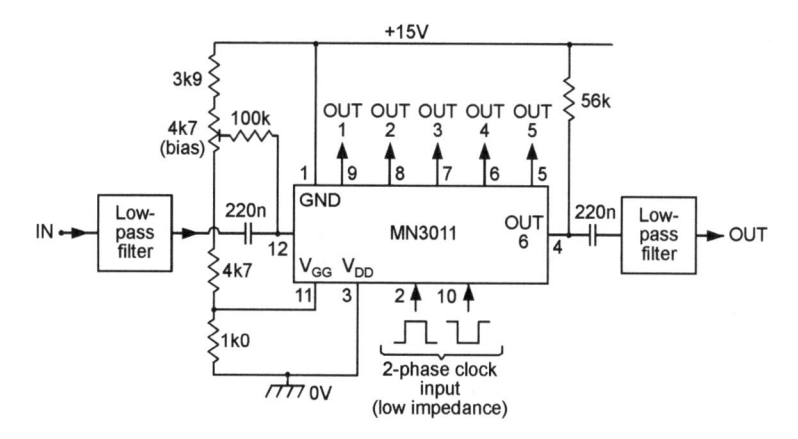

**Figure 8.33** *Delay line using the MN3011*

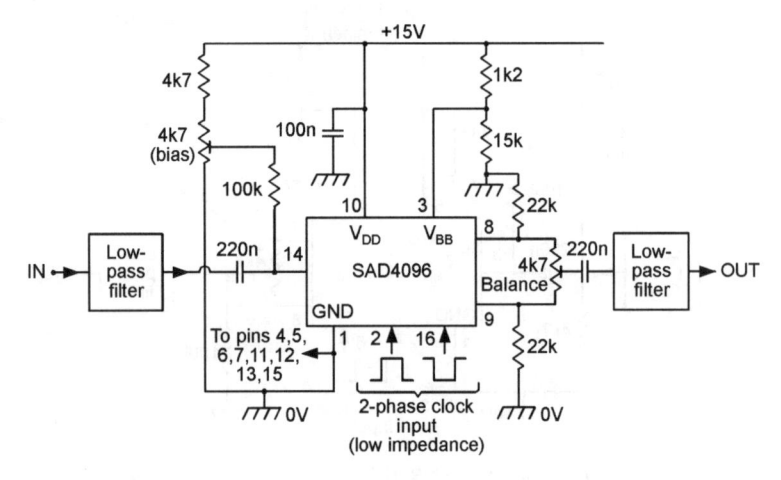

**Figure 8.34** *Delay line using the SAD4096*

input of delay-line B; the unused pin-6 output of delay-line A is disabled by shorting it to pin 7.

Note in the *Figure 8.33* circuit that the MN3011 is shown used as a simple delay line, with its 6th (longest delay) output fed to the outside world via a low-pass filter but with each of its other five outputs externally accessible. If only a single output is required from this circuit, but is wanted from a different output, simply remove the 56k resistor and the 220n capacitor junction from pin 4 (the 6th output point) and reconnect it to the output pin that gives the desired delay. If several independent outputs are required from the IC, fit each one with its own 56k resistor and 220n capacitor and low-pass filter, in the same way as shown used on the 6th output. If the IC is to be used in its primary mode, as an ambience synthesizer or acoustic 'room expander', in which several outputs are mixed together, refer to *Figure 8.45* and its associated text.

### Clock generator circuits

Most BBD delay line ICs need clean squarewave clock signals that switch fully between the supply rail voltages. Clock generator design for the SAD512D delay line is very easy, since this IC incorporates a divider stage on its clock input line that lets it accept non-symmetrical single-phase clock signals, but most other BBDs require good 2-phase clock signals.

Clock generator design for the MN3004 and MN3011 delay lines is made easy via a special 2-phase clock generator/driver IC, the MN3101. *Figure 8.35* shows the outline and pin notation of this 8-pin device. The MN3101 houses a 2-stage input amplifier that drives a frequency divider/buffer stage

*Audio delay-line systems and circuits* 225

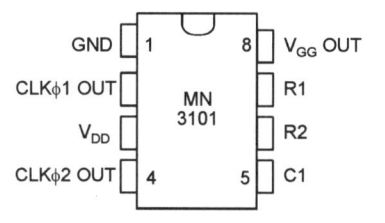

**Figure 8.35** *Outline and pin notation of the MN3101 2-phase clock generator*

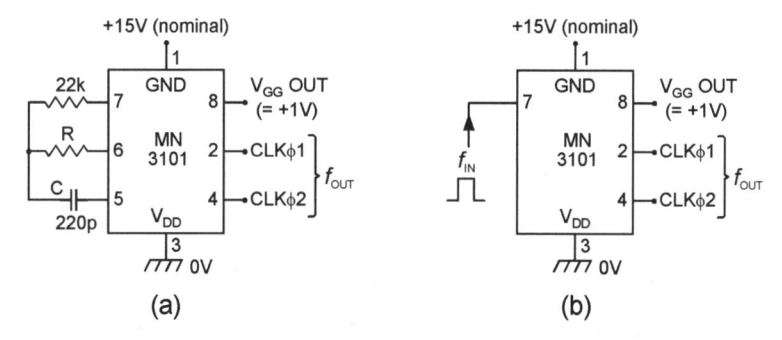

**Figure 8.36** *Basic ways of using the MN3101 as (a) a self-contained or (b) externally driven 2-phase clock waveform generator*

that, when driven by a single-phase oscillator input, generates a high-quality 2-phase low-impedance squarewave output that can be fed directly to the clock terminals of the delay line IC (up to 8192 bucket brigade stages can be driven by a single MN3101). The MN3101 also provides a $V_{GG}$ output bias voltage that can be fed directly to the $V_{GG}$ terminal of the MN3004 or MN3011 IC.

*Figure 8.36* shows two alternative ways of using the MN3101 as a 2-phase clock waveform generator. In *Figure 8.36(a)*, the IC's input amplifier is used as a free-running oscillator, and the circuit acts as a self-contained clock generator in which the output clock frequency is inversely proportional to the *R* value and equals 125kHz at 5k0 or 620Hz at 1M0. In *Figure 8.36(b)* the IC is used as a simple divider/buffer in which an external single-phase 'clock' signal is applied to input pin 7; in this case the output clock frequency is half that of the input.

Note that the MN3101 can be used as a clock generator with any of the eight delay line ICs mentioned in the previous *Delay-line circuits* section, with the single exception of the MN3207 low-voltage BBD IC, for which a special clock generator IC, the MN3102, is available. *Figure 8.37* shows the outline and pin notation of the MN3102, which is internally similar to the MN3101, and *Figure 8.38* shows the basic way of connecting the IC as a self-contained clock generator that can operate from any supply in the range +4V

226  Audio delay-line systems and circuits

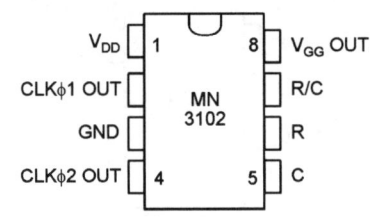

**Figure 8.37**  Outline and pin notation of the MN3102 low-voltage 2-phase clock generator

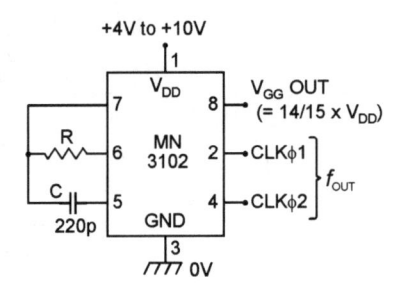

**Figure 8.38**  Basic ways of using the MN3102 as a self-contained 2-phase clock waveform generator

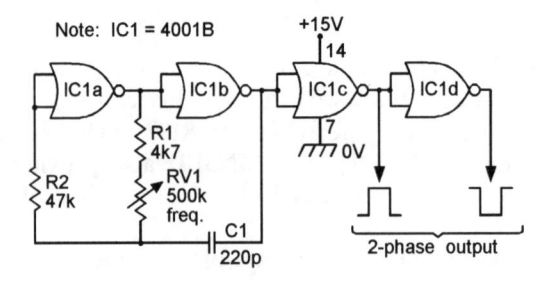

**Figure 8.39**  Variable-frequency general-purpose 2-phase CMOS clock generator

to +10V and can drive up to 4096 bucket brigade stages. Resistor $R$ can be given any value in the range 22k to 1M0.

In most cases, BBD delay line ICs can be adequately clocked by generators designed around readily-available low-cost CMOS digital ICs, and *Figures 8.39* to *8.41* show some practical examples of such circuits. The simple 2-phase generator of *Figure 8.39* is based on a 4001B IC, and can be used in most applications where a fixed or manually variable frequency is needed; the frequency can be swept over a 100:1 range via RV1, and the centre frequency can be altered by changing the C1 value.

The high-performance 2-phase generator of *Figure 8.40* is based on the voltage-controlled oscillator (VCO) section of a 4046B phase-locked loop IC,

*Audio delay-line systems and circuits* 227

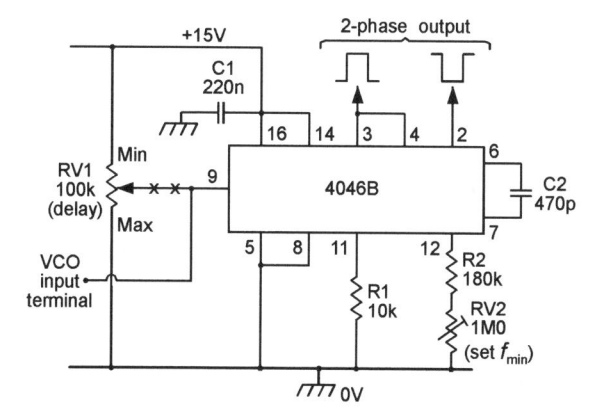

**Figure 8.40** *High-performance voltage-controlled 2-phase CMOS clock generator*

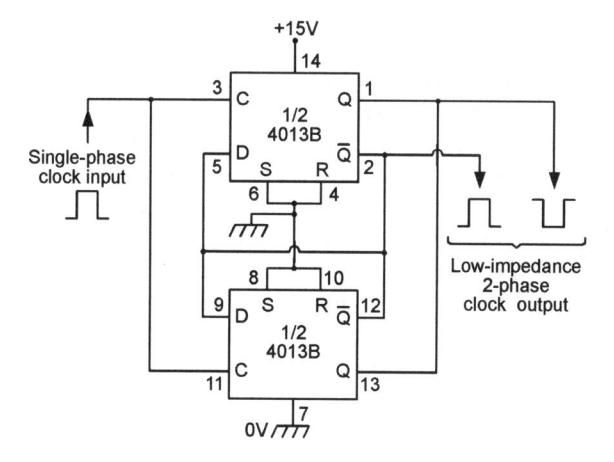

**Figure 8.41** *Single-phase to 2-phase converter, with low-impedance output*

and is useful in applications where the frequency needs to be swept over a very wide range, or needs to be voltage controlled. The frequency is controlled by the voltage on pin 9, being at maximum (minimum delay) when pin 9 is high, and minimum (maximum delay) when pin 9 is low. Maximum frequency is determined by the C2–R1 values, and minimum frequency by the value of C2 and the series values of R2–RV2. The frequency (delay) can be manually controlled via pot RV1, or can be controlled by an external voltage by breaking the RV1 connection at the 'x–x' points and feeding the control voltage to pin 9 as shown.

The *Figure 8.39* and *8.40* circuits can be used to directly clock all BBD delay lines except the MN3011 and SAD4096, which have clock terminal capacitances of 1000pF or more and need a low-impedance clock drive that is best provided by the *Figure 8.41* circuit, which uses both halves of a 4013B divider wired in parallel to give the required low-impedance 2-phase output; the circuit is driven by a single-phase clock signal, which can be obtained from either of the *Figure 8.39* or *8.40* circuits.

## Filter and mixer circuits

In most applications, a low-pass filter must be inserted between the actual input signal and the input of the BBD delay line, to prevent aliasing problems, and another in series with the output of the line, to provide clock-signal rejection and integration of the composite 'sample' signals. For maximum bandwidth, both filters usually have a cut-off frequency that is one third (or less) of the maximum used clock frequency; the input filter usually has a 1st-order or better response, and the output filter has a 2nd-order or better response.

*Figure 8.42* shows the practical circuit of a 25kHz 2nd-order low-pass filter with ac-coupled input and output. The non-inverting terminal of the op-amp is biased at half-supply volts, usually by a simple potential divider network. The cut-off frequency can be varied by giving C1 and C2 alternative values, but in the same ratio as shown in the diagram; e.g. cut-off can be reduced to 12.5kHz by giving C1 and C2 values of 1n0 and 6n0, respectively.

Most delay lines suffer from a certain amount of insertion loss; typically, if 100mV is put in at the front of a delay line, only 70mV or so appears at the output. Often, the output low-pass filter is given a degree of compensatory gain, to give zero overall signal loss. *Figure 8.43* shows such a circuit, which has a nominal cut-off frequency of about 12kHz, depending on the setting of the *gain balance* control.

In most BBD applications, at least one multi-input analogue signal mixer (adder) is incorporated somewhere in the system. *Figure 8.44* shows how a 2-input unity-gain mixer (adder) can also be made to act as a 1st-order low-pass filter by simply wiring a roll-off capacitor (C3) between the output and the inverting terminal of the op-amp. This type of circuit is often used at the front end of BBD flanger and reverberation designs.

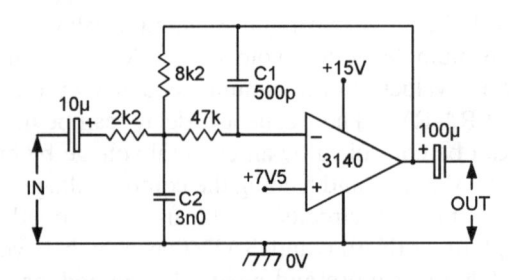

**Figure 8.42** *25kHz 2nd-order maximally-flat low-pass filter*

*Audio delay-line systems and circuits* 229

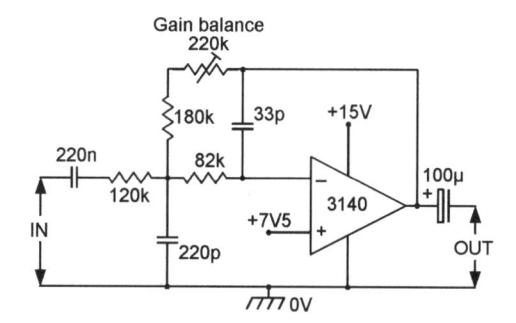

**Figure 8.43** *Adjustable-gain 2nd-order low-pass output filter*

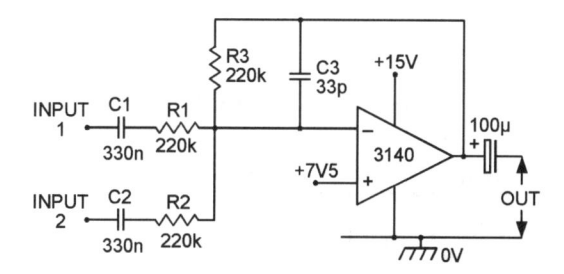

**Figure 8.44** *Combined 2-input mixer/1st-order low-pass filter*

Simple echo-reverb circuits of the basic type shown in *Figure 8.4* simulate the kind of reverberation that occurs when a sound bounces back and forth along a single fixed path between two hard surfaces (such as walls). Ambience synthesizers (acoustic 'room expanders'), on the other hand, try to simulate the kind of multi-path reverberation that occurs in real-life buildings such as churches and cathedrals, where sounds ricochet along a near-infinite numbers of paths between various walls and the hard floor and ceiling or inner-dome of the building. An adequately close approximation to this simulation can actually be achieved by wiring several of the basic *Figure 8.4* circuits (with independently adjustable delay times) in parallel and adding their outputs together in an audio mixer (in the manner shown in the simple 'two paths' synthesizer circuit of *Figure 8.19*), but such a system would obviously be very expensive. The 6-output multiple-delay MN3011 BBD IC (see *Figures 8.23, 8.24, 8.26,* and *8.33*) offers an easy and reasonably priced solution to this problem, as shown in the circuits of *Figures 8.45* and *8.46*.

*Figure 8.45* shows the basic circuit of a practical MN3011 ambience synthesizer. Here, each of the IC's six delayed (echo) outputs has its own volume control, and the outputs of these controls are mixed together, along with the original input signal, and passed on to the outside world via a simple low-pass filter. The output with the longest delay (Output 6) is also coupled back

230  Audio delay-line systems and circuits

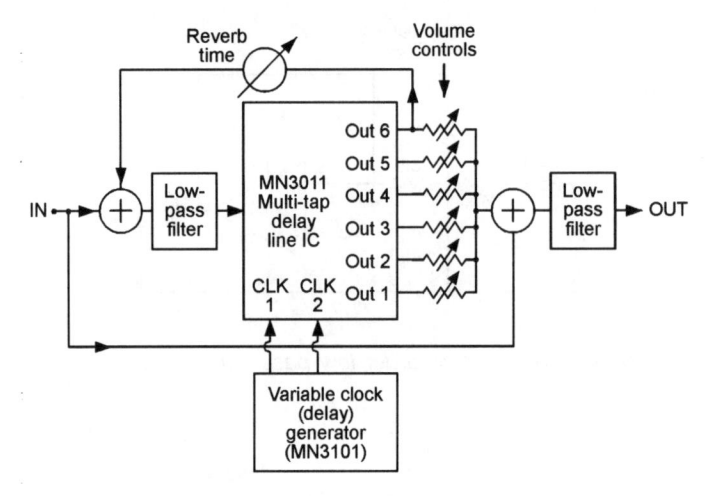

**Figure 8.45**  Basic MN3011 ambience synthesizer circuit

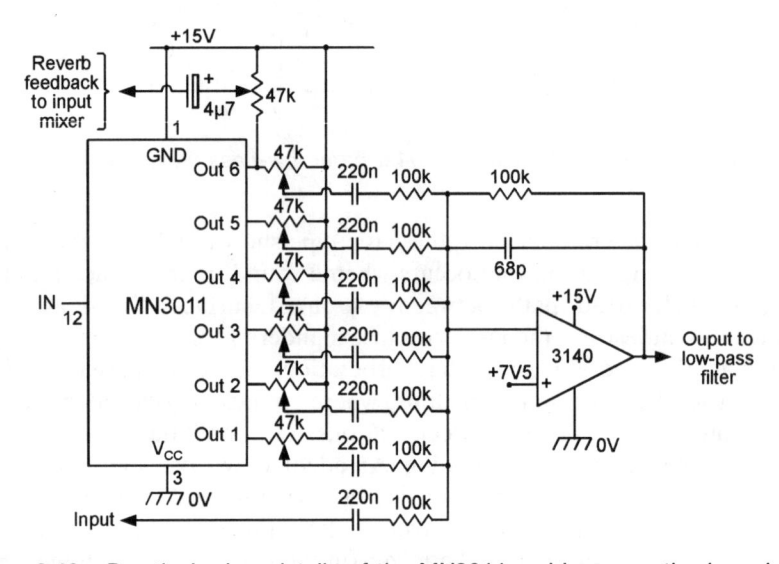

**Figure 8.46**  Practical mixer details of the MN3011 ambience synthesiser circuit

to the delay line's input via a mixer (adder) and low-pass filter, to provide the system with a realistic reverberation effect.

*Figure 8.46* shows practical details of the output mixer of the above MN3011 ambience synthesizer circuit. Here, each of the six MN3011 delayed outputs has its own volume control in the form of a 47k pot wired between the output and the +15V supply rail, and the outputs of each of these, plus the original undelayed input signal, are added together in a simple 7-input unity-gain mixer of the basic *Figure 8.44* type and then passed on to the outside world via a low-pass filter. This MN3011 circuit gives an excellent ambience synthesizer performance.

## Digital delay-line basics

Electronic delay-line systems can be built using either analogue bucket brigade delay-line (BBD) ICs or digital ICs. Each system has its own particular set of advantages and disadvantages. *Figure 8.47* shows, in its very simplest form, the basic elements of a digital delay-line system, which is best described here in terms of its 'input' and 'output' sections.

On the 'input' side of this system, the analogue input signal is first applied to an $n$-bit analogue-to-digital converter (ADC) unit which repeatedly takes high-speed samples of the instantaneous input voltage and, after converting each sample into an $n$-bit digital form, then latches the resulting digital word into a unique address in a digital random access memory (RAM) storage unit; the RAM address is incremented one step upwards on each successive ADC conversion, returning to the first address after the RAM's final address is used.

The major element on the 'output' side of the system is an $n$-bit digital-to-analogue converter (DAC) unit, which can convert each of the stored digital words (values) of the RAM back into its original analogue form. The operation of the DAC is synchronized to that of the ADC by the system's timing generator, but its RAM address selector follows $x$-places behind that of the ADC. Thus, if the system is clocked at a 10kHz (100µs) rate and the ADC's '$x$-places' value is 2000, the system's analogue output signals actually appear 200ms after they are applied to the system's input, and the system thus functions as a useful audio-frequency delay line.

The *Figure 8.47* circuit is greatly oversimplified. *Figure 8.48* shows a slightly more realistic version of a digital delay-line system. Here, the

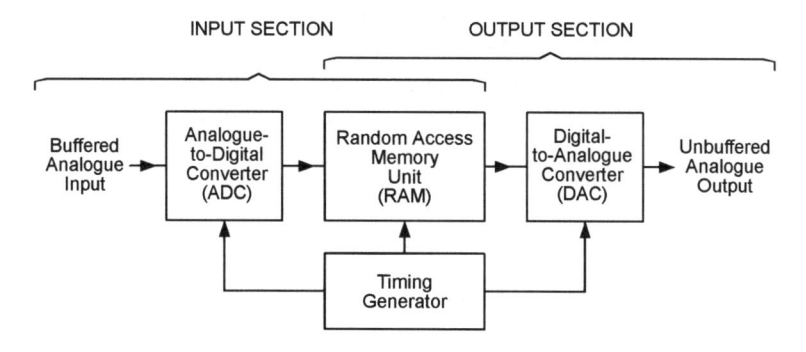

**Figure 8.47** *The basic elements of a digital delay-line system*

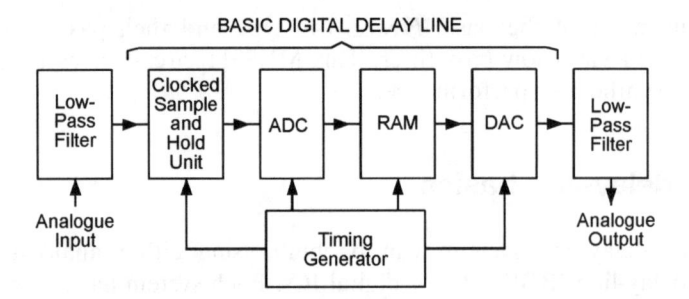

232  Audio delay-line systems and circuits

**BASIC DIGITAL DELAY LINE**

Low-Pass Filter → Clocked Sample and Hold Unit → ADC → RAM → DAC → Low-Pass Filter

Analogue Input ... Timing Generator ... Analogue Output

**Figure 8.48** *A more realistic version of the basic digital delay-line system*

analogue input signal is first passed through a low-pass filter, to eliminate signal 'aliasing' problems, and is then applied to a sample-and-hold unit that is clocked in synchrony with the rest of the basic digital delay line. The sample-and-hold unit effectively takes a high-speed snapshot of the instantaneous analogue input voltage each time it is clocked, thus presenting the unit's ADC with a stable input voltage sample while it performs its $n$-bit analogue-to-digital conversion. The results of the A-to-D conversion are then latched into the RAM, where they are later accessed and converted back into analogue form by the DAC, which passes them on to the outside world via another low-pass filter that eliminates any clock break-through signals and simultaneously converts the quantized output of the DAC back to the clean analogue form of the original input signal. Details of the individual elements of this system are as follows.

## The input filter

In BBD and digital delay line systems, the input and clocking (sampling) signals inevitably interact and produce unwanted outputs equal to the sum and the difference frequencies of the two signals. Thus, if a delay line is fed with a 3kHz input and is clocked at 10kHz, the line will produce output signals of 3kHz, 7kHz, 10kHz and 13kHz. If the line's output filter has a sharp cut-off at 6kHz, only the 3kHz signal will be passed by the system, and there will be no adverse output effects. If, on the other hand, a 4.9kHz input is applied to the same system, the line will produce output signals at 4.9kHz, 5.1kHz, 10kHz and 14.9kHz, and the second of these signals (which is known as an aliasing signal) will pass easily through the output filter and will also interact with the 4.9kHz signal to produce an annoying 200Hz 'difference' tone. It is the task of the unit's input filter to act as an anti-aliasing device, by eliminating all input signals above a designed frequency limit.

A basic law of electronic delay-line design states that the system's clock frequency must be at least *double* the upper cut-off frequency of the input filter, but in practice this ratio should be made as large as possible, to help minimize signal aliasing problems. Thus, if the above mentioned delay line had a simple low-pass 5kHz input filter and a 20kHz clock, an input of 4.9kHz would produce output signals of 4.9kHz, 15.1kHz, 20kHz and 24.9kHz, and all but the 4.9kHz signal would fall well beyond the pass range of the system's filter units.

## The sample-and-hold unit

The sample-and-hold unit's task is to take a sequence of high-speed snapshots of the instantaneous analogue input voltage and hold each one steady for a few dozen microseconds while it is processed by the system's ADC unit. *Figure 8.49* shows a practical sample-and-hold circuit in which IC2 is an IGFET-type op-amp. Both op-amps are wired as unity-gain voltage followers, with IC1 acting as an input buffer. IC2 has a 1n0 capacitor wired between its non-inverting input pin and ground, and has its input connected to IC1's output via a clocked CMOS bilateral switch. When the switch is closed (during the *sample* period), IC2's output follows the input signal, but when the switch opens (at the start of the *hold* period) the prevailing instantaneous input voltage is stored in the 1n0 capacitor and appears at IC2's output as a stable voltage for the duration of the 'hold' period. Note that, to give IC2 an adequately high input impedance (at least 10 000 Megohms), the circuit's PCB must provide IC2's input pin with a printed guard ring that is bootstrapped directly from IC2's output.

*Figure 8.50* shows the basic sample-and-hold waveforms obtained from a 1kHz sinewave that is centred on +1.00V and is clocked at a 13.4kHz rate. During each clock cycle, the output waveform (which is shaded grey) follows that of the input during the 'sample' period, but locks it into a steady state during the 'hold' period. Note here that the output of the delay line's DAC has a shape similar to that shown, but is converted back to the original sinewave input shape by the line's low-pass output filter.

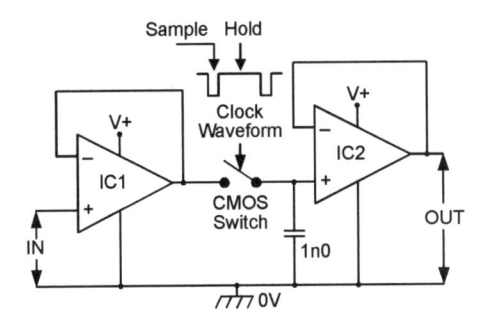

**Figure 8.49** *A practical sample-and-hold circuit using an IGFET op-amp*

234  Audio delay-line systems and circuits

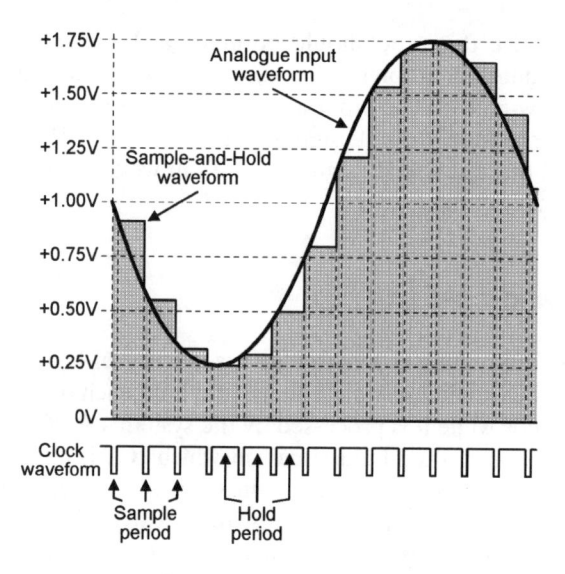

**Figure 8.50**  Sample-and-hold waveforms obtained from a 1kHz sinewave at a 13.4kHz clock frequency

### The analogue-to-digital converter (ADC)

The ADC's task is to convert the steady output voltage of the sample-and-hold unit into a representative $n$-bit digital (binary) code word that can then be latched into the delay line's RAM. *Figure 8.51* shows the basic elements of a typical ADC IC. The IC usually incorporates a precision band-gap voltage reference (normally 2.55V in an 8-bit ADC), which can be used to set the IC's full-scale voltage reading value. Most modern ADCs use a fast

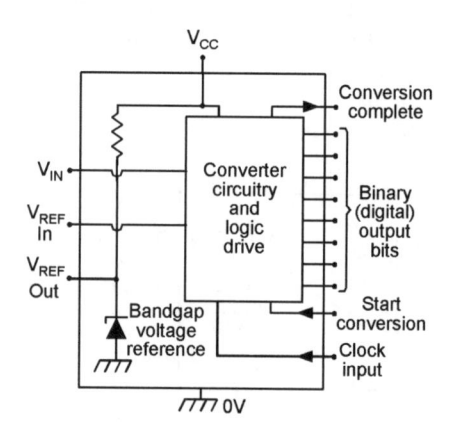

**Figure 8.51**  Basic elements of a typical analogue-to-digital converter IC

logic-driven successive approximation method of A-to-D conversion, in which the converter is driven by a fast (about 1MHz) clock waveform and the conversion process is started by applying a pulse to a *Start conversion* terminal (in some ADCs, the clock generator and/or the sample-and-hold unit is/are built into the IC). The conversion process usually takes one clock cycle per bit plus another two processing cycles, i.e. 10 cycles on an 8-bit ADC, 14 cycles on a 12-bit ADC. When the conversion is complete a *Conversion complete* terminal changes state and the $n$-bit digital conversion code is locked into the IC's output terminals.

The most important basic parameter of any fast ADC is its bit-size. Most modern ADCs have 8-bit or 12-bit outputs; a few have a 10-bit output, and the more expensive types have a 16-bit output. The ICs signal-to-noise (S/N) ratio, useful dynamic operating range and signal resolution are determined by its bit size. An 8-bit ADC can generate 256 ($2^8$) different 8-bit binary codes, and if set to read 2.55V full-scale will generate a '11111111' code with an input of 2.55V or '00000000' with an input of zero volts, and will have a discrimination (the analogue voltage difference between adjacent code steps) of 10mV. Thus, in this example, the converter has a discrimination of 0.4% of reading at 2.5V input, 4.0% at 0.25V (at –20dB relative to full scale), and 40% at 0.025V (at –40dB).

The matter of the ADC's 'useful dynamic range' in *ac*-signal processing applications is best explained as follows, using the above-mentioned 8-bit ADC as a working example. This 8-bit ADC has a discrimination of 10mV and can handle absolute maximum signal swings of 2.55V peak-to-peak, which it can record in terms of 256 distinct step-voltage levels (including 0V). If this input voltage is reduced by 20dB (to 255mV pk–pk), the ADC can record the signal in terms of only 26 step levels, and if it is reduced by 40dB (to 25.5mV) it can record it in terms of only 3 step levels, which is the absolute minimum number of levels needed to enable a moderately accurate (but highly distorted) version of the original signal to be reconstructed via a DAC and filter network. Consequently, if the input signal is reduced to below the 20mV level, the ADC will be unable to record it in a useful (greater than 2 step levels) form. This ADC thus has an absolute maximum useful *ac* dynamic range of 2.55V/0.02V, which translates into 42dB.

In practice, the maximum useful *ac* dynamic range of an ADC is directly proportional to its bit size, in the ratio 6dB per bit, minus 6dB, and can be simply calculated from the equation:

useful dynamic range = $6 \times (n - 1)$ dB

where $n$ is the ADC's bit size. Thus, an 8-bit ADC has a *maximum* useful *ac* dynamic range of 42dB, a 12-bit ADC has a range of 66dB, and a 16-bit ADC has a range of 90dB.

*Figure 8.52* lists the above data in tabular form, as applicable to 8-bit, 10-bit, 12-bit and 16-bit ADCs; exactly the same data also applies to DACs.

236  Audio delay-line systems and circuits

| PARAMETER | 8-BIT | 10-BIT | 12-BIT | 16-BIT |
|---|---|---|---|---|
| Conversion steps | 256 | 1024 | 4056 | 65,536 |
| Step size at 2.56V full scale reading | 10mV | 2.5mV | 0.625mV | 39μV |
| Discrimination at 2.5V (= 0dB) | 0.4% | 0.1% | 0.025% | 0.0016% |
| Discrimination at 0.25V (= -20dB) | 4.0% | 1.0% | 0.25% | 0.0156% |
| Discrimination at 25mV (= -40dB) | 40% | 10% | 2.5% | 0.156% |
| Discrimination at 25mV (= -60dB) | n.a | n.a | 25% | 1.56% |
| S/N-ratio (dB) | 48dB | 60dB | 72dB | 96dB |
| Useful dynamic range (maximum) | 42dB | 54dB | 66dB | 90dB |

**Figure 8.52**  *Table listing the basic characteristics of modern 8-bit, 10-bit, 12-bit and 16-bit analogue-to-digital converters (ADCs); the same data is applicable to digital-to-analogue converters (DACs)*

Note, however, that the listed 'useful dynamic range' figures are the absolute maximum attainable, and must be reduced by a further 6dB if a good low-distortion low-level performance is required from the ADC.

### The random access memory (RAM)

The RAM's task is to act as a temporary store for the digital data that is generated by the ADC. RAMs are available as either *Static* (SRAM) or *Dynamic* (DRAM) devices. Each bit of data that is fed into a SRAM is stored in an individual flip-flop cell, and all data is retained until it is overwritten or the SRAM's power supply is interrupted. Each bit of data that is fed into a DRAM is stored as an electrical charge in a small capacitor cell, and decays rapidly (in a few milliseconds) unless refreshed at frequent intervals. DRAM is easier to construct than SRAM, and is thus significantly cheaper than SRAM in large-memory-sized ICs.

In most 'audio' digital delay line applications, relatively little RAM space is required. SRAM and DRAM ICs are available in standard memory sizes of 4096-bit (4K), 16 384-bit (16K), 65 536-bit (64K), 262 144-bit (256K), and 1 048 576-bit (1M), and the maximum delay (in seconds) given by a digital delay line is given by the formula:

Delay = memory size/($f_{CLK}$ × bit-size)

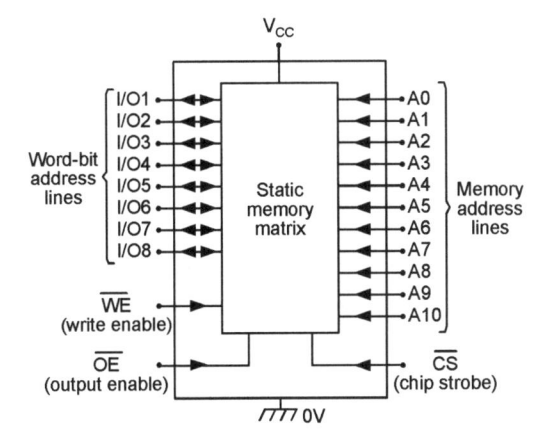

**Figure 8.53** *Basic usage elements of a typical static RAM (SRAM) IC*

where 'bit-size' refers to the system's ADC. Hence, a 12-bit system that uses a 20kHz clock gives a maximum delay of 68ms from a 16k RAM, 273ms from 64k, 1092ms from 256k, or 4.37 seconds from a 1M RAM. Most of these delay times are greatly in excess of normal needs, so the easiest and cheapest way to design a modern digital delay line system is to simply use a clock frequency that is at least four times greater than the required system bandwidth, thus minimizing the system's filtering requirements, and to use lots of easy-to-use and modestly priced static RAM (SRAM).

*Figure 8.53* shows the basic usage elements of a typical SRAM. This particular diagram applies specifically to a 16K device that is organized as a 2048 × 8-bit memory, i.e. it has 2048 addresses (each with its own 11-bit address code) that can each hold one 8-bit word that is applied or accessed (written or read) via eight bidirectional input/output (I/O) lines; the lines have 3-state outputs that are controlled via the NOT-OE terminal. To write an 8-bit data word to the SRAM, the desired memory location is selected via its 11-bit address code, the data word is applied to the eight I/O lines, the NOT-WE pin is tied low, and a negative-going clock pulse is applied to the NOT-CS terminal. To read a stored data word, its memory location is selected via its 11-bit address code, NOT-WE is tied high and NOT-OE is tied low, and a negative-going clock pulse is applied to the NOT-CS terminal, at which point the data word appears on the I/O lines.

### *The digital-to-analogue converter (DAC)*

The DAC's task is to sequentially read $n$-bit digital (binary) data words from the RAM and convert them back into their original analogue form. The DAC's characteristics must match those of the systems ADC in terms of bit-size and

238   Audio delay-line systems and circuits

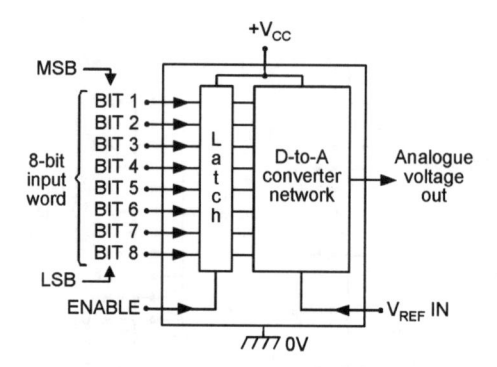

**Figure 8.54**  Basic usage elements of a latched 8-bit parallel-input, voltage-output DAC

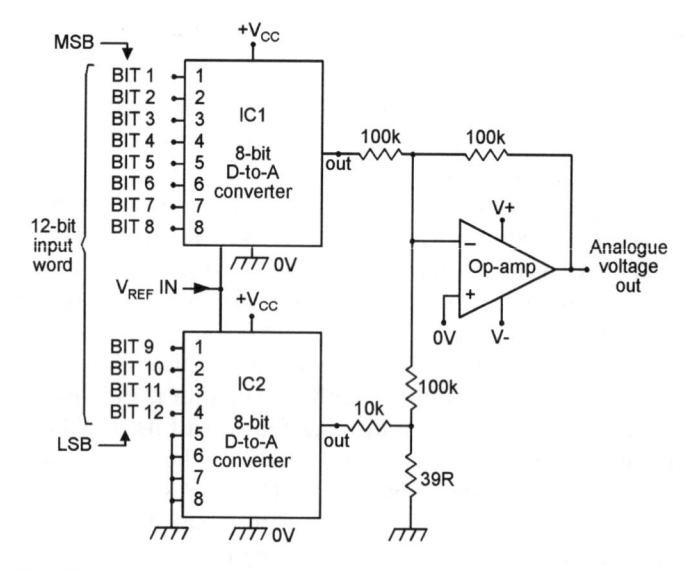

**Figure 8.55**  Way of connecting two 8-bit DACs to give 12-bit D-to-A conversion, with an inverted analogue output

scaling, and the basic data listed in *Figure 8.52* is applicable to both DACs and ADCs. DACs come in a variety of basic types and bit sizes. Their analogue outputs may take the form of a voltage or a current (that can be converted to a proportional voltage via an external op-amp), their $n$-bit inputs may be applied in parallel form (to $n$ input pins, which may or may not be connect to internal latches) or in serial form (to a single input pin), and the DAC may or may not be microprocessor compatible.

The ideal DAC for use in individually designed digital delay line systems is the simple voltage output type that has parallel inputs with internal latches. *Figure 8.54* shows the basic usage elements of a typical 8-bit DAC of this type. The IC requires a $V_{REF}$ input with a value equal to the desired full-scale analogue output voltage (normally 2.55 volts in an 8-bit DAC). Usually, the IC's input latch is transparent when the ENABLE pin is low (i.e. the IC gives a direct analogue output voltage equivalent of the applied 8-bit input word under this condition), and latches and holds the data when the ENABLE pin is taken high.

It is possible to wire two 8-bit DACs so that they act as a crude but useful DAC that will accept any word size in the range 9 to 16 bits. *Figure 8.55* shows two non-latched voltage-output 8-bit DACs and an op-amp adder used to make a simple 12-bit DAC. The eight most significant bits (bits 1 to 8) of the input word are applied to IC1, and the remaining lower-order bits (bits 9 to 12) are applied to IC2's four most significant inputs; IC2's four unused (least significant) input terminals are grounded. The analogue output of IC1 is fed directly to one input of the 2-input unity-gain op-amp adder, and the output of IC2 is fed to the other input of the adder via a 256:1 attenuator made from a 10k and 39R resistor. Note that this converter gives an inverted output, and that it is far less accurate than a dedicated 12-bit DAC but is sufficiently accurate for use in all practical delay-line applications.

## The output filter

The output of the DAC is a quantized version of the original analogue input signal, and has a step-like form similar to that shown in *Figure 8.50*. The output filter's task is to restore the waveform's original smooth shape by filtering out its high-frequency 'step' components, and to remove any residual clock break-through signals from the system's final output. If the clock frequency is at least four times greater than the system's designed bandwidth, the output filter can be a simple low-pass type. If the clock frequency is only two times greater than the system bandwidth, the output filter may have to be a well designed 4th-order or better low-pass type.

## The timing generator

The timing generator's task is to synchronize the operations of the delay line's main units. *Figure 8.56* shows the basic elements of the timing generator system. It consists of a fast clock generator that drives the system's main timing generator, which in turn controls the actions of the sample-and-hold unit and the ADC and also produces the clock signals for the RAM address register and the DAC. The main timing system and the RAM address register can be reset to zero (at initial switch-on or on request) via a logic unit. *Figure 8.57* shows the basic form of the main set of timing waveforms. In this particular case the ADC is an 8-bit unit and the 20kHz RAM clock is derived

240  Audio delay-line systems and circuits

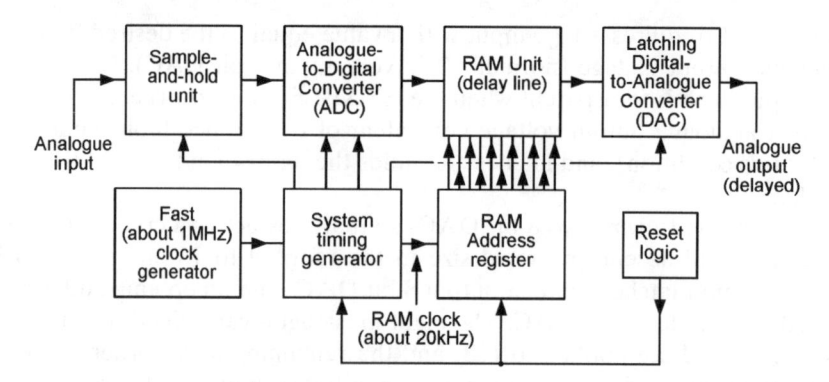

**Figure 8.56** Basic elements of a digital delay line's timing generator system

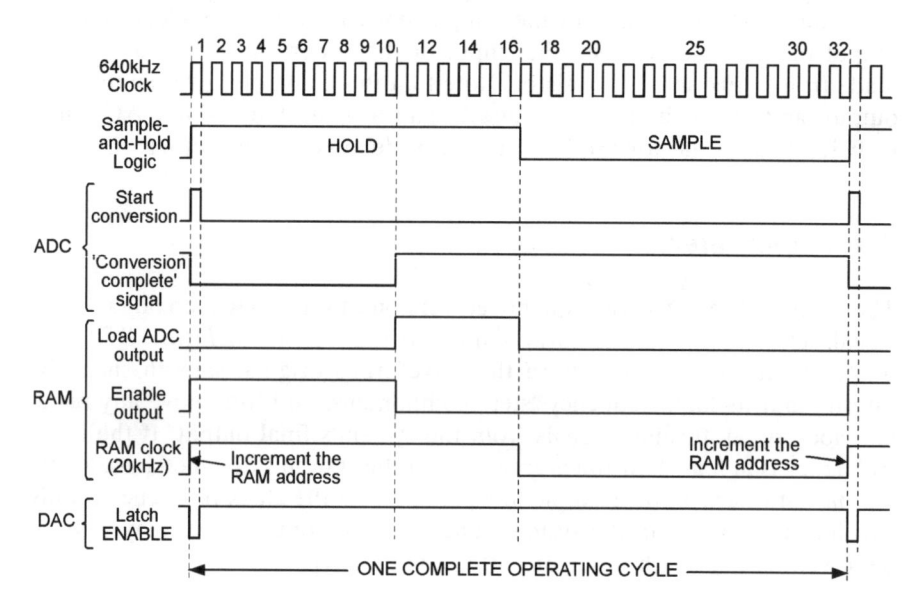

**Figure 8.57** Basic form of the main set of timing waveforms when using an 8-bit ADC and a 20kHz RAM clock

from a 640kHz master clock (via a divide-by-16 counter); the two clocks are thus synchronized. This delay-line system's basic sequence of operations is as follows.

At the start (on the rising edge) of each new 20kHz RAM clock cycle, the RAM address is incremented one step upwards, and its contents are inspected and latched by the DAC, which converts them into an analogue output. Simultaneously, the sample-and-hold unit is switched into the HOLD mode, and the ADC starts its A-to-D conversion operation which (in the case of an 8-bit ADC) takes ten 640kHz clock cycles; at the end of these ten

*Audio delay-line systems and circuits* 241

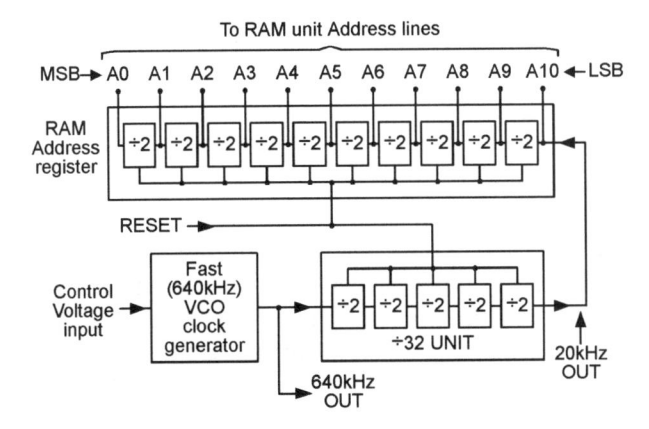

**Figure 8.58** *Block diagram of the clock generator system and the RAM address register*

cycles the ADC produces a *Conversion complete* signal, and the ADC's 8-bit output is then latched into the RAM's current address; a few 640kHz clock cycles later, the sample-and-hold unit is switched into the SAMPLE mode, completing the basic sequence of operations, which is repeated in each 20kHz RAM clock cycle.

Note from the above description that, if the RAM has an *effective* length of '$x$' addresses, the DAC will initially find each address empty until it reaches its $x+1$ RAM-clock cycle, at which point it will find the data that was written to that address $x$ clock cycles earlier. The system thus acts as an $x/f$ delay line, where $f$ is the RAM clock cycle frequency.

*Figure 8.58* shows the basic form of the clock generator system and the RAM address register. Note that the fast (640kHz) clock generator is a VCO type such as that used in a 4046B CMOS IC, thus allowing the delay line to be used in vibrato, phasing and flanging applications. The 640kHz signal is reduced to 20kHz via a 5-stage ripple counter, which drives the RAM address register, which in this case is a 10-stage ripple counter and thus produces an 11-bit output that can select a maximum of 2048 addresses. The maximum number of selectable RAM addresses doubles each time a new ripple stage is added to the address register, so the address limit can be raised to 32 768 by simply adding four extra ripple stages.

## Digital delay-line variations

The last few sections have presented a fairly comprehensive description of modern digital delay-line circuit principles, together with sufficient information to enable most experienced electronics enthusiasts and professionals to

242  Audio delay-line systems and circuits

design their own personal systems. There are, however, a few special and potentially useful circuit variations that have not yet been mentioned, as follows.

## Delay-time variation

A digital delay line's delay time can be varied by altering either its clock frequency or its RAM's *effective* address 'length'. If the 'variable clock frequency' technique is used, it is important to remember that the system may run into severe aliasing problems if the system's *clock frequency* to low-pass input-filter *cut-off frequency* ratio falls below the critical 2:1 value. One way of avoiding this problem is to 'gang' the system's *clock* and *cut-off* frequencies by making the low-pass input and output filters from clocked switched-capacitor active filter ICs (such as the MF10), driven by the system's clock in the way shown in *Figure 8.59* and given a 2.5:1 *clock* to *cut-off* ratio. This system gives a 3kHz bandwidth at a 7.5kHz clock frequency, rising to 20kHz at a 50kHz clock frequency.

The simplest way of varying the RAM's *effective* address length (and thus the delay time) is to use the technique shown in *Figure 8.60*, where the RAM Address register can generate up to thirteen address bits and can thus select a maximum of 8192 addresses from the 64k RAM, but can have its *effective* address range reduced (in divide-by-2 steps) by opening appropriate switches in the A0 to A4 address lines in the ways shown in the table. Thus, with only the A0 switch open, the effective address range is reduced to 4096, and with all five switches open it falls to 256 addresses. In practice, the address switches are usually electronic (CMOS) types, controlled by simple logic networks, and the circuit is used in conjunction with a limited range (up to 2.5:1) variable-clock system of the *Figure 8.59* type, thus making the delay fully variable over an 80:1 range.

*Figure 8.61* shows a rather more sophisticated way of varying the RAM's effective address length. Here, the 13-bit output of the RAM's clocked

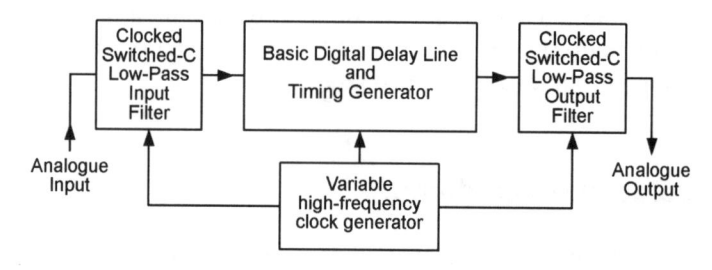

**Figure 8.59**  *Basic variable-delay system using 'ganged' clock generator and input and output low-pass filters*

*Audio delay-line systems and circuits* 243

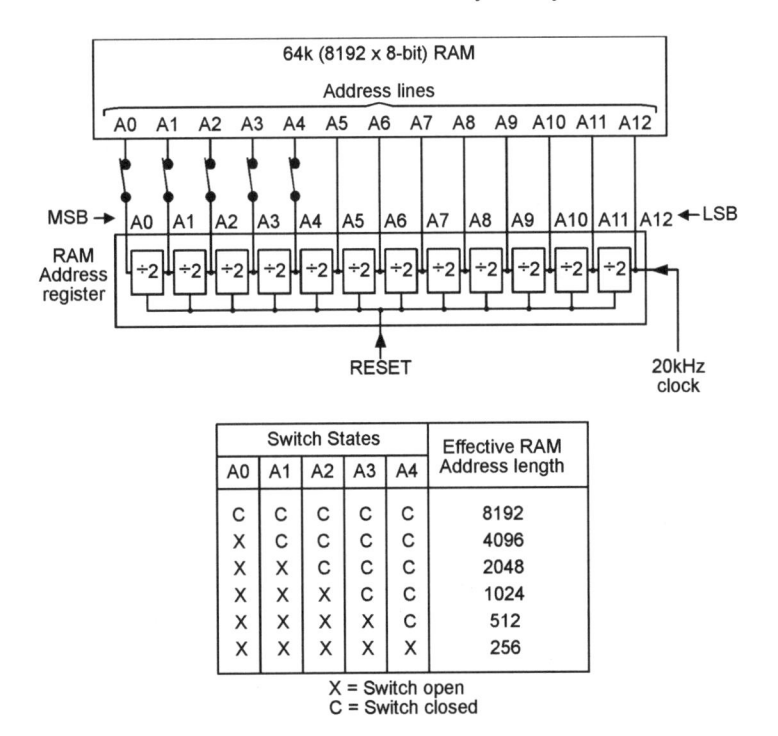

**Figure 8.60** *Simple method of varying the RAM's effective address length (and thus the delay time) via selector switches*

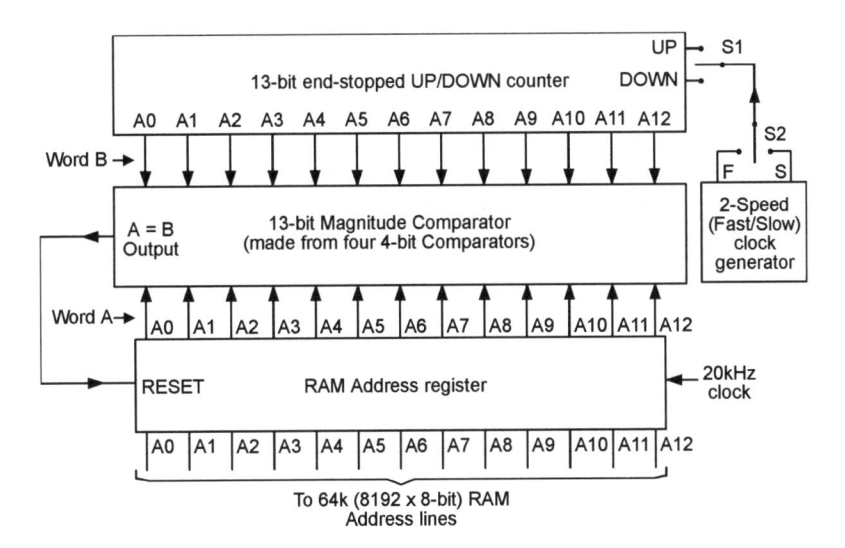

| Switch States | | | | | Effective RAM |
|---|---|---|---|---|---|
| A0 | A1 | A2 | A3 | A4 | Address length |
| C | C | C | C | C | 8192 |
| X | C | C | C | C | 4096 |
| X | X | C | C | C | 2048 |
| X | X | X | C | C | 1024 |
| X | X | X | X | C | 512 |
| X | X | X | X | X | 256 |

X = Switch open
C = Switch closed

**Figure 8.61** *Method of varying the RAM's effective address length over its full range, in 1-digit steps*

Address register is applied to both the RAM and the 'Word A' input of a 13-bit magnitude comparator (made from four 74LS85 4-bit comparator ICs), which has its 'Word B' input supplied by a 13-bit UP/DOWN counter which can have its direction and clock speed controlled via switches S1 and S2 and which gives 'end-stopped' operation (i.e. it cannot be clocked above '1111111111111' in the UP direction or below '0000000000000' in the DOWN direction, but can be set to any value within this range). The circuit's basic action is such that, in each full operating sequence, the Address register's output sequentially steps upwards from the 'all-zeros' value until its 13-bit output code coincides with that set in the UP/DOWN counter, at which point the 'A = B' output of the magnitude comparator goes high and resets the Address register's output back to the 'all-zeros' value, and the sequence starts to repeat again.

Note that the effective address length of the *Figure 8.61* circuit is fully variable from zero to 8192 in 1-digit steps via the UP/DOWN counter controls, and that it allows the delay line's delay-time to be varied over a very wide range without altering its band-pass values or causing aliasing problems.

## *Signal pre-emphasis*

When used in ac-signal processing applications, ADCs of less than 12-bit size have inherently poor dynamic range performances, due to their digital 'step size' limitations, and these ranges cannot be improved by using 'companding' techniques of the type that are often used in linear audio circuit designs. Their *effective* dynamic ranges can, however, be improved by about 12dB by using conventional audio pre-emphasis and de-emphasis techniques, as shown in the basic digital delay line system of *Figure 8.62*.

The amplitudes of most voice and music signals are dominated by bass frequencies, with the amplitudes of the higher 'treble' frequency signals, which are mainly harmonics of the bass signals, falling off at a 6dB/octave rate. When these signals are fed through a delay-line system with a poor dynamic range, the higher-frequency signals are – because of their relatively

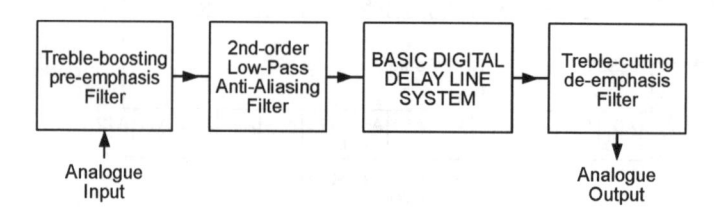

**Figure 8.62** *Adding pre-emphasis to a digital delay line can boost its effective dynamic range by 12dB*

Audio delay-line systems and circuits 245

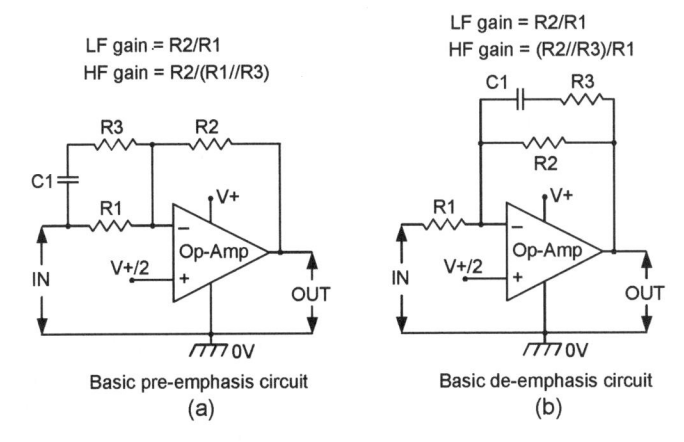

**Figure 8.63** Basic pre-emphasis (treble-boost) and de-emphasis (treble-cut) filter circuits

|  | 8-BIT System | 10-BIT System | 12-BIT System | 16-BIT System |
|---|---|---|---|---|
| Useful dynamic range, basic delay-line system | 33dB | 45dB | 57dB | 81dB |
| Effective dynamic range, using pre-emphasis | 45dB | 57dB | 69dB | 93dB |

**Figure 8.64** Table showing typical 'music signal' dynamic ranges of four basic digital delay-line systems, with and without pre-emphasis

low amplitudes – the first to be lost, and the resulting audio output sounds flat and unpleasant. This problem can be overcome by feeding the input signals to the delay line via a pre-emphasis filter that boosts the treble frequency amplitudes by 6dB/octave, and – after they leave the end of the delay line – then restoring the signals to their original form via a matching de-emphasis filter that cuts the treble frequency amplitudes by 6dB/octave. *Figure 8.63* shows the basic forms of the two filters, which normally have their maximum boost or cut limited to about 20dB by their resistance values, and have their turnover frequency (usually about 800Hz) set via C1.

The table of *Figure 8.64* lists the typical 'music signal' dynamic ranges of 8-bit, 10-bit, 12-bit and 16-bit digital delay-line systems, with and without pre-emphasis. Note first that the 'music signal' (low distortion) dynamic range of any ADC is about 6dB less than its absolute maximum dynamic range, as listed in *Figure 8.52*, and that an additional 3dB of dynamic range is lost when the ADC's digital output is converted back to analogue form by the DAC

246  *Audio delay-line systems and circuits*

at the end of the digital delay line. Also note that a delay-line system's dynamic range increases by 6dB for each 1-bit increase in its ADC/DAC size; thus, a 12-bit system's performance is 12dB better than that of a 10-bit system. Finally, note that the use of pre-emphasis increases the dynamic range of any digital delay line by about 12dB, and thus effectively raises the systems 'bit' size by 2 bits.

## Which delay-line system is best?

The 'best' delay-line system is the one that is the most cost-effective for use in a specified application, and the table of *Figure 8.65* lists relevant details of nine different systems. In this table, the 'maximum delay' is specified at a clock frequency of 10kHz, at which the line's signal bandwidth is typically 4kHz, and the 'comparative chip-set cost' (the total cost of the system's ICs)

| Delay Line Type | Dynamic range (music, without pre-emphasis) | Max delay at 10kHz clock frequency | Comparative chip-set cost |
|---|---|---|---|
| 512-stage BBD (MN3004, with MN3101 clock) | 85dB | 25.6mS | 5.1 units |
| 1024-stage BBD (MN3207, with MN3102 clock) | 73dB | 51.2mS | 5.6 units |
| 3072-stage BBD (3 x MN3207, with MN3102 clock) | 70dB | 153.6mS | 13.6 units |
| 3328-stage 6-tap BBD (MN3011, with MN3101 clock) | 76dB | 166.4mS | 16.0 units |
| 8-BIT Digital System, with 64k (8192 x 8-bits) RAM (6264) | 33dB | 819.2mS | 12.6 units |
| 10-BIT Digital System (HT8955A), with 64k DRAM | 45dB | 655.4mS | 5.5 units |
| 10-BIT Digital System (HT8955A), with 256k DRAM | 45dB | 2621.6mS | 5.1 units |
| 12-BIT Digital System, with 64k (8192 x 8-bits) RAM (6264) | 57dB | 409.6mS | 23.5 units |
| 16-BIT Digital System, with 64k (8192 x 8-bits) RAM (6264) | 81dB | 409.6mS | 43.5 units |

Notes:
Analogue BBD system = BBD IC + Clock IC + 2 op-amp filters
Digital system = ADC, DAC, clocks, Address Register, etc., + RAM + 2 op-amp filters.

Use of pre-emphasis increases the effective dynamic ranges of all systems by 12dB (equal to a 2-bit upgrade on digital systems) at an additional chip-set cost of 0.53 units.

In the 'comparative chip-set cost' column, 1 unit equals the typical cost of a monthly electronics magazine (= £2.25 in the UK, $3.50 in the USA at late 1996 prices).

**Figure 8.65** *Table listing basic performance details of nine different analogue and digital delay-line systems*

is specified in 'units', where 1 unit equals the typical cost of a monthly electronics magazine (the actual costs can thus easily be worked out and index-linked over a reasonable period of time and in various parts of the world). From this table, note the following.

If a dynamic range of at least 70dB is required, at delays up to about 200ms, analogue bucket-brigade delay-line systems are best. If a dynamic 'music' range of 45dB basic or 57dB with pre-emphasis is acceptable, and very long (greater than 50ms) delays are required, a 10-bit digital system based on the Holtek HT8955A 'voice echo' IC is best (this IC is described in the next section of this chapter). If a dynamic range of 81 to 93dB is required, at delays greater than 25ms, a 16-bit digital delay system will be needed. Note that the most expensive ICs in a digital system's chip-set are the ADC and the DAC, and that the RAM (which determines the system's maximum delay time) is comparatively inexpensive. Thus, 64k and 256k SRAM ICs typically cost only 1.5 and 3.5 units respectively, and 64k and 256k DRAM types costs 1.95 and (surprisingly) 1.55 units, respectively.

## The Holtek HT8955A delay-line IC

The Holtek HT8955A is a low-cost but fairly sophisticated 24-pin IC that, when used in conjunction with an external dynamic RAM, acts as a complete 10-bit digital delay line system that – when operated at a 25kHz sampling clock rate – can generate delays of up to 200ms when using a 64k DRAM or 800ms when using a 256k DRAM. The IC is a CMOS type designed to operate from 5V supply lines, and incorporates a 10-bit ADC and a 10-bit DAC and full control and DRAM-interfacing circuitry, plus a built-in analogue pre-amplifier. The device is intended for use in popular applications such as cheap voice echo units, low-cost karaoke systems, and simple sound effects generators.

*Figures 8.66* to *8.68* show the internal block diagram of the HT8955A, together with its outline and pin notation details, and *Figure 8.69* lists its claimed basic operating specification. Note in *Figure 8.69* that the manufacturer's claims regarding the unit's S/N ratio and THD are in fact suspiciously optimistic for a mere 10-bit unit, and may not be attained in reality. In actual fact, the HT8955A functions as an excellent 'cheapo' delay-line IC that gives exactly the kind of performance that would be expected from such a unit. Its output signals are noisy and badly distorted, but its range of delay times is excellent. In short, the IC offers an excellent low-cost introduction to the practicalities of modern digital delay-line usage. The IC's basic operation is as follows:

The HT8955A has two built-in oscillators, a 'fast' one (controlled via pins 6 and 7) that controls (via the unit's timebase generator) the main ADC and DAC circuitry, and a 'slow' one (controlled via pins 8 and 9) that

248  *Audio delay-line systems and circuits*

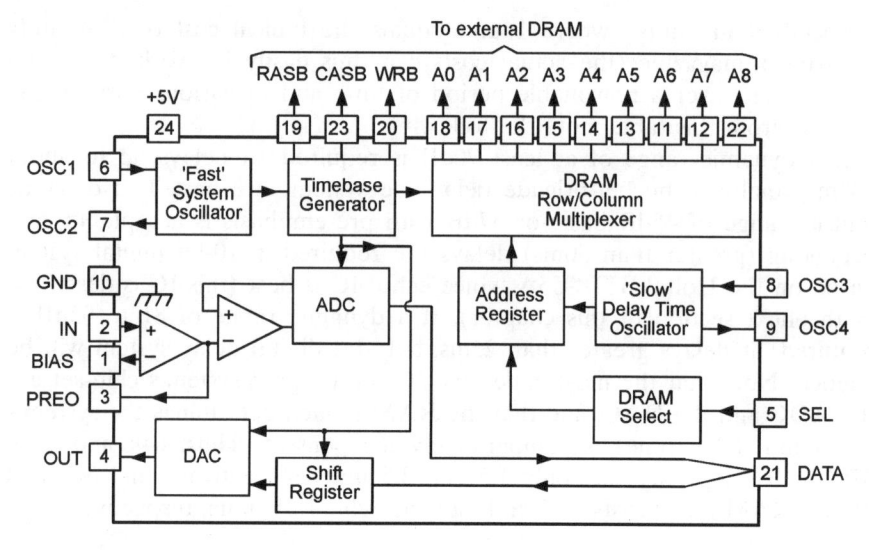

**Figure 8.66**  *Functional block diagram of the Holtek HT8955A digital delay-line IC*

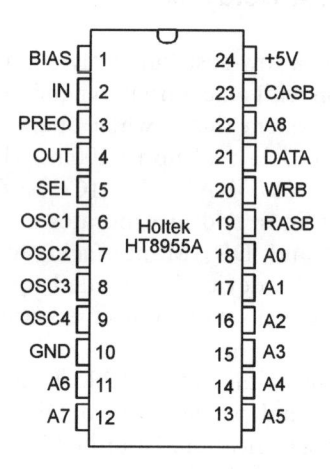

**Figure 8.67**  *Outline and pin notation of the HT8955A IC*

controls the system's Address register and thus exercises control over the delay line's actual delay time. The IC's analogue input signals are applied to the built-in pre-amplifier via pin 2 and are then passed on to the IC's ADC, which sequentially samples them and converts each sample into a 10-bit digital word, which is made available, when required, in serial form on a bidirectional data bus connected to pin 21, ready to be passed on to the external dynamic RAM (DRAM). The stored 10-bit data words of the DRAM are – when required – accessed (via a shift register) by the IC's

Audio delay-line systems and circuits  249

| Pin No. | Pin Name | Description |
|---|---|---|
| 1 | BIAS | Bias of internal pre-amp; connect to decoupler 'C'. |
| 2 | IN | Audio signal input pin (inverting) to pre-amp. |
| 3 | PREO | Pre-amp output pin. |
| 4 | OUT | Delayed audio signal (from DAC) output pin. |
| 5 | SEL | Delay-time select pin (open = 64k DRAM size, 0V (GND) = 256k DRAM size). |
| 6 | OSC1 | 'Fast' System Oscillator input (timing) pin. |
| 7 | OSC2 | 'Fast' System Oscillator output (timing) pin. |
| 8 | OSC3 | 'Slow' Delay Time Oscillator input (timing) pin. |
| 9 | OSC4 | 'Slow' Delay Time Oscillator output (timing) pin. |
| 10 | GND | Power supply ground (0V). |
| 11 | A6 | Connect to external DRAM 'A6' Address pin. |
| 12 | A7 | Connect to external DRAM 'A7' Address pin. |
| 13 | A5 | Connect to external DRAM 'A5' Address pin. |
| 14 | A4 | Connect to external DRAM 'A4' Address pin. |
| 15 | A3 | Connect to external DRAM 'A3' Address pin. |
| 16 | A2 | Connect to external DRAM 'A2' Address pin. |
| 17 | A1 | Connect to external DRAM 'A1' Address pin. |
| 18 | A0 | Connect to external DRAM 'A0' Address pin. |
| 19 | RASB | Connect to external DRAM 'RASB' Control pin. |
| 20 | WRB | Connect to external DRAM 'WRB' Control pin. |
| 21 | DATA | Data I/O pin to and from the external DRAM. |
| 22 | A8 | Connect to external DRAM 'A8' Address pin. |
| 23 | CASB | Connect to external DRAM 'CASB' Control pin. |
| 24 | +5V | Connect to positive (+5V) supply rail. |

**Figure 8.68**  *Table listing the HT8955A's pin descriptions*

| Characteristic | Test condition | Min. | Typ. | Max. | Unit. |
|---|---|---|---|---|---|
| Operating voltage | – | 4.5 | 5.0 | 5.5 | V |
| Operating current | No load | – | 2.5 | 8 | mA |
| Pre-amp open-loop voltage gain (Av) | $R_L > 100k$ | – | 2000 | – | V/V |
| Input voltage range, with +5V supply | – | 1.5 | – | 3.5 | V |
| Maximum output volts | $R_L > 470k$ | 1.0 | 1.5 | – | V |
| Maximum delay time, with 64k DRAM. | SEL = o/c 25kHz sampling | 150 | 200 | – | mS |
| Maximum delay time, with 256k DRAM. | SEL = +5V 25kHz sampling | 600 | 800 | – | mS |
| Signal-to-noise (S/N) ratio. | Vout = 1V at 400Hz, Bandwidth = 10kHz | – | 55 | – | dB |
| Total harmonic distortion (THD) | Vout = 1V at 400Hz, Bandwidth = 7kHz | – | 0.5 | – | % |

**Figure 8.69**  *Table detailing the basic working specification of the HT8955A IC*

250  *Audio delay-line systems and circuits*

DAC via the pin-21 data bus and are then made available, in time-delayed analogue form, on pin 4 of the IC.

The most complex part of the HT8955A IC is the section that controls the flow of data bits to and from the external DRAM. Each one of these individual bits (*from* the IC's ADC or *to* its DAC) is clocked into (or from) a unique address in the DRAM via the IC's Address register and Row/Column Multiplexer, which can select any one of up to 262 144 addresses in a 256k DRAM. This operation requires the use of an 18-bit address code, and this is applied to the DRAM in the form of one 9-bit (A0 to A8) *Row* word (controlled via the Row Address Select or 'RASB' pin) followed by one 9-bit (A0 to A8) *Column* word (controlled via the Column Address Select or 'CASB' pin). The *direction* of the data flow (to or from the DRAM) is controlled via the IC's Write/Read or 'WRB' pin. The actual data (*from* the ADC or *to* the DAC) takes the form of a 10-bit word that flows from or to the pin 21 DATA terminal of the HT8955A in *serial* form (one bit at a time). Each one of these 10-bit words thus occupies a 'field' of 10-bits of DRAM space.

Thus, in each 'slow' operating cycle of the HT8955A, the IC goes through the following operating sequence. First, it executes an ADC conversion operation, then opens up a new 10-bit DRAM field and accesses each individual part of its 10-bit data word, sequentially transferring each existing bit to the IC's DAC and replacing it with a corresponding new bit from the ADC until, at the end of the sequence, the 'old' 10-bit data word has appeared in time-delayed analogue form at the IC's pin-4 output terminal, and has been replaced in the DRAM field with a new 10-bit data word derived from the IC's pin-2 analogue input terminal. The IC then moves on to the next operating sequence, during which it carries out similar operations on the DRAM's next multi-bit data field, and so on.

## Basic usage data

The Holtek HT8955A is designed to be very easy to use, and is specifically intended for use with 4164 (64k) or 41256 (256k) dynamic RAM ICs. These

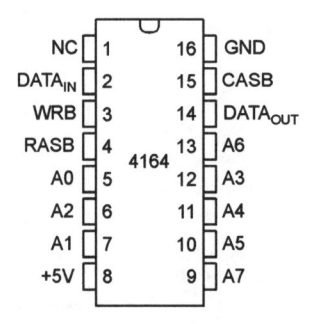

```
      NC  [ 1      16 ] GND
   DATA_IN [ 2      15 ] CASB
     WRB  [ 3      14 ] DATA_OUT
     RASB [ 4      13 ] A6
              4164
      A0  [ 5      12 ] A3
      A2  [ 6      11 ] A4
      A1  [ 7      10 ] A5
     +5V  [ 8       9 ] A7
```

**Figure 8.70**  *Outline and pin notation of the 4164 64k dynamic RAM IC*

*Audio delay-line systems and circuits* 251

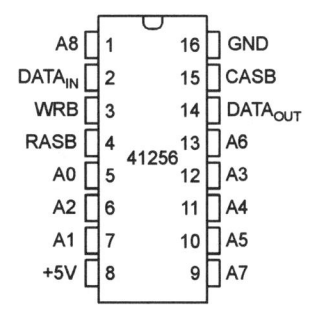

**Figure 8.71** *Outline and pin notation of the 41256 256k dynamic RAM IC*

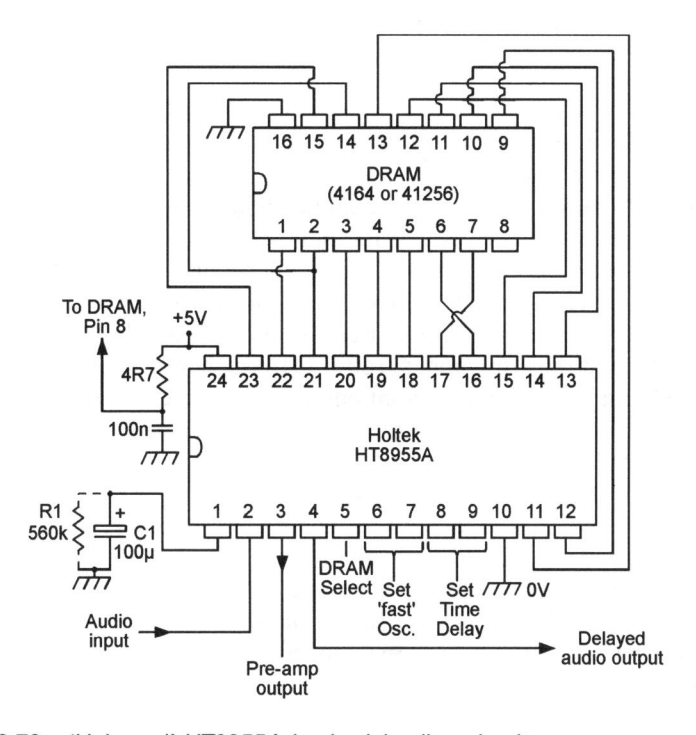

**Figure 8.72** *'Universal' HT8955A basic delay-line circuit*

ICs are 16-pin types, with very similar pin functions, as shown in *Figures 8.70* and *8.71*, which show the normal pin notation modified to conform with those used on the HT8955A IC. The only significant difference between the two DRAMs (from the user's point of view) is that the 41256 uses a basic 9-bit (A0 to A8) address system, with the A8 bit going to pin 1, and the 4164 uses a basic 8-bit (A0 to A7) address system, with the IC's pin-1 terminal internally unconnected. These facts enable the HT8955A and either type of DRAM to be used in the 'universal' basic delay-line circuit of *Figure 8.72*.

252  *Audio delay-line systems and circuits*

The basic *Figure 8.72* circuit must be powered from a well regulated 5V supply, and consumes up to 45mA when used with a 4164 DRAM or up to 70mA with a 41256 DRAM. If a 4164 DRAM is used pin 5 of the HT8955A must be left unused, and if a 41256 DRAM is used pin 5 must be tied to ground. To use this basic circuit in practical applications, the user must first connect pin 5 in the appropriate mode, connect a resistor between pins 6 and 7 to set the IC's 'fast' oscillator frequency, and connect a fixed and a variable resistor between pins 8 and 9 to set the desired 'delay' time of the system. C1 is used to decouple the internal pre-amp's built-in bias network, and R1 is an optional biasing resistor who's function is explained in the next few paragraphs. To use the pre-amp, an appropriate feedback network must be connected between pins 2 and 3 (the pre-amp's input and output pins), and an audio input signal applied to pin 2 will then produce a time-delayed audio output signal on pin 4.

The best way to learn about the HT8955A is to use it in a simple test circuit, and *Figure 8.73* shows the connections needed to make the 'universal' HT8955A circuit act as a basic test unit that uses a 64k (4164) DRAM, and *Figure 8.74* shows the alternative connections for use with a 256k (41256) DRAM. Note in these circuits that the pin 6–7 resistor sets the frequency of the IC's 'fast' oscillator, the pin 8–9 components control the 'slow' oscillator, and the pin 5 connection selects '64k' or '256k' DRAM operation. In both cases, the IC's internal pre-amp is used as an audio amplifier that gives ×2 voltage gain and has its upper roll-off frequency set at 5kHz by the 100k//330pF pin 2–3 *R* and *C* component values; the roll-off frequency is inversely proportional to the *C* value, and can be doubled by reducing the *C* value to 165pF.

To use the *Figure 8.73* and *8.74* circuits, simply connect an audio signal (from an AF signal generator or from an entertainment source) to the circuit's input and then monitor (with an oscilloscope or a hi-fi system) the circuit's three audio output points, first at the pin-3 pre-amp output, then at the pin-4 'Delay OUT' point, and finally at the 'Delay output' point at the junction of the 10k resistor and 3n3 capacitor. During these tests, increase the input signal amplitude until clipping occurs on the Delay output, then see if the performance can be improved significantly by fitting various values (not less than 100k) of R1 between pin 1 and either ground or the +5V supply rail. During these tests you will probably note that the delay outputs are rather noisy and have a very limited useful dynamic range, and that the performance can be adversely affected by poor circuit layout.

When you have finished with the *Figure 8.73* or *8.74* test circuits, you can move on and convert them into simple low-cost echo/reverb units by using the basic connections shown in *Figure 8.75*. This particular diagram shows the connections for use with a 256k (41256) DRAM, but those for use with a 64k (4164) DRAM are very similar. In both cases, the IC's internal pre-amp is used as a 1st-order low-pass filter with a 5kHz break frequency, and

*Audio delay-line systems and circuits* 253

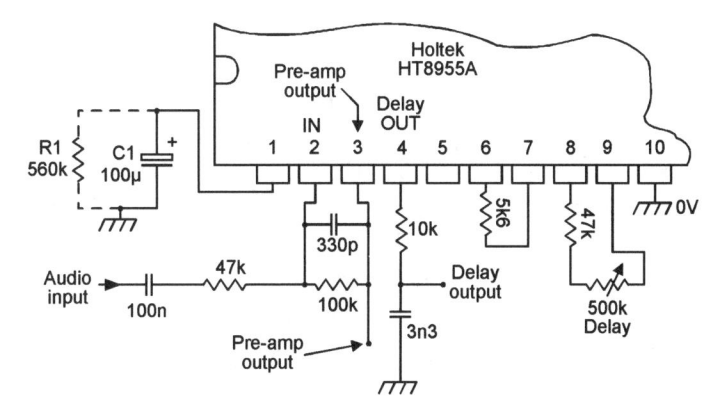

**Figure 8.73** *The 'universal' HT8955A circuit wired as a basic test unit with a 64k (4164) DRAM*

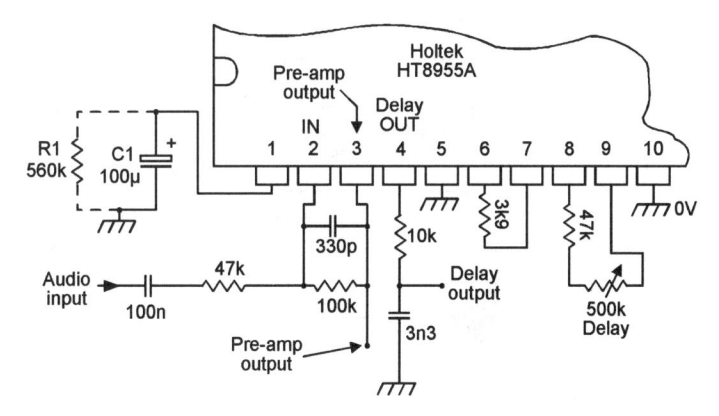

**Figure 8.74** *The 'universal' HT8955A circuit wired as a basic test unit with a 256k (41256) DRAM*

also as an audio mixer that gives ×2 voltage gain to the audio input signal and a ×0.17 to ×1.2 voltage gain to a 'reverb' feedback signal from the delay line's pin-4 output. Consequently, the output of the pre-amp consists of the original audio input signal plus 'reverberating echo' signals from the delay line output, and is made available via a 20k volume control. The echo/reverb sounds are particularly impressive when used with voice inputs. *Figure 8.76* shows a simplified equivalent functional diagram of the low-cost echo/reverb circuit, together with its basic waveforms.

*Figures 8.77* to *8.79* shows three useful ancilliary circuits that can be used in conjunction with any HT8955A delay-line system. The *Figure 8.77* circuit is that of a simple voltage regulator that can be used to supply the IC's +5V regulated supply (at load currents up to 100mA); the circuit is powered from an unregulated input of about +12V, and is designed around a 78L05 IC that

254  Audio delay-line systems and circuits

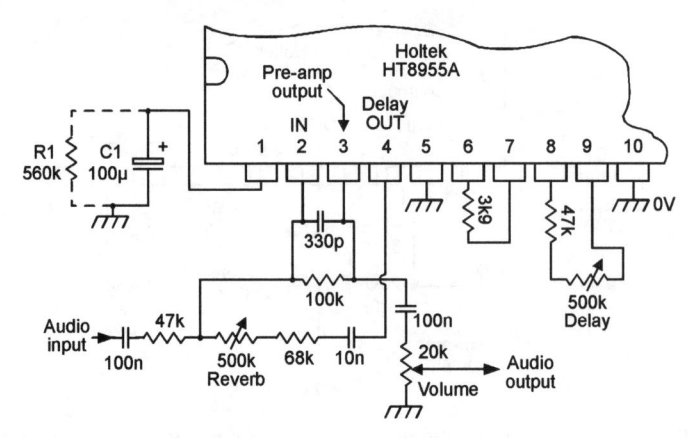

**Figure 8.75**  The 'universal' HT8955A circuit adapted as a low-cost echo/reverb unit with a 256k (41256) DRAM

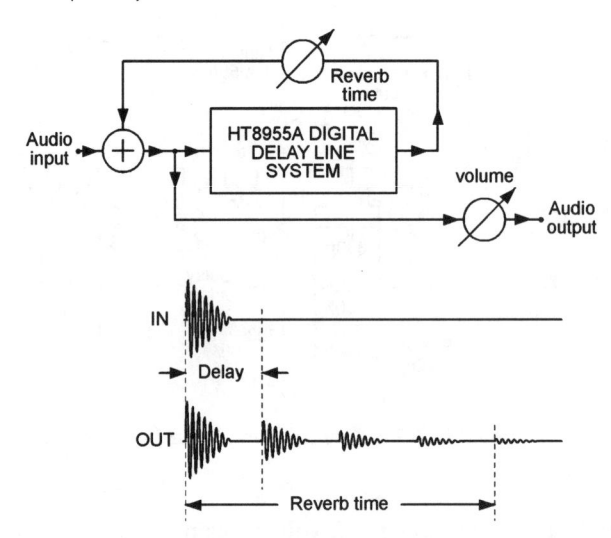

**Figure 8.76**  Simplified equivalent of the low-cost echo/reverb circuit

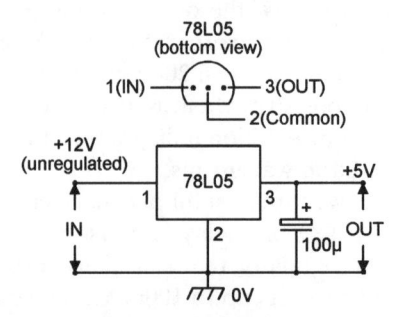

**Figure 8.77**  Simple +5V, 0–100mA regulated power supply circuit

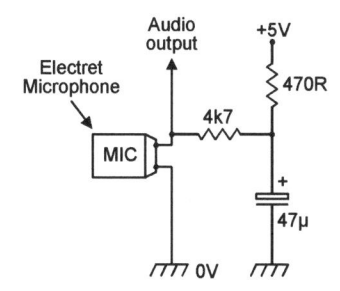

**Figure 8.78** *Way of powering an electret microphone from the +5V supply*

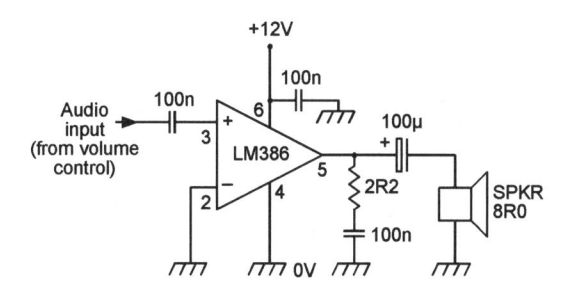

**Figure 8.79** *Low-cost power amplifier that can generate several hundred milliwatts in an 8R0 speaker*

is housed in a 3-pin TO-92 plastic package and normally uses the pin connections shown in the diagram. Note, however, that a few manufacturer's versions of this IC have their IN and OUT connections reversed, so if this circuit fails to work, try swapping the IC's IN and OUT connections.

*Figure 8.78* shows a simple way of connecting an electret microphone directly to the input of the *Figure 8.75* circuit and powering it from the circuit's +5V supply. This type of microphone has a built-in FET amplifier, and in the diagram the 4k7 resistor is used as the FET's drain load, thus making the microphone's output directly available.

Finally, *Figure 8.79* shows a simple low-cost power amplifier that can be powered from an unregulated +12V supply (which is also used for powering the HT8955A system's +5V voltage regulator) and can be driven directly from the delay line's volume control, and which can generate an output of several hundred milliwatts in an 8R0 speaker. The IC used is an LM386 type, which is housed in an 8-pin plastic package.

## A karaoke circuit

A major application area of the HT8955A is in simple karaoke systems, in which the voices of one or more amateur singers are fed through an echo-reverb

256  Audio delay-line systems and circuits

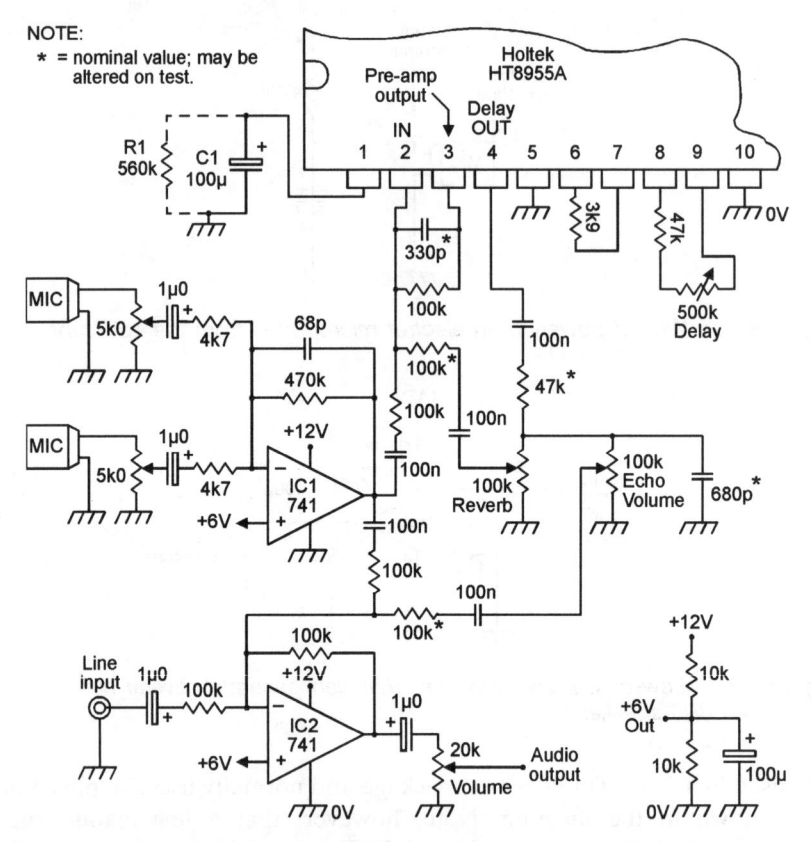

**Figure 8.80**  *The 'universal' HT8955A circuit adapted as a simple karaoke system with a 256k DRAM*

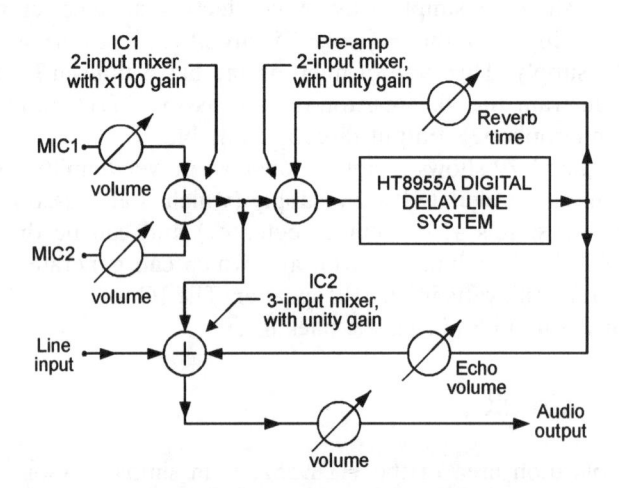

**Figure 8.81**  *Equivalent circuit of the simple karaoke system*

unit and are mixed with an unmodified music signal. To conclude this look at the HT8955A delay-line IC, *Figure 8.80* shows how the 'universal' HT8955A circuit of *Figure 8.72* can be adapted as a low-cost karaoke unit with a 256k (41256) DRAM, and *Figure 8.81* shows, in block diagram form, the karaoke unit's equivalent circuit.

The karaoke unit can accept voice inputs from two dynamic (moving coil) microphones, plus a single 'line input' music signal. Each microphone input has its own volume control, and the outputs of these are mixed together in IC1, which gives a ×100 voltage gain to low-frequency signals; the 68pF capacitor wired across the 470k resistor causes the gain to fall off at a 6dB/octave rate above 5kHz. IC1's output is tapped off in two directions; one output signal is fed to the input of the HT8955A's pre-amp, where it is mixed with part of the delay line's output to give a 'reverb' effect, and the other is fed to the input of IC2, where it is mixed with 'line input' music signal and with a fraction of the delay line's output to give a final composite audio output signal.

Note in *Figure 8.80* that the non-inverting input pin of each 741 op-amp is biased at half of the op-amps +12V supply voltage via a decoupled divider made of two 10k resistors. Also note that the values of the components marked with a star (*) may be altered on test, to give a modified circuit performance, to suit individual preferences. Thus, the values of the three marked resistors affect the voltage gain in various parts of the circuit, and the values of the two marked capacitors affect the frequency response. The circuit thus offers plenty of scope for experiment.

# 9

# Power supply circuits

Two frequent tasks facing the audio-IC user are those of designing basic power supplies to enable the audio equipment to operate from AC power lines, and of designing voltage regulators to power individual circuit sections at precise DC voltage values over wide ranges of load current variations. Both of these design tasks are fairly simple. Basic power supply circuits consist of little more than a transformer–rectifier–filter combination, so the designer merely has to select the circuit values (using a few simple rules) to suit his/her own particular design requirements.

Voltage regulator circuits can vary from simple zener networks, designed to provide load currents up to only a few mA, to fixed-voltage high-current units using dedicated voltage regulator ICs and designed for driving hi-fi stereo power amplifiers, etc. Practical examples of all of these circuits are described in this chapter.

## Power supply circuits

Basic power supply circuits are used to enable the audio equipment to safely operate from the AC power lines (rather than from batteries), and consist of little more than a transformer that converts the AC line voltage into an electrically isolated and more useful AC value, and a rectifier–filter combination that converts this new AC voltage into smooth DC of the desired voltage value.

*Figures 9.1* to *9.4* show the four most useful basic power supply circuits that the reader will probably ever need. The *Figure 9.1* circuit provides a single-ended DC supply from a single-ended transformer and bridge rectifier combination, and gives a performance that is virtually identical to that of the centre-tapped transformer circuit of *Figure 9.2*. The *Figure 9.3* and *9.4* circuits each provide 'split' or dual DC supplies with nearly identical performances. The rules for designing these four circuits are very simple, as follows.

Power supply circuits 259

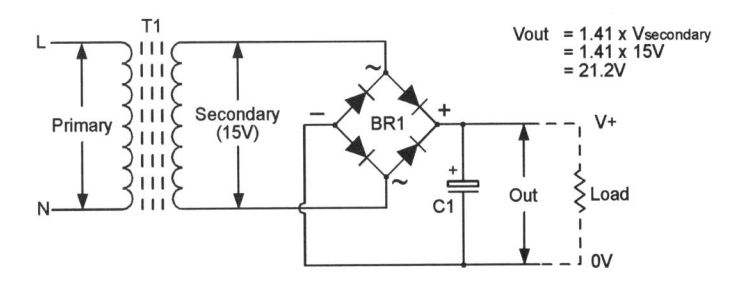

**Figure 9.1** *Basic single-ended power supply using a single-ended transformer and bridge rectifier*

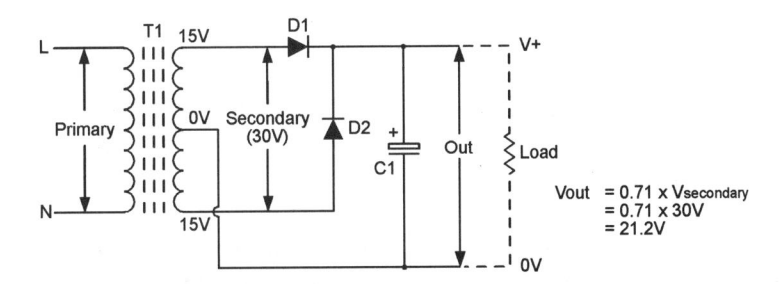

**Figure 9.2** *Basic single-ended power supply using a centre-tapped transformer and two rectifiers*

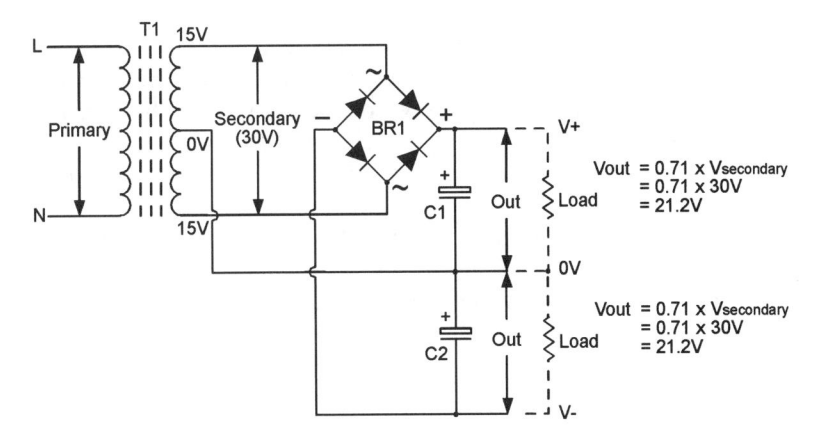

**Figure 9.3** *Basic split or dual power supply using a centre-tapped transformer and bridge rectifier*

## 260 Power supply circuits

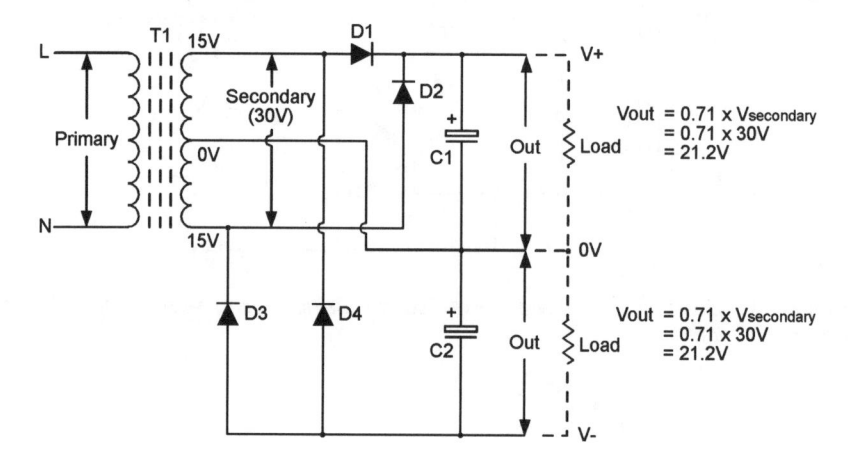

**Figure 9.4** *Basic split or dual power supply using a centre-tapped transformer and individual rectifiers*

### Transformer–rectifier selection

The three most important parameters of a transformer are its secondary voltage, its power rating, and its regulation factor. The secondary voltage is always quoted in r.m.s. (root mean square) terms at full rated power load, and the power load is quoted in terms of VA (volt–amps) or watts (W). Thus, a 15V 20VA transformer gives a secondary voltage of 15Vrms when its output is loaded by 20W. When the load is removed (reduced to zero) the secondary voltage rises by an amount implied by the *regulation factor*. Thus, the output of a 15V transformer with a 10% regulation factor (a typical value) rises to 16.5V when the output is unloaded.

Note here that the r.m.s. output voltage of the transformer secondary is *not* the same as the DC output voltage of the complete full-wave rectified power supply which, as shown in *Figure 9.5*, is in fact 1.41 times greater than that of a single-ended transformer, or 0.71 times that of a centre-tapped transformer (ignoring rectifier losses). Thus, a single-ended 15Vrms transformer with 10% regulation gives an output of about 21V at full rated load (just under 1A at 20VA rating) and 23.1V at zero load. When rectifier losses are taken into account the output voltages are slightly lower than shown in the graph. In the 'two-rectifier' circuits of *Figures 9.2* and *9.4* the losses are about 600mV, and in the bridge circuits of *Figures 9.1* and *9.3* they are about 1.2V. For maximum safety, the rectifiers should have current ratings at least equal to the DC output currents.

Thus, the procedure for selecting a transformer for a particular task is quite simple. First, decide the DC output voltage and current that is needed;

*Power supply circuits* 261

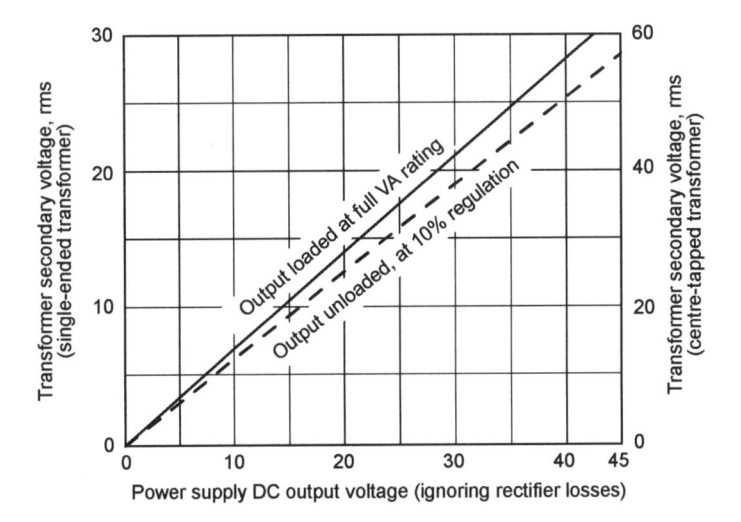

**Figure 9.5** *Transformer selection chart. To use, decide on the required loaded DC output voltage (say 21V), then read across to find the corresponding transformer secondary voltage (15V single-ended or 30V centre-tapped)*

the product of these values gives the minimum VA rating of the transformer. Finally, consult the graph of *Figure 9.5* to find the transformer secondary r.m.s. voltage that corresponds to the required DC voltage. The transformer may be a conventional 'square' type or a toroidal type. Toroidal types offer smaller size, weight and magnetic radiation than conventional types, but draw fairly high initial surge currents at switch-on, and should thus be provided with an anti-surge fuse on their AC power-connection side.

## The filter capacitor

The purpose of the filter capacitor is to convert the full-wave output of the rectifier into a smooth DC output voltage; its two most important parameters are its working voltage, which must be greater than the off-load output value of the power supply, and its capacitance value, which determines the amount of ripple voltage that will appear on the DC output when current is drawn from the circuit.

As a rule of thumb, in a full-wave rectified power supply operating from a 50Hz to 60Hz power line, an output load current of 100mA will cause a ripple waveform of about 700mV pk-to-pk to be developed on a 1000µF filter capacitor, the amount of ripple being directly proportional to the load current and inversely proportional to the capacitance value, as shown in the design guide of *Figure 9.6*. In most practical applications, the ripple should be kept below

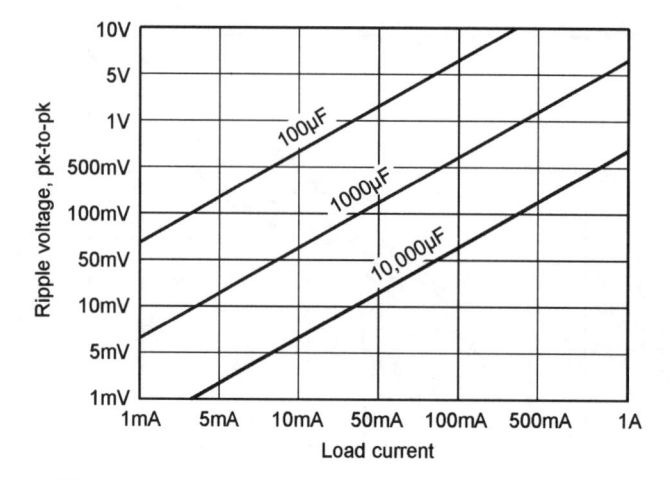

**Figure 9.6** *Filter capacitor selection chart, relating capacitor size to ripple voltage and load current in a full-wave rectified 50–60Hz powered circuit*

1.5V pk-to-pk under full load conditions. If very low ripple is needed, the basic power supply can be used to feed a 3-terminal voltage regulator IC, which can easily reduce the ripple by a factor of 60dB or so at low cost.

## Voltage regulator circuits

Practical voltage regulators may vary from simple zener diode circuits designed to provide load currents up to only a few milliamps, to fixed- or variable-voltage high-current circuits designed around dedicated 3-terminal voltage regulator ICs. Circuits of all these types are described in the remainder of this chapter.

### Zener-based circuits

*Figure 9.7* shows how a zener diode can be used to generate a fixed reference voltage by passing a current of about 5mA through it from the supply line via limiting resistor $R$. In practice, the output reference voltage is not greatly influenced by sensible variations in the diode current value, and these may be caused by variations in the values of $R$ or the supply voltage, or by drawing current from the output of the circuit. Consequently, this basic circuit can be made to function as a simple voltage regulator, generating output load currents up to a few tens of mA, by merely selecting the $R$ value as shown in *Figure 9.8*.

Here, the value of $R$ is selected so that it passes the maximum desired output current plus 5mA. Consequently, when the specified maximum output load current is being drawn the zener passes only 5mA, but when zero load

Power supply circuits 263

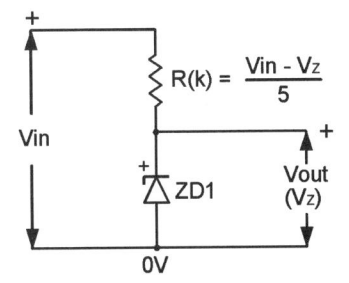

**Figure 9.7** *This basic zener reference circuit is biased at about 5mA*

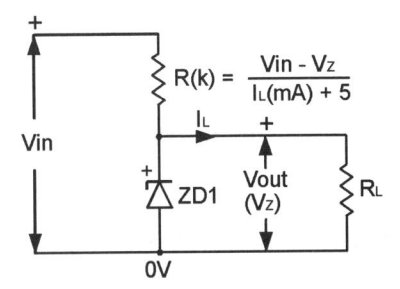

$$R(k) = \frac{V_{in} - V_z}{5}$$

**Figure 9.8** *This basic zener regulator circuit can supply load currents of a few tens of mA*

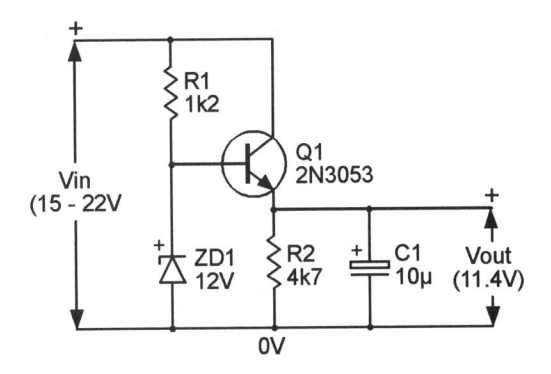

$$R(k) = \frac{V_{in} - V_z}{I_L(mA) + 5}$$

**Figure 9.9** *This series-pass zener based regulator circuit gives an output of 11.4V and can supply load currents up to about 100mA*

current is being drawn it passes all of the *R* current, and the zener thus dissipates maximum power. Note that the power rating of the zener must not be exceeded under this 'no load' condition.

The available output current of a zener regulator can easily be increased by wiring a current-boosting voltage follower circuit into its output, as shown in the series-pass voltage regulator circuits of *Figures 9.9* and *9.10*. In *Figure*

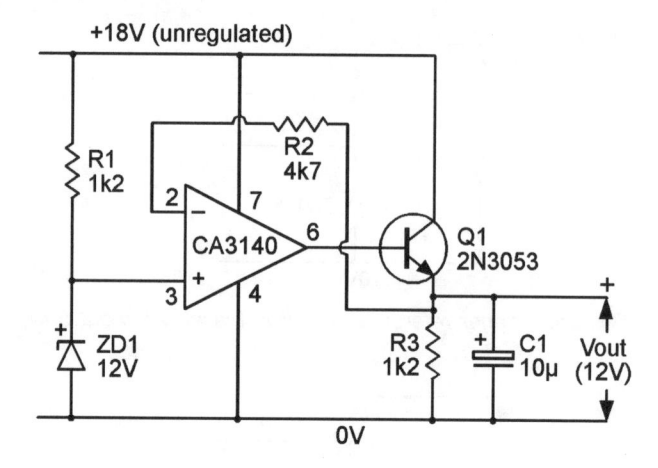

**Figure 9.10** *This op-amp based regulator gives an output of 12V at load currents up to 100mA and gives excellent regulation*

9.9, emitter follower Q1 acts as the voltage following current booster, and gives an output voltage that is 600mV below the zener value under all load conditions; this circuit gives reasonably good regulation. In *Figure 9.10*, Q1 and the CA3140 op-amp form a precision current-boosting voltage follower that gives an output equal to the zener value under all load conditions; this circuit gives excellent voltage regulation. Note that the output load current of each of these circuits is limited to about 100mA by the power rating of Q1; higher currents can be obtained by replacing Q1 with a power Darlington transistor.

### Fixed 3-terminal regulator circuits

Fixed-voltage regulator design has been greatly simplified in the past decade or two by the introduction of 3-terminal regulator ICs such as the 78xxx series of positive regulators and the 79xxx series of negative regulators, which incorporate features such as built-in fold-back current limiting and thermal protection, etc. These ICs are available with a variety of current and output voltage ratings, as indicated by the 'xxx' suffix; current ratings are indicated by the first part of the suffix (L = 100mA, blank = 1A, S = 2A), and the voltage ratings by the last two parts of the suffix (standard values are 5V, 12V, 15V and 24V). Thus, a 7805 device gives a 5V positive output at a 1A rating, and a 79L15 device gives a 15V negative output at a 100mA rating.

Three-terminal regulators are very easy to use, as shown in the basic circuits of *Figures 9.11* to *9.13*, which show the connections for making positive, negative and dual regulator circuits, respectively. The ICs shown are 12V types with 1A

Power supply circuits 265

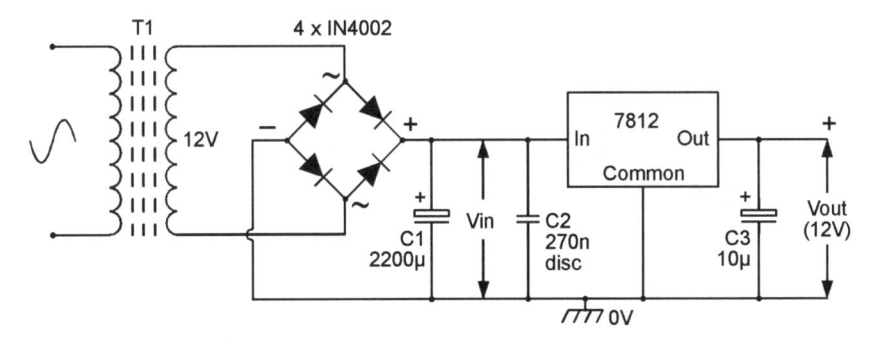

**Figure 9.11** *Connections for using a 3-terminal positive regulator, in this case a 12V 1A '78' type*

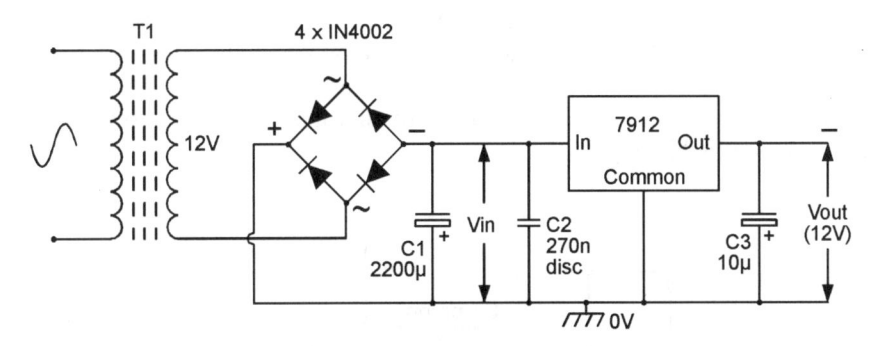

**Figure 9.12** *Connections for using a 3-terminal negative regulator, in this case a 12V 1A '79' type*

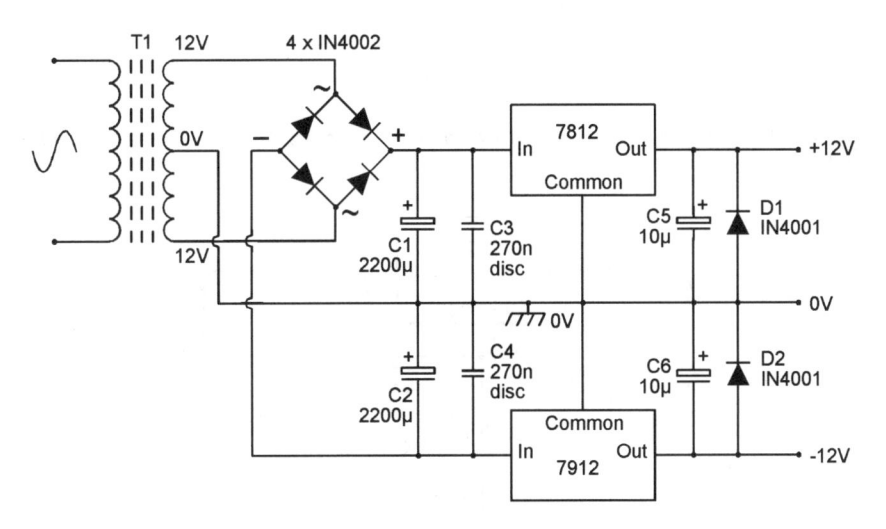

**Figure 9.13** *Complete circuit of a 12V 1A dual power supply using 3-terminal regulator ICs*

266  Power supply circuits

ratings, but the basic circuits are valid for all other voltage values, provided that the unregulated input is at least 3V greater than the desired output voltage. Note that a 270nF or greater disc (ceramic) capacitor must be wired close to the input terminal of the IC, and a 10μF or greater electrolytic is connected across the output. The regulator ICs typically give about 60dB of ripple rejection, so 1V of input ripple appears as a mere 1mV of ripple on the regulated output.

## Voltage variation

The output voltage of a 3-terminal regulator IC is actually referenced to the IC's 'common' terminal, which is normally (but not necessarily) grounded. Most regulator ICs draw quiescent currents of only a few mA, which flow to ground via this commom terminal, and the IC's regulated output voltage can thus easily be raised above the designed value by simply biasing the common terminal with a suitable voltage, making it easy to obtain odd-ball output voltage values from these 'fixed voltage' regulators. *Figures 9.14* to *9.16* show three ways of achieving this.

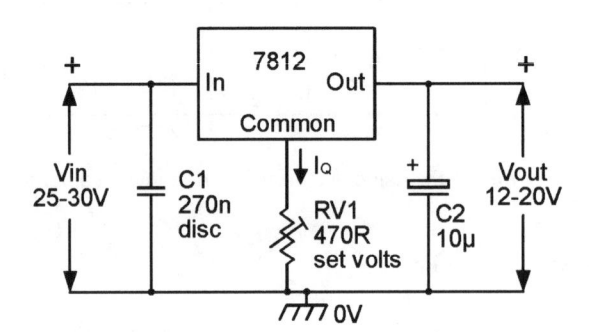

**Figure 9.14** *Very simple method of varying the output voltage of a 3-terminal regulator*

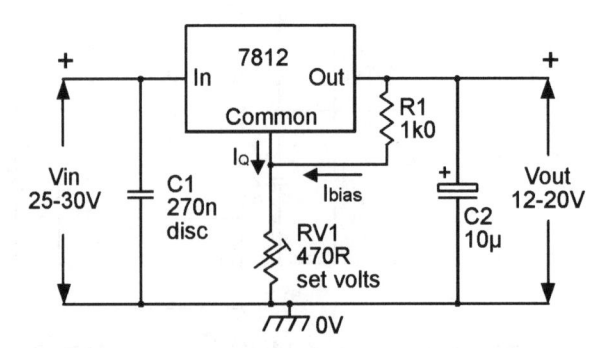

**Figure 9.15** *An improved method of varying the output of a 3-terminal regulator*

*Power supply circuits* 267

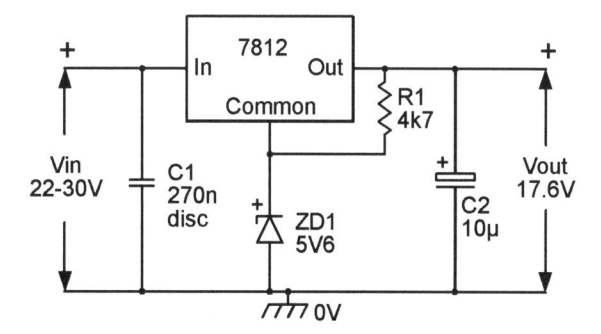

**Figure 9.16** *The output voltage of a 3-terminal regulator can be increased by a fixed amount by wiring a suitable zener diode in series with the common terminal*

In *Figure 9.14* the bias voltage is obtained by passing the IC's quiescent current (typically about 8mA) to ground via RV1. This design is adequate for many applications, although the output voltage shifts slightly with changes in quiescent current. The effects of such changes can be minimized by using the circuit of *Figure 9.15*, in which the RV1 bias voltage is determined by the sum of the quiescent current and the bias current set by R1 (12mA in this example). If a fixed output with a value other than the designed voltage is required, it can be obtained by wiring a zener diode in series with the common terminal as shown in *Figure 9.16*, the output voltage then being equal to the sum of the zener and regulator voltages.

## Current boosting

The output current capability of a 3-terminal regulator can be increased by using the circuit of *Figure 9.17*, in which current boosting can be obtained

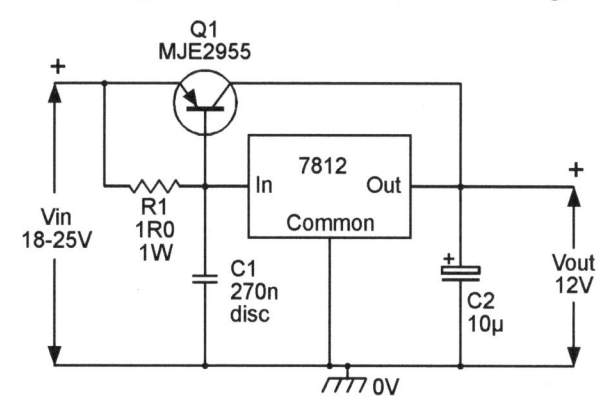

**Figure 9.17** *The output current capacity of a 3-terminal regulator can be boosted via an external transistor. This circuit can supply 5A at a regulated 12V*

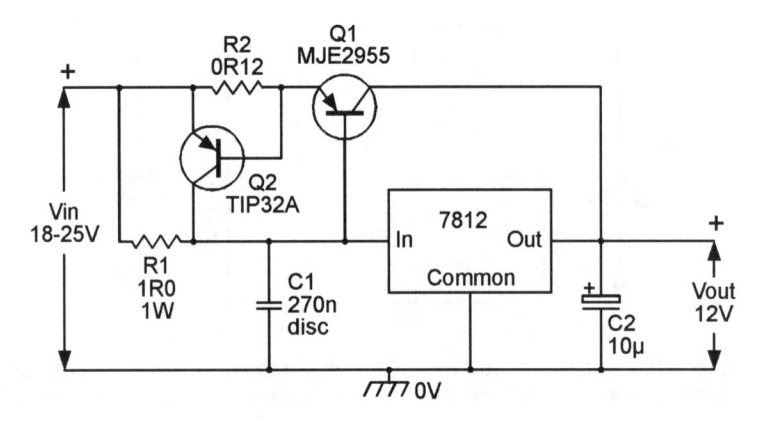

**Figure 9.18** *This version of the 5A regulator has overload protection provided via Q2*

via by-pass transistor Q1. Note that R1 is wired in series with the regulator IC; at low currents insufficient voltage is developed across R1 to turn Q1 on, so all the load current is provided by the IC. At currents of 600mA or greater sufficient voltage (600mV) is developed across R1 to turn Q1 on, so Q1 provides all current in excess of 600mA.

*Figure 9.18* shows how the above circuit can be modified to provide the by-pass transistor with overload current limiting via 0R12 current-sensing resistor R2 and turn-off transistor Q2, which automatically limit the output current to about 5A.

### Variable 3-terminal regulator circuits

The 78xxx and 79xxx range of 3-terminal regulator ICs are designed for use in fixed-value output voltage applications, although their outputs can, as already shown, in fact be varied over limited ranges. If the reader needs regulated output voltages that are variable over very wide ranges, however, they can easily be obtained from 3-terminal 'variable' regulator ICs such as the 317K or 338K types.

*Figure 9.19* shows the outline, basic data and the basic variable regulator circuit that is applicable to these two devices, which each have built-in fold-back current limiting and thermal protection and are housed in TO3 steel packages. The major difference between the devices is that the 317K has a 1.5A current rating compared to the 5A rating of the 338K. Major features of both devices are that their output terminals are always 1.25V above their 'adjust' terminals, and their quiescent or adjust-terminal currents are a mere 50μA or so.

Thus, in the *Figure 9.19* circuit, the 1.25V difference between the 'adjust' and output terminals makes several mA flow to ground via RV1, thus causing a variable adjust voltage to be developed across RV1 and applied to the adjust

*Power supply circuits* 269

| Parameter | 317K | 338K |
|---|---|---|
| Input voltage range | 4-40V | 4-40V |
| Output voltage range | 1.25-37V | 1.25-32V |
| Output current rating | 1.5A | 5A |
| Line regulation | 0.02% | 0.02% |
| Load regulation | 0.1% | 0.1% |
| Ripple rejection | 65dB | 60dB |

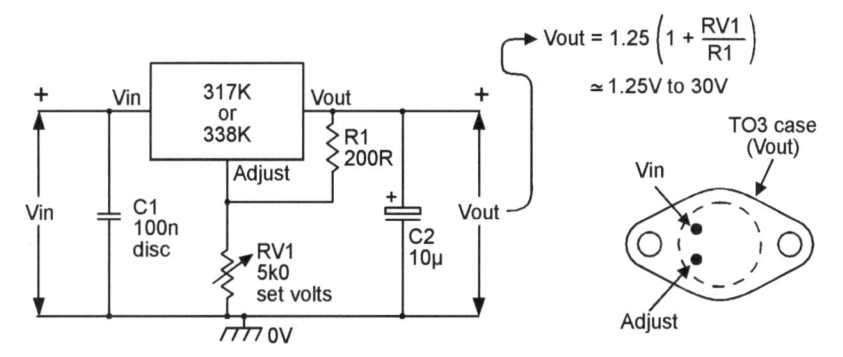

$$V_{out} = 1.25\left(1 + \frac{RV1}{R1}\right)$$
$$\approx 1.25V \text{ to } 30V$$

**Figure 9.19** *Outline, basic data and application circuit of the 317K and 338K variable-voltage 3-terminal regulators*

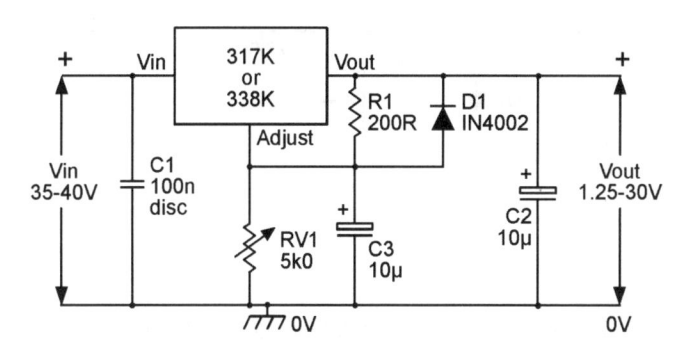

**Figure 9.20** *This version of the variable-voltage regulator has 80dB of ripple rejection*

terminal. In practice the output of this circuit can be varied from 1.25V to 33V via RV1, provided that the unregulated input voltage is at least 3V greater than the output. Alternative voltage ranges can be obtained by using other values of R1 and/or RV1, but for best stability the R1 current should be at least 3.5mA.

The basic *Figure 9.19* circuit can be usefully modified in a number of ways; its ripple rejection factor, for example, is about 65dB, but this can be increased to 80dB by wiring a 10μF bypass capacitor across RV1, as shown in *Figure 9.20*, together with a protection diode that stops the capacitor discharging into the IC if its output is short-circuited.

270  Power supply circuits

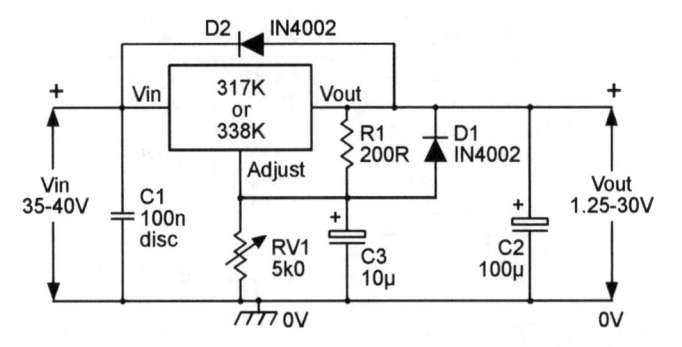

**Figure 9.21**  *This version of the regulator has 80dB ripple rejection, a low impedance transient response, and full input and output short-circuit protection*

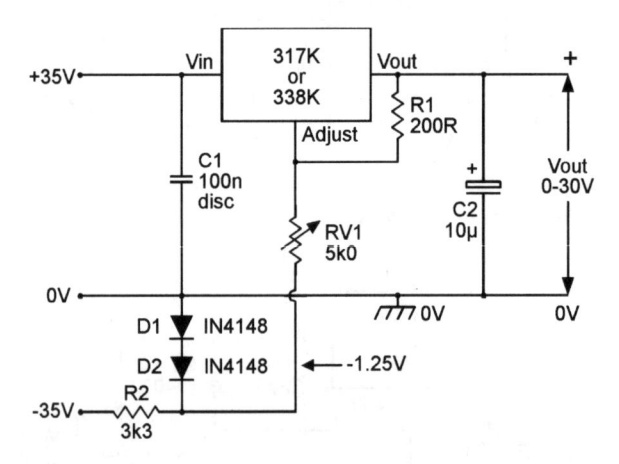

**Figure 9.22**  *The output of this version of the regulator is fully variable from zero to 30V*

*Figure 9.21* shows a further modification of the *Figure 9.20* circuit; here, the transient output impedance of the regulator is reduced by increasing the C2 value to 100µF and using diode D2 to protect the IC against damage from the stored energy of this capacitor if an input short occurs.

Finally, *Figure 9.22* shows how the circuit can be modified so that its output is variable all the way down to zero volts, rather than to the 1.25V of the earlier designs. This is achieved by using a 35V negative rail and a pair of series-connected diodes that clamp the low end of RV1 to minus 1.25V.

# 10
# *End notes*

Chapters 1 to 9 of this handbook have each dealt with a specific aspect of audio IC technology or with a particular range or class of audio ICs. Between them, these various chapters have covered almost the full spectrum of matters relating to the practical aspects of audio IC usage. There are, however, still a few reasonably important details of audio power amplifier IC usage that have not yet been covered in this handbook, and included among them are the matters of loudspeaker selection, power amplifier sensitivity, power supply parameter selection, IC power dissipation, heatsink selection, and practical audio power amplifier design techniques. All of these subjects are dealt with in this final chapter.

## Loudspeaker selection

The loudspeaker is the final and perhaps weakest link in the audio signal communication chain. In each channel of a hi-fi system the speaker unit is normally connected (directly or via an isolating capacitor) between the amplifiers output and the ground (common) line. The speaker unit may take the form of a single full-range loudspeaker, as shown in the basic 'single-ended power supply' amplifier circuit of *Figure 10.1*, but more often takes the form of two or more speakers that are driven via a passive cross-over filter, as shown in the simple example of *Figure 10.2*. In *Figure 10.2*, one speaker handles the bass frequencies and the other (called a tweeter) handles the treble tones; the diagram shows typical component values for a 6dB/octave filter that drives a pair or 8R0 speakers and has a cross-over frequency of 5kHz. Note that the unit is designed to give an input impedance equal to that of a single speaker (8R0 in this case), and thus appears to the amplifier as a single speaker load.

The most important basic parameter of a loudspeaker is (ignoring its frequency response and its power handling capacity) its input or coil impedance.

272    *End notes*

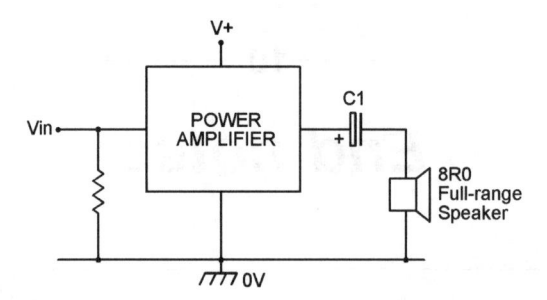

**Figure 10.1**  *A power amplifier, operated from a single-ended supply, driving a single full-range loudspeaker*

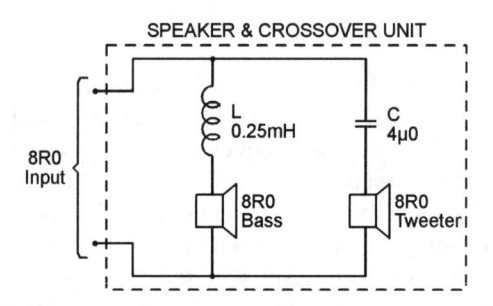

**Figure 10.2**  *A speaker unit consisting of two loudspeakers and a simple (6dB) cross-over filter unit*

Modern loudspeakers have standard impedances of 8R0, 6R0 or 16R (but sometimes 64R in very-low-power types) in most domestic or music-system applications, or 4R0 in low-voltage units such as in-car entertainment systems (which operate at a nominal 14.4V under actual running conditions). Often, in in-car systems, two 4R0 speakers are wired in parallel to form a high-power 2R0 unit. To obtain the maximum possible output power from an audio system, it is necessary to correctly match the effective speaker load impedance to the operating characteristics of the audio power amplifier.

All audio power amplifier ICs have specific supply-voltage and load-current handling and power dissipation limits, and these greatly influence the maximum power that can be safely fed to speakers with particular values of load impedance. The following formulas define the relationships between speaker load impedance, voltage, current and power:

(1) $I = \sqrt{\dfrac{W}{R}}$    (2) $E = \sqrt{RW}$    (3) $W = \dfrac{E^2}{R}$    (4) $W = I^2 R$

were $E$ = load voltage in r.m.s. volts, $I$ = load current in r.m.s. amps, $R$ = load impedance in ohms, $W$ = load power dissipation in r.m.s. watts. Thus, from (1) and (2) it can be deduced that an 8R0 load operating at a 10W power level consumes 1.12A at 8.94V, from (3) that an amplifier that produces an output of 10V r.m.s. generates 12.5W in an 8R0 load, and from (4) that an amplifier that can produce a maximum output of 1.2A can generate a maximum of 11.52W in an 8R0 speaker.

In practice, IC drive-current limits are usually specified in terms of $I_{peak}$, and their output voltage swing limits (which are typically about 10% less than their supply voltage values) are specified in $V_{pk}$ terms in split-supply applications or in $V_{pk-pk}$ ($V_{pp}$) in single-ended-supply applications; the following formulas show the relationships between r.m.s. (rms), peak (pk) and peak-to-peak (pp) voltage or current values:

(5) $I_{pk} = I_{rms}\sqrt{2}$  (6) $I_{rms} = \dfrac{I_{pk}}{\sqrt{2}}$  (7) $V_{rms} = \dfrac{V_{pk}}{\sqrt{2}} = \dfrac{V_{pp}}{2\sqrt{2}}$

(8) $V_{pk} = V_{rms}\sqrt{2}$  (9) $V_{pp} = 2V_{rms}\sqrt{2}$

Thus, from (6), (4), (2) and (9) it can be deduced that an IC that can supply an $I_{pk}$ of 3.5A can supply a maximum $I_{rms}$ of 2.47A and generate a maximum of 49W in an 8R0 speaker at a drive voltage of 19.8$V_{rms}$ or 56$V_{pp}$, thus requiring a typical supply voltage of at least 62V from a single-ended power unit or ±31V from a split supply unit. The same IC could generate a maximum of only 24.5W (at 9.9$V_{rms}$) in a 4R0 speaker, but a pair of ICs could generate 98W in a 16R speaker (or a pair of series-wired 8R0 types) when connected in the bridge-drive mode (in which they effectively double the speaker's drive voltage).

Thus, the first step in designing an IC audio power amplifier system for a particular application is to select a suitable IC and speaker, using the basic data given above. *Figure 10.3*, for example, lists the actual voltage and current values of five different speakers when they are operating at 10W power levels. At low operating voltages, low impedance (2R0 or 4R0) speakers must be

| Speaker impedance | $V_{rms}$ | $I_{rms}$ | $V_{pk-pk}$ | $I_{pk}$ |
|---|---|---|---|---|
| 16R | 12.65V | 0.79A | 35.8V | 1.12A |
| 8R0 | 8.94V | 1.12A | 25.3V | 1.58A |
| 6R0 | 7.75V | 1.29A | 21.9V | 1.82A |
| 4R0 | 6.32V | 1.58A | 17.9V | 2.23A |
| 2R0 | 4.47V | 2.24A | 12.6V | 3.16A |

**Figure 10.3** *Voltage and current values of various speakers at 10W power levels*

used to obtain a good power output. At reasonable high voltages, 8R0 or 16R speakers can be used.

Note, when making calculations of the types described in this section, that if you do not have access to precise data concerning an IC's $I_{pk}$ limit, the data can often be inferred from the manufacturer's application circuits, which are usually designed to show the product operating close to its performance limits. Thus, the LM3886 circuit of *Figure 6.43*, which can generate 68W in a 4R0 load, implies (using formulas (2) and (5)) that the IC can comfortably generate output currents of at least 4.12$A_{rms}$ or 5.8$A_{pk}$.

## Power amplifier sensitivity

A power amplifier's 'sensitivity' value defines the magnitude of r.m.s. input signal voltage needed to produce a particular output power level from the amplifier. Commercial hi-fi units usually specify the sensitivity at both a standard 1W output level and at the full rated output power level. The 'standard' (1W) sensitivities of commercial units vary between roughly 25mV and 200mV; a good compromise value is 100mV. The power amplifier's output power level is proportional to the square of the input voltage, thus, if the amplifier has a basic 100mV sensitivity, it intrinsically gives an output of 4W at 200mV, 16W at 400mV, 36W at 600mV, and 40W at 632mV. The basic formulas that link an amplifier's output power, standard sensitivity, and actual sensitivity are:

(10) $W_{out} = \left(\dfrac{V_{input}}{V_{sens}}\right)^2$   (11) $V_{input} = V_{sens}\sqrt{W_{out}}$

where $V_{sens}$ and $V_{input}$ are in millivolts and $W_{out}$ is in watts.

The amplifier's sensitivity is determined by its voltage gain ($A_V$), speaker impedance ($R$), and output power level ($W$), using the formulas:

(12) $A_V = \dfrac{V_{out}}{V_{input}} = \dfrac{\sqrt{RW}}{V_{sens}}$

In practice, 4R0 and 8R0 speaker loads absorb 2.0V and 2.83V of drive, respectively, at 1W power levels. Thus, to give a standard sensitivity value of 100mV, the amplifier needs an $A_V$ of ×20 with a 4R0 load or ×28.3 with an 8R0 load. Most modern audio power amplifier ICs function like high-power op-amps that are used in either the inverting or non-inverting mode, and their $A_V$ (and hence sensitivity) values can thus be set to any desired values by selecting their R1 and R2 feedback resistor values in the basic ways shown in *Figures 10.4* and *10.5*.

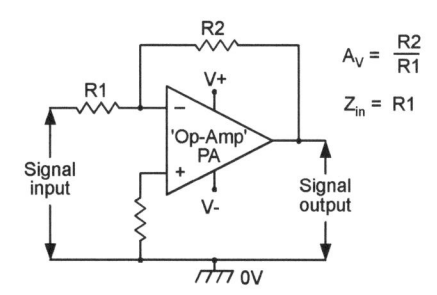

**Figure 10.4** *Basic circuit and formula for setting the $A_V$ (and hence sensitivity) value of a modern inverting 'op-amp' type of power amplifier (PA) circuit*

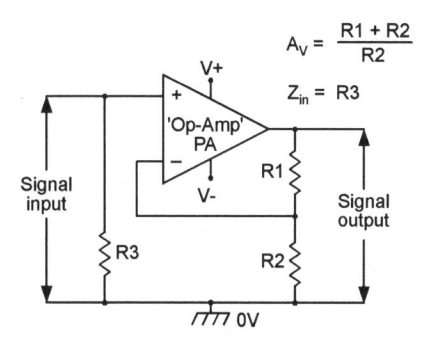

**Figure 10.5** *Basic circuit and formula for setting the $A_V$ (and hence sensitivity) value of a modern non-inverting 'op-amp' type of power amplifier (PA) circuit*

## Power supply requirements

An audio power amplifier IC's power supply is a critical part of the audio system, and if poorly specified or designed can have a highly detrimental affect on the audio power amplifier IC's operating efficiency and power dissipation. *Figure 10.6* illustrates some important audio power amplifier IC characteristics that are relevant to power supply design. The diagram shows a basic amplifier that is powered from a single-ended supply, together with its maximum peak-to-peak sinewave output waveform superimposed on the supply's half-voltage line. The following formulas are relevant to this diagram and to the next few paragraphs of text:

(13) $I_{DC} = 0.45 \times I_{rms}$   (14) $P_{INPUT} = V_{SUPPLY} \times I_{DC}$

(15) Efficiency $= \dfrac{P_{OUT}}{P_{INPUT}}$   (16) $P_{IC} = P_{INPUT} - P_{OUT}$

Regarding these formulas, which all assume that the IC is a single (rather than a dual) type, formula (13) defines the IC's power supply current ($I_{DC}$)

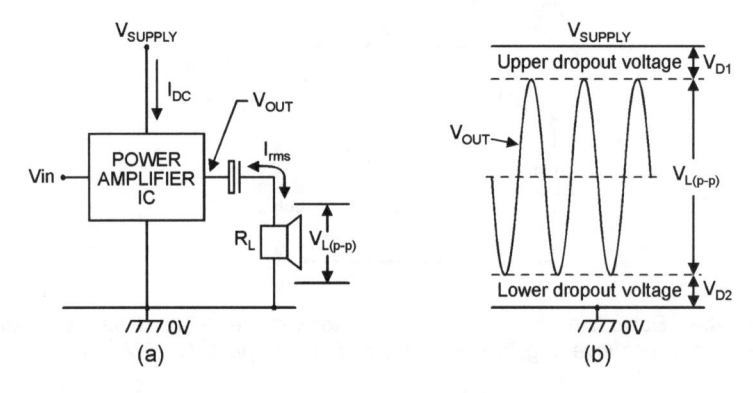

**Figure 10.6** *Basic details of an audio power amplifier IC circuit operated from a single-ended supply (a), and its sinewave output waveform when driving maximum power into a speaker load (b)*

as a function of the IC's load (speaker) current, $I_{rms}$ (see formula (1)), but ignores the effects of the IC's quiescent currents, which are usually relatively small. Formula (14) defines the total power input into the IC, in term of supply voltage and current ($I_{DC}$). Formula (15) defines the power amplifier IC's power conversion efficiency in terms of $P_{OUT}$ (to the speaker load) versus $P_{INPUT}$. Finally, formula (16) defines the actual power dissipated across the IC as $P_{INPUT}$ minus $P_{OUT}$.

Returning now to *Figure 10.6*, note that in an ideal class-AB power amplifier the load's sinewave output voltage would be able to swing fully between the two supply rail levels (0V and $V_{SUPPLY}$) without clipping. Such an amplifier would, at maximum power output, operate with an efficiency of 78.5%. If this ideal amplifier is driving 10W into an 8R0 load at its maximum output level, $V_{SUPPLY}$ will have a value of 25.28V (the load's $V_L$ peak-to-peak voltage value), the IC will consume a supply current ($I_{DC}$) of 0.504A (= 0.45 times the load's 1.12A r.m.s. signal current), a total of 12.74W (= 25.28V × 0.504A) will be consumed from the power supply, and 2.74W of this will be developed across the IC and 10W across the load.

In a real-life class-AB power amplifier IC, significant voltage losses occur in the IC's output driver stages, and in *Figure 10.6* these are represented by two 'dropout' voltages, $V_{D1}$ and $V_{D2}$, and to compensate for these the above 10W amplifier needs a minimum supply voltage of $V_L+V_{D1}+V_{D2}$. Suppose that this combination calls for a minimum $V_{SUPPLY}$ value of 30V. In this case a total of 15.12W (= 30V × 0.504A) will be consumed from the supply at 10W output, and 5.12W of this is generated across the IC, which thus operates with an efficiency of only 66.1% (= $P_{OUT}/P_{IN}$). (Note that dropout voltage values are often given in manufacturer's data sheets, but if this information is not readily available it can, if the IC is fitted with an adequate heatsink, easily be ascertained from practical power amplifier tests with the aid of an oscilloscope.)

Suppose now that, in the above 10W amplifier circuit, the power supply is a badly designed unregulated type that, at the nominal AC power line voltage, gives an actual output of 35V at full load. In this case a total of 17.64W (= 35V × 0.504A) will be consumed from the supply at 10W output, and 7.64W of this is developed across the IC, which thus operates with an efficiency of only 56.7% at full output. If the AC power line voltage subsequently rises to its permitted 'nominal+10%' upper limit, the IC power dissipation will rise to 9.4W at 10W output, and the IC will operate with an efficiency of only 51.5%. Good power supply design is thus vital.

From the above, it can be seen that the ideal power supply should be a low ripple type that, at the designed maximum audio power output level, produces a per-channel output current of $I_{DC}$ (see formula (13)) and a voltage of $V_L + V_{D1} + V_{D2}$ if the supply is a single-ended type or ± half of this value if it is a split type. Note that if the supply is an unregulated type, the supply voltage will (depending on the supply's actual regulation factor) typically rise by 15% under very-low-load condition and by a further 10% if the AC power line voltage rises to its maximum permitted value. This worst-case voltage must not exceed the IC's supply voltage limit. If the IC is operating close to its critical supply voltage or power dissipation levels, a regulated power supply should be used.

## IC power dissipation

The previous few paragraphs outlined some basic principles concerning power dissipation in audio power amplifier ICs. A good understanding of this subject is important in practical applications, since IC power dissipation generates heat in the IC and must, if the IC is to function correctly, be kept under control (usually with the aid of a suitable heatsink). This section provides additional information on the subject, as follows.

If the sinewave input signal voltage to an IC audio power amplifier circuit is progressively increased from zero to maximum, the circuit's load (speaker) voltage and current both increase proportionately, and the load power thus increases in proportion to the square of the input voltage. The power dissipation of the actual IC, however, varies in a rather different way, since any increase in input voltage causes an increase in the IC's supply current but a decrease in the volt drop across the IC's power-driving output stages. Initially, the IC's power dissipation increases as the input voltage is increased, but eventually a point is reached where the power dissipation reaches a peak value, and any further increase in input voltage results in a *decrease* in the IC's power dissipation.

278  *End notes*

The IC's peak power dissipation point ($P_{IC(MAX)}$) occurs when $V_L$ (the load's peak-to-peak signal voltage) equals 63.7% of $V_S$ (the IC's supply voltage), at which point $P_{IC(MAX)}$ has, in theory, a value of:

(16) $P_{IC(MAX)} = \dfrac{(V_S)^2}{20R_L}$

Note, however, that this widely quoted but slightly simplified theoretical figure assumes the use of an electronically perfect IC that (among other things) is perfectly linear and generates zero signal distortion. In practice, actual $P_{IC(MAX)}$ figures may be as much as 20% higher than given in formula (16), particularly if very low impedance (2R0 or 4R0) speaker loads are used. Thus, in the well designed 10W, 30V, 8R0 amplifier mentioned in the previous section of this chapter, the IC has a theoretical $P_{IC(MAX)}$ value of 5.62W, but in practice has a probable value of about 6.2W.

If the IC's sinewave input signal voltage is increased beyond the $P_{IC(MAX)}$ point, a level is eventually reached where the load is operating at its designed maximum power level, and any further increase in input voltage generates a clipped (square shaped if severely clipped) output signal that has a greater power content than a simple sinewave; the IC's power dissipation decreases sharply under this condition.

As a consequence of the various factors described above, *all* audio power amplifier ICs generate a $P_{IC}$ versus $P_L$ curve that follows the basic form shown in *Figure 10.7*, in which – as the $P_L$ value is increased from zero to maximum – the $P_{IC}$ value rises exponentially from zero to a peak value that occurs at the $P_{IC(MAX)}$ point, and then starts to fall again as the $P_L$ value heads towards the signal-clipping point.

Note that the above $P_{IC(MAX)}$ figures apply to 'single' ICs. In dual (2-channel) ICs, or in single types that form one half of a bridge-type amplifier, the above $P_{IC(MAX)}$ figures must be doubled.

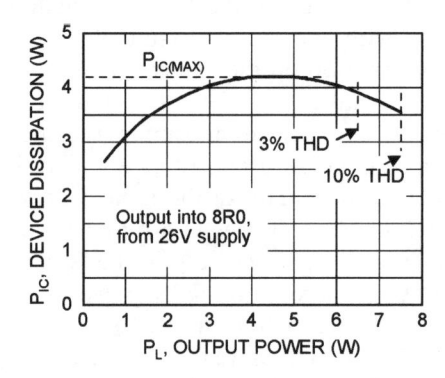

**Figure 10.7** *Typical $P_{IC}$ versus $P_L$ curve of an LM384-based 6.5W amplifier driving an 8R0 load from a 26V supply*

## Heatsink calculations

The heart of any IC is its solid-state chip, and it is this that becomes heated when power is generated within the IC. By industry-wide convention, an IC chip's recommended safe maximum power handling capacity is normally specified at a chip temperature of 25°C, and – in commercial ICs that use a plastic package – the chip's recommended maximum allowable operating temperature is limited to 150°C, at which temperature its power handling capacity is deemed to be zero. The chip's power handling capacity is inversely proportional to the chip temperature, and is 40% of maximum at 100°C. Most modern audio power amplifier IC chips incorporate thermal protection circuitry that automatically cuts off the chip's power feed at a chip temperature of 165°C, thus protecting the IC against catastrophic power-induced over-temperature damage.

*Figure 10.8* presents the above data in graphic form, with the chip's power handling capacity presented as a percentage of the maximum figure attainable at the 25°C value. Thus, if the chip can handle a maximum of 10W at 25°C, its temperature must be kept below 100°C if it is to be used in an application where its maximum power dissipation ($P_{IC(MAX)}$) is expected to reach 4W. This temperature control is normally exercised via an external heatsink, which helps the internally power-generated heat to flow away and be absorbed by the ambient conditions.

Heat flow is best modelled in terms of thermal resistance, where a 'thermal resistance' represents the thermal conductor's input-to-output temperature drop divided by the power dissipated by the conductor, in °C/W. The symbol used to represent thermal resistance is θ (theta), and is usually followed by a subscript that denotes the direction of heat flow. Thus, the symbol $θ_{JC}$ denotes the thermal resistance that exists between the IC's junction (chip) and case.

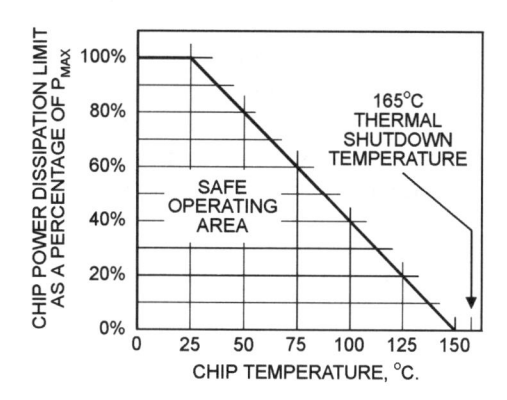

**Figure 10.8** *Graph showing the relationship between chip temperature and chip power dissipation limits*

280   End notes

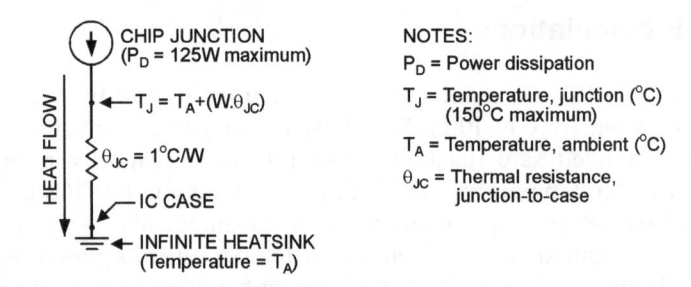

**Figure 10.9**  *Simple thermal resistance model of a power amplifier IC that has its case bonded to an infinite heatsink*

*Figure 10.9* shows a simple thermal resistance model of an audio power amplifier IC that has a power dissipation limit of 125W at a chip temperature of 25°C, has a $\theta_{JC}$ value of 1°C/W, and has its case bonded to an infinite heatsink that is operating at an ambient temperature of $T_A$°C. It is fairly obvious from this model that, at a chip power dissipation value of $W$, the chip's junction temperature ($T_J$) equals $T_A + (W \cdot \theta_{JC})$. Thus, if $T_A$ is 25°C, the $T_J$ value works out at 35°C at 10W, 75°C at 50W, or 150°C at 125W, and if $T_A$ is 45°C the $T_J$ value works out at 55°C at 10W, 95°C at 50W, or 150°C at 105W, and so on.

In the real world, the point of union between the case and heatsink inevitably has a certain thermal resistance value, and this is given the symbol $\theta_{CS}$ (case-to-sink) and has a typical value of 0.2°C/W if the union is a bolted type made via a good *modern* heat transfer compound. Also, the heatsink is inevitably finite, and is represented by the symbol $\theta_{SA}$ (sink-to-ambient), and has a value of $X$°C/W. The *total* thermal resistance between the chip junction and the ambient temperature is represented by the symbol $\theta_{JA}$ (junction-to-ambient) and is equal to the sum of these three thermal resistance value. Thus, $\theta_{JA} = \theta_{JC} + \theta_{CS} + \theta_{SA}$.

*Figure 10.10* shows a thermal resistance model of a real-world version of the *Figure 10.9* audio power amplifier IC, bolted to a finite heatsink that has a $\theta_{SA}$ value of 2°C/W and is operating at an ambient temperature of 25°C. The IC again has a power dissipation limit of 125W at a chip temperature of 25°C, and has a $\theta_{JC}$ value of 1°C/W. The total thermal resistance, $\theta_{JA}$, between the chip junction and ambient thus equals 1°C/W + 0.2°C/W + 2°C/W, i.e. 3.2°C/W. Consequently, in this example, the IC's $T_J$ value equals $T_A + (W \cdot \theta_{JA})$, and works out at 57°C at 10W, or 121°C at 30W.

In practice, the two most valuable pieces of data that can be gleaned from a real-world thermal model of the *Figure 10.10* type are, first, the maximum power dissipation limit ($P_{IC(MAX)}$) of an IC at a given $\theta_{JA}$ value, and second, the *minimum* heatsink value ($\theta_{SA}$) needed to dissipate a known value of IC

```
                    ↓  CHIP JUNCTION              NOTES:
                       (P_D = 125W maximum)       P_D = Power dissipation
                    ←—T_J = T_A+(W.θ_JA)          T_J = Temperature, junction (°C)
                                                       (150°C maximum)
         ⎰  θ_JC = 1°C/W  ⎤                       T_A = Temperature, ambient (°C)
    H    ⎱               ⎥
    E    ⎰  θ_CS = 0.2°C/W ⎬ θ_JA = 3.2°C/W      θ_JC = Thermal resistance,
    A    ⎱               ⎥                              junction-to-case
    T                    ⎥
    F    ⎰               ⎥                       θ_CS = Thermal resistance,
    L    ⎱  θ_SA = 2°C/W  ⎦                             case-to-sink
    O
    W    ←— Ambient temperature (T_A)            θ_SA = Thermal resistance,
    ↓   ═   = 25°C                                       sink-to-ambient

                                                  θ_JA = Thermal resistance,
                                                        junction-to-ambient
```

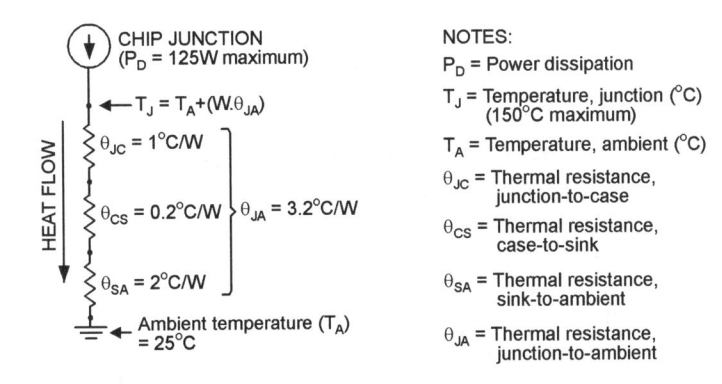

**Figure 10.10** *Real-world thermal model of a power amplifier IC that has its case bolted to a finite heatsink*

power, $P_{IC}$. The following two formulas, in which $T_{J(MAX)}$ has a value of 150°C in normal commercial ICs, define this data.

(17) $\quad P_{IC(MAX)} = \dfrac{T_{J(MAX)} - T_A}{\theta_{JA}} \qquad$ (18) $\quad \theta_{SA} = \dfrac{T_{J(MAX)} - T_A}{P_{IC}} - (\theta_{JC} + \theta_{CS})$

Thus, from (17) it can be seen that, when using the $\theta_{JA}$ value of 3.2°C/W shown in *Figure 10.10*, the IC can dissipate a maximum of 39W, and, from (18), that, for the IC to dissipate 50W, the heatsink needs a *maximum* $\theta_{SA}$ value of 1.3°C/W.

## Heatsink practicalities

When using formula (18) it is always wise to use a realistic (rather than optimistic) worst-case $T_A$ figure, such as 30°C rather than 25°C. The vital $\theta_{JC}$ value used in the formula is usually given in the IC's data sheet. It is important to note that formula (18) indicates the theoretical *absolute maximum* size of heatsink $\theta_{SA}$ value that should be used in a particular application, and that in practice a heatsink with a somewhat lower $\theta_{SA}$ value should actually be used. In the case of the *Figure 10.10* 50W amplifier given above, for example, where the formula gives a maximum $\theta_{SA}$ value of 1.3°C/W, a heatsink with an actual value of 1°C/W should be used. Note, however, that the physical size (and cost) of a heatsinks is inversely proportional to its $\theta_{SA}$ value, and a 1°C/W heatsink is thus physically larger and more expensive than a 1.3°C/W type. Heatsinks are readily available in a wide range of $\theta_{SA}$ values.

When selecting an actual heatsink size after completing the formula (18) calculations, some thought should be given to two conflicting theories in the subject. These can be called the 'real-life audio' and the 'IC reliability' theories. The first of these theories states that real-life audio systems deal primarily with music or speech signals, and these, at full volume, have a mean power output that is typically less than one third of the maximum continuous sinewave value used in the formula (18) calculation. This theory concludes that in hi-fi types of audio application, there is no practical advantage in using a heatsink that is physically larger than the minimum value indicated by formula (18).

The alternative 'IC reliability' theory is based on IC failure rate studies that show that failure rates are exponentially proportion to the IC's junction temperature, and under worst-case conditions increase by a factor of three for every 10°C rise in junction temperature. When these failure rate studies are applied to the above-mentioned 50W audio amplifier, they show that when the IC is operating under worst-case conditions at full (50W) sinewave power, the IC's failure rate is reduced by a factor of five by using a heatsink with a $\theta_{SA}$ value of 1°C/W, rather than the 1.3°C/W value indicated by formula (14), and that under typical full-volume (16.6W average output power) music conditions the failure rate is almost halved by using the 1°C/W heatsink.

## Basic audio power amplifier design

To conclude this book, this final section shows how the information so far presented in this chapter can be used to design a complete and reliable basic audio power amplifier system. The first step in creating such a design is to draw up a specification that includes per-channel details such as (a) maximum power output, (b) output load impedance, (c) input impedance, (d) input sensitivity, and (e) a basic amplifier description, and to then select a specific IC for use in the design. Suppose that this basic specification is as follows:

(a) Maximum power output = 30W.
(b) Load (speaker) impedance = 8R0.
(c) Input impedance = 47k.
(d) Input sensitivity (at 1W output) = 100mV.
(e) Amplifier type = non-inverting; powered from a split supply.

The precedure for designing such an amplifier is as follows.

Regarding specifications (a) and (b), formulas (2), (8), (1) and (5) in the Loudspeaker selection section of this chapter show that the power amplifier must supply the 8R0 load with a peak voltage of 21.9V and a peak current of 2.74A at the 30W power level. These values are well within the capabilities of the LM3875 IC (see Chapter 6). This IC's data sheet shows that the

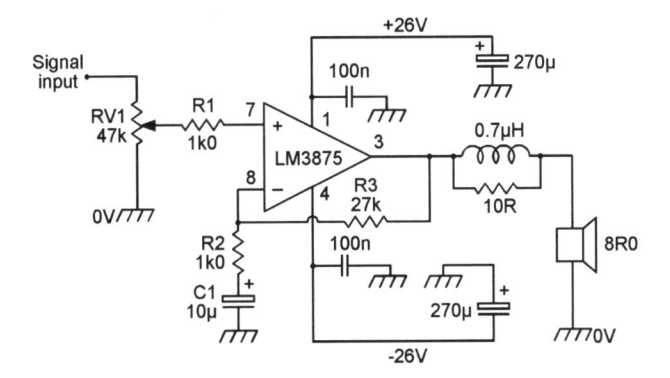

**Figure 10.11** *Design example of a 30W non-inverting amplifier powered from a split (dual) supply (see text for design precedure)*

IC's upper and lower dropout voltage each have values of 4V under the specified maximum load driving conditions, and the split (dual) power supply must thus generate an output of ±26V at (from formulas (1) and (13)) 0.87A under full load conditions. The worst-case off-load voltage, assuming a 15% regulation factor in the power supply and a +10% voltage rise in the AC supply line, is thus ±32.5V, and is well within the operating limits of the LM3875.

Regarding specification (e), this calls for a basic amplifier design of the type shown in *Figure 10.11* (which is based on the *Figure 6.39* circuit shown in Chapter 6). In this design, specification (c) can be met by giving volume control RV1 a value of 47k. Specification (d) can, as described in the Power amplifier sensitivity section of this chapter, be met by giving the amplifier an $A_V$ value of (ideally) ×28.3, and in *Figure 10.11* this is achieved by giving R3 a value of 27k.

The final step in the amplifier design exercise is that of selecting a suitable heatsink. From formula (16), it can be seen that in this circuit the IC's maximum power dissipation is (remembering that the IC is using a dual, rather than single-ended, supply) 16.9W. The LM3875 IC has a $\theta_{JA}$ value of 1°C/W, and the $\theta_{CS}$ value is typically 0.2°C/W. Consequently, formula (18) shows that, at an ambient temperature of 30°C, the heatsink needs a maximum $\theta_{SA}$ value of 5.9°C/W. In practice, a heatsink with a somewhat lower $\theta_{SA}$ value should be used.

# Index

*Individual types of integrated circuits (ICs) mentioned in this Manual are listed in a separate section at the end of the Index.*

Acoustic room expanders (ambience synthesizer) 214-15, 229-31
Active filters
 2nd-order Butterworth 113
 10kHz low-pass 31-2
 100Hz high-pass 32-3
 300Hz to 3.4kHz speech 33-4
 basics 30-1
 NAB tape equalization 107
 RIAA phono equalization 107
 variable 34-5
 variable rumble/scratch/speech 34-5
ADC (analogue-to-digital converter) 15-16
 for digital delay-lines 234-6
 dynamic range 235-6
AGC (automatic gain control) amplifiers 56-7
Aliasing signals 232
AM (amplitude modulation) 53-5
AM-radio audio power amp 133-8
Ambience synthesizers (acoustic room expander) 214-15, 229-31
Amplifiers
 AGC (automatic gain control) 56-7
 closed-loop 23-4
 constant-volume 45-6
 fuzz effect 44
 high input impedance 26, 27
 linear 25-7
 as mixers 29
 non-linear 43-4

RIAA phono pre-amp 41-3, 107-8, 111-12, 148
 variable gain/voltage controlled 50-3
 *see also* Op-amps; OTAs (operational transconductance amplifiers); Power amplifiers; Pre-amplifiers
Analogue switches
 basics 73-6
 channel separation 75-6
 practical circuits 76-83
 THD (total harmonic distortion) 75
Audio delay lines *see* Delay-lines
Audio signals
 audio range 1
 hearing 1
Audio systems
 basic elements 2-3
 definition 2-3

Bar-graph displays *see* LED (light emitting diode) bar-graph displays
BBD (Bucket Brigade Delay) line
 clock generator circuits 224-8
 delay time available 209-10
 filters for 228-31
 practical circuits 218-24
 practical ICs 216-18
 principles 205-9
 usage connections 208, 209
 *see also* Psycho-acoustics
Biamplification 175-7

## Index

Bipolar op-amps 24-5
Breathing/pumping, compandors 14-15

Capacitor filter, for power supplies 261-2
Cassette tape *see* Tape recorders
CCD (Charge-Coupled Device) delay line *see* BBD (Bucket Brigade Delay) line
CD (Compact Discs), dynamic range 15-17
Channel separation, analogue switches 75-6
Chorus generator with delay-lines 211-12
Clipping distortion 5
Comb filter circuits with delay-lines 213-14
Compandors (compressors/expanders)
 breathing/pumping 14, 67
 circuit description 60-2
 compressor circuit 66
 and digital delay lines 244-6
 for hi-fi applications 69-70
 noise in 67
 principle 11-14
 rectifier bias current cancellation 68-9
 theory 65
Constant-volume amplifiers 45-6, 62-3
Cross-over distortion 5-6
Cross-over filters 176

DACs (Digital-to-Analogue Converters) 15-16
 digital delay-lines 237-9
Darlington power output stages 121-3
De-emphasis/pre-emphasis 10-11, 244-6
Delay equalization with delay-lines 216-17
Delay-lines
 basics 201-5
 chorus generator 211-12
 comb filter circuits 213-14
 continuous-loop recorder method 204
 for delay equalization 216-17
 double tracking effect 201-2

echo/reverb units 202-5, 214-15, 229
flangers 213-14
mini-chorus effect 211-12
mini-echo/mini-chorus effect 201-2
musical effects 211-12
phasers 213
practical circuits 218-28
practical ICs 216-18
predictive switching with 215-16
reverb time 203
selecting best 246-7
spring and tape system 204-5
vibrato with 211-12
*see also* BBD (Bucket Brigade Delay) line; Digital delay-lines; Holtek HT8955A delay-line; Psycho-acoustics
Destination points 2-3
Differential amplifiers 21-2
Digital audio, basics 15-17
Digital delay-lines
 ADCs (analogue-to-digital converters) 234-6
 basics 231-41
 DACs (Digital-to-Analogue Converters) 237-9
 delay-time variation 242-4
 input filter 232-3
 output filter 239
 and pre-emphasis/de-emphasis 244-6
 principle 231-2
 RAM (Random Access Memory) 236-7
 sample-and-hold unit 233-4
 selecting best 246-7
 timing generator 239-41
 *see also* Holtek HT8955A
Digitally controlled ICs
 principle 85-7
 for tone/volume control 87-94
Distortion *see* Frequency response distortion; Harmonic signal distortion
DNL (dynamic noise limiter) 70-3
DNR (dynamic noise reduction) 70-3
Dolby systems 15
Double tracking effect 201-2
DRAM (Dynamic RAM) 236

Index 287

Ducking (voice-over) unit 63-5
Dynamic range
    ADCs (Analogue-to-digital
        Converters) 235-6
    cassette tapes 10
    CDs 15-17
    hybrid control 14-15
    and noise 8-9
    stereo phonograph 10
    *see also* Compandors; Pre-emphasis

Echo/reverb units 202-5, 214-15, 229
Efficiency, power amplifiers 275-7
Electronic attenuator 58-60
Equalizers 39-41
    *see also* RIAA phono pre-
        amplifiers/amplifiers/ equalizers
Expanders *see* Compandors
    (compressors/expanders)

Fidelity 3-4
Filters
    for BBD delay-lines 228-31
    for digital delay-lines 232-3, 239
    for frequency response distortion 7-8
    *see also* Active filters; Cross-over
        filters; Switch capacitor filter ICs
Flangers with delay-lines 213-14
Frequency response, of op-amps 22-3
Frequency response distortion 6-8
Fuzz effect amplifier 44

Graphic equalizers 39-41
Guard rings 28-9

Harmonic signal distortion
    clipping distortion 5-6
    cross-over distortion 5
    definition 4
    non-linear distortion 6
Hass effect 211
Hearing
    age effect 1-2
    unimpaired 1

Heatsinks
    bonding 280
    and IC reliability 282
    for power amplifiers 150
    practicalities 281-2
    and thermal models 279-81
    and thermal resistance 279-81
Hi-fi systems 3
    typical form 17-19
Holtek HT8955A delay-line
    functional block diagram 248
    functional description 247-50
    karaoke circuit for 255-7
    pin notations and descriptions 248, 249
    usage data 250-5

IC (Integrated Circuit) power
    dissipation 277-8

JFET op-amps 25

Karaoke circuit, using Holtek HT8955A
    delay-line 255-7

LED (light emitting diode) bar-graph
    displays
    10 LED linear meter 188
    bar-mode voltmeters 193-6
    basics 177-81, 187-90
    dot-mode 178
    dot-mode voltmeters 190-3
    linear scaled meter 181-2
    log displays 198-200
    log and linear displays 179, 187
    and over-range detectors 178, 185-7
    power meter applications, log-scaled
        187, 198-200
    practical circuits 181-7, 190-200
    reference voltage sources 187, 190
    semi-log scale, VU meters 187, 198-200
    supply voltage for 181
    voltmeter applications, linear 187
    VU (volume unit) meters 198

288  Index

Loudspeakers 18-19
   and biamplification 175-6
   coil impedance 271-4
   impedance/voltage/current/power
      relationships 272-4
   for in-car entertainment 272
   selection 271-4
   two with cross-over filters 271-2
Low-fi systems 4

Magnetic tape recorders *see* Tape
   recorders
Medium-fi systems 3-4
Meters, LED linear scale 181-2
Mini-echo/mini-chorus effect 201-2, 211-12
Mixers 29
Modulation
   amplitude 53-5
   ring 54, 55-6
MOSFET op-amps 24-5
Musical effects with delay lines 211-12

NAB tape playback equalization 107, 111-12
Negative feedback 23-4
Noise
   and dynamic range 8-9
   S/N-ratio 8-9
Non-linear amplifiers 43-4
Non-linear distortion 6

Octave equalizers 39-41
Op-amps
   basics 20-3
   bipolar 24, 25
   as differential amplifiers 21-2
   frequency response 22-3
   inverting/non-inverting 21-2, 23-4, 25-6
   JFET 25
   MOSFET 24, 25
   output voltage 21
   as voltage followers 27-9
   *see also* Amplifiers

OTAs (operational transconductance
   amplifiers)
   as AGC amplifiers 56-7
   applications 46
   basic circuits 47-50
   practical devices 46-7
   with ring modulators 56
   as a voltage controlled amplifier 58-60
Over-range detectors 178, 185-7

Pan pot circuit 114
Phasers with delay-lines 213
Phono pre-amplifier/amplifiers with
   RIAA equalization 41-3, 107-8, 111-12, 148
Phonograph, dynamic range 10
Power amplifiers
   basic design summary 282-3
   biamplification 175-7
   boosted-output op-amps 130-1
   class-A basics 116-17
   class-AB
      with auto-bias 121-2
      basics 118-21
      with complementary pairs 123
      with Darlington output stages 121-3
      with quasi-complementary output
         stages 121-2
      transformerless type 119-21
   class-B basics 117-18
   complete power amplifier IC 127-8
   driver circuits
      bootstrapping 124-5
      with dc and ac feedback 126-7
      decoupled parallel dc feedback 126
   efficiency 275-7
   and heat sinks 150
   high power 6W to 17W 150-63
   high power 18W to 68W 163-75
   low power op-amps 128-9
   peak available output power 132-3
   power supply requirements 275-7
   sensitivity 274-5
   temperature/dissipation considerations 150-1
   *see also* Power ICs

Power dissipation
  ICs 277-8
  see also Heatsinks
Power ICs
  1.2W at 3V to 15V 140
  1.5W at 4V to 12V 133-8
  1.5W at 4V to 15V 141-3
  2.0W at 3V to 16V 143-5
  2.0W/5.5W at 8V to 26V 147
  5W/6W at 5V to 20V 133-8
  6W at 4V to 20V for automobiles 158-9
  9.5W at 6V to 18V for automobiles 160-1
  12W at 6V to 15V split power supply 162-3
  18W at 6V to 18V split power supply 164-5
  20W at 6V to 18V for automobiles 165
  22W at 3V to 20V split power supply 165-7
  25W at 20V to 60V 167-8
  32W at 9V to 50V 168-70
  40W at 7.5V to 30V split power supply 170-1
  56W at 20V to 84V split power supply 171-3
  56W at 24V to 84V split power supply 171-2, 173-4
  68W at 20V to 84V split power supply 171-2, 174-5
  325mW at 4V to 15V 133-8
  dual 1.0W at 1.8V to 15V for headphones 143, 144
  dual 2.0W at 6V to 26V 145-7
  dual 4W/5W/8W at 6V to 32V 151-8
  dual, 220mW per channel at 1.8V to 6V 138-40
Power supplies
  basic circuits 258-60
  filter capacitor 261-2
  regulation factor 260
  requirements of power amplifiers 275-7
  ripple 261-2
  transformer-rectifier selection 260-1
  see also Voltage regulator circuits

Pre-amplifiers
  basics 101-2
  differential biasing 105
  as filter amplifiers 107-9
  inverting/non-inverting 106, 111-12
  practical circuits 102-15
Pre-emphasis/de-emphasis 10-11, 244-6
Predictive switching with delay-lines 215-16
Psycho-acoustics
  Hass effect 211
  laws of 210-11
Public address delay equalization 216-17
Pumping/breathing, compandors 14-15

RAM (Random Access Memory), digital delay-lines 236-7
Rectifiers, for power supplies 260-1
Regulation factor, power supplies 260
Reverb time 203
RF pickup/instability 110-11
RIAA phono pre-amplifiers/amplifiers/equalizers 41-3, 107-8, 111-12, 148
Ring modulation 54, 55-6
Ripple, power supplies 261-2, 266
Room expanders, acoustic 214-15
Rumble/scratch/speech active filter 34-5, 108

S/N-ratio
  basics 8-9
  CDs 15-17
Sample-and-hold unit, digital delay-lines 233-4
Scratch/rumble/speech active filter 34-5
  with pre-amplifiers 108-9
Signal compression see Compandors
Source points 2-3
SRAM (Static RAM) 236-7
Stereo phonograph, dynamic range 10
Subjective impressions of sound 19
Switch capacitor filter ICs
  basics 94-6
  practical circuits 96-100
Switches see Analogue switches

Tape recorders
  and compandors 65
  dynamic range 10
  pre-amplifier with NAB equalization 107-8, 111-12
  with predictive switching 215-16
THD (total harmonic distortion)
  analogue switches 75
  with compandors 67
Timing generator, digital delay-lines 239-41
Tone control
  with pre-amplifiers 108, 114
  with voltage-controlled ICs 83-5, 87-94
Tone control circuits
  active 37-9
  basics 35-7
Transformers
  for power amplifiers 117
  for power supplies 260-1

Variable-gain circuits
  with CA3080 50-1
  with LM13600 52-3

VCAs (voltage controlled amplifiers) 50-3, 58-60, 62
Vibrato with delay-lines 211-12
VOGAD (stereo constant-volume amplifier) 62-3
Voice-over (ducking) unit 63-5
Voltage followers 27-9
Voltage regulator circuits
  fixed 3-terminal
    current boosting 267-8
    principle and usage 264-6
    ripple rejection 266
    variable 267-8
    voltage variation 266-7
  variable 3-terminal 268-70
  zener-based 262-4
Voltmeters
  20-LED 196-7
  bar-mode 193-6
  dot-mode 190-3
Volume control, with voltage-controlled ICs 83-5, 87-94
VU (volume unit) meters 198

Zener-based voltage regulator circuits 262-4

# Integrated circuits (ICs) by type number

| | | |
|---|---|---|
| 317K | Voltage regulators | Ch. 9 |
| 338K | Voltage regulators | Ch. 9 |
| 4013B | Low impedance clock driving | Ch. 8 |
| 4016/4066 family | Quad bilateral analogue switches | Ch. 3 |
| 4046B | VCO with phase-locked loop as clock generator | Ch. 8 |
| 741 | Bipolar op-amp | Ch. 2/5 |
| 74HC4052 | Dual 4-channel analogue multiplexer switch | Ch. 3 |
| 78xxx series | Voltage regulators | Ch. 9 |
| 79xxx series | Voltage regulators | Ch. 9 |
| CA3080 | Operational transconductance amplifier | Ch. 2 |
| CA3130E | MOSFET op-amp | Ch. 2 |
| CA3140E | MOSFET op-amp | Ch. 2/9 |
| HT8955A | Digital delay line | Ch. 8 |
| LF13741 | JFET op-amp | Ch. 2 |
| LF351 | JFET op-amp | Ch. 2 |
| LF411 | JFET op-amp | Ch. 2 |
| LF441 | JFET op-amp | Ch. 2 |
| LM1036 | Dual dc-controlled tone/volume/balance | Ch. 3 |
| LM1037 | Dual 4-channel analogue switch | Ch. 3 |
| LM13600 | Operational transconductance amplifier | Ch. 2 |
| LM13700 | Operational transconductance amplifier | Ch. 2 |
| LM1875 | Audio power amplifier | Ch. 6 |
| LM1877 | Dual protected output power amplifier | Ch. 5 |
| LM1894 | Dynamic noise reduction IC | Ch. 3 |
| LM2877/2878/2879 | Dual high power audio amplifiers | Ch. 6 |
| LM380/LM384 | Audio power amplifier | Ch. 5 |
| LM383 (TDA2003) | Audio power amplifier | Ch. 6 |
| LM386 | Audio power amplifier | Ch. 5 |
| LM387 | Dual pre-amp | Ch. 4 |
| LM3875 | Audio power amplifier | Ch. 6 |
| LM3876 | Audio power amplifier | Ch. 6 |
| LM388 | Audio power amplifier | Ch. 5 |
| LM3886 | Audio power amplifier | Ch. 6 |
| LM3914/15/16 | Dot/bar-graph driver ICs | Ch. 7 |
| LM831 | Dual power amplifier | Ch. 5 |
| LM833 | Dual audio op-amp | Ch. 4 |
| LMC1983 | Digitally controlled 3-channel stereo selector and tone/volume control | Ch. 3 |
| MC3340P | Voltage controlled amplifier | Ch. 3 |
| MF10C | Switched-capacitor filter IC | Ch. 3 |
| MN3004 | 512 stage high-performance delay line | Ch. 8 |
| MN3011 | 3328 stage delay line | Ch. 8 |
| MN3101 | Delay-line clock generator | Ch. 8 |
| MN3102 | Delay-line low voltage clock generator | Ch. 8 |
| MN3207 | 1024 stage low-voltage delay line | Ch. 8 |

*Integrated circuits (ICs) by type number*

| | | |
|---|---|---|
| NE531 | Bipolar op-amp | Ch. 2 |
| NE570/571 | Dual compandor | Ch. 3 |
| SAD1024 | Dual 512 stage delay line | Ch. 8 |
| SAD4096 | 4096 stage delay line | Ch. 8 |
| SAD512D | 512 stage delay line with clock divider | Ch. 8 |
| SSM2120 | Hi-fi compandor | Ch. 3 |
| TBA810P | Audio power amplifier | Ch. 6 |
| TBA820M | Audio power amplifier | Ch. 5 |
| TDA1020 | Audio power amplifier | Ch. 6 |
| TDA1022 | Low cost 512 stage delay line | Ch. 8 |
| TDA1097 | 1536 stage delay line | Ch. 8 |
| TDA1514A | Audio power amplifier | Ch. 6 |
| TDA2005M | Audio power amplifier | Ch. 6 |
| TDA2006 | Audio power amplifier | Ch. 6 |
| TDA2030 | Audio power amplifier | Ch. 6 |
| TDA2040 | Audio power amplifier | Ch. 6 |
| TDA2050 | Audio power amplifier | Ch. 6 |
| TDA2822 | Dual audio power amplifier | Ch. 6 |
| TDA7052 | Audio power amplifier | Ch. 5 |
| U237/47/57/67 | Bar-graph drivers | Ch. 7 |